BIRKHÄUSER

Trends in Mathematics is a series devoted to the publication of volumes arising from conferences and lecture series focusing on a particular topic from any area of mathematics. Its aim is to make current developments available to the community as rapidly as possible without compromise to quality and to archive these for reference.

Proposals for volumes can be sent to the Mathematics Editor at either

Birkhäuser Verlag
P.O. Box 133
CH-4010 Basel
Switzerland

or

Birkhäuser Boston Inc.
675 Massachusetts Avenue
Cambridge, MA 02139
USA

Material submitted for publication must be screened and prepared as follows:

All contributions should undergo a reviewing process similar to that carried out by journals and be checked for correct use of language which, as a rule, is English. Articles without proofs, or which do not contain any significantly new results, should be rejected. High quality survey papers, however, are welcome.

We expect the organizers to deliver manuscripts in a form that is essentially ready for direct reproduction. Any version of TeX is acceptable, but the entire collection of files must be in one particular dialect of TeX and unified according to simple instructions available from Birkhäuser.

Furthermore, in order to guarantee the timely appearance of the proceedings it is essential that the final version of the entire material be submitted no later than one year after the conference. The total number of pages should not exceed 350. The first-mentioned author of each article will receive 25 free offprints. To the participants of the congress the book will be offered at a special rate.

Hypercomplex Analysis

Irene Sabadini
Michael Shapiro
Frank Sommen
Editors

Birkhäuser
Basel · Boston · Berlin

Editors:

Irene Sabadini
Dipartimento di Matematica
Politecnico di Milano
Via Bonardi 9
20133 Milano
Italy
e-mail: irene.sabadini@polimi.it

Frank Sommen
Clifford Research Group
Department of Mathematical Analysis
Universiteit Gent
Galglaan 2
9000 Gent
Belgium
e-mail: fs@cage.ugent.be

Michael Shapiro
Escuela Superior de Física y Matemáticas
Instituto Politécnico Nacional
Avenida IPN s/n
07338 México, D.F.
México
e-mail: shapiro@esfm.ipn.mx

2000 Mathematical Subject Classification: 30G35

Library of Congress Control Number: 2008942605

Bibliographic information published by Die Deutsche Bibliothek. Die Deutsche Bibliothek lists this publication in the Deutsche Nationalbibliografie; detailed bibliographic data is available in the Internet at http://dnb.ddb.de

ISBN 978-3-7643-9892-7 Birkhäuser Verlag AG, Basel - Boston - Berlin

Basel · Boston · Berlin
P.O. Box 133, CH-4010 Basel, Switzerland
Part of Springer Science+Business Media
Printed on acid-free paper produced from chlorine-free pulp. TCF ∞
Cover Design: Alexander Faust, Basel, Switzerland
Printed in Germany

ISBN 978-3-7643-9892-7 e-ISBN 978-3-7643-9893-4

9 8 7 6 5 4 3 2 1 www.birkhauser.ch

Contents

Preface

This volume contains some papers written by the participants to the Session "Quaternionic and Clifford Analysis" of the 6th ISAAC Conference (held in Ankara, Turkey, in August 2007) and some invited contributions. The contents cover several different aspects of the hypercomplex analysis. All contributed papers represent the most recent achievements in the area as well as "state-of-the art" expositions.

The Editors are grateful to the contributors to this volume, as their works show how the topic of hypercomplex analysis is lively and fertile, and to the referees, for their painstaking and careful work. The Editors also thank professor M.W. Wong, President of the ISAAC, for his support which made this volume possible.

October 2008,

Irene Sabadini
Michael Shapiro
Frank Sommen

Quaternionic and Clifford Analysis
Trends in Mathematics, 1–9

An Extension Theorem for Biregular Functions in Clifford Analysis

Ricardo Abreu Blaya and Juan Bory Reyes

Abstract. In this contribution we are interested in finding necessary and sufficient conditions for the two-sided biregular extendibility of functions defined on a surface of $\mathbb{R}^{2n}$, but the latter without imposing any smoothness requirement.

Mathematics Subject Classification (2000). Primary 30E20, 30E25; Secondary 30G20.

Keywords. Clifford analysis, biregular functions, Bochner-Martinelli formulae, extension theorems.

1. Introduction and preliminary facts

The study of two-sided biregular Clifford-valued functions goes back to [14]. These functions, which are of two higher-dimensional variables in Euclidean spaces, are a non trivial generalization of monogenic functions of one Clifford variable, i.e., nullsolutions of the Dirac operator in Euclidean space. The latter being called Clifford analysis, see [4].

It is proved that many important properties of holomorphic functions of one complex variable may be extended for the class of biregular functions in the framework of Clifford analysis, see [10] for more details.

In a classical formulation, the characterization of the two-sided biregular extension of functions defined on the boundary of a domain of $\mathbb{R}^{2n}$, $n \geq 2$, is tied up with certain a priori smoothness restrictions on the boundary in order to ensure the existence of a pointwise normal vector on it. However, these restrictions may be entirely avoided, if the normal vector is replaced by the exterior normal in Federer's sense, see [9], p. 477.

The natural question arises whether it is possible to extend two-sided biregularly a merely continuous function defined on the boundary of a domain, after making the assumption that the boundary is an Ahlfors David regular and rectifiable surface. New results in this direction to be described in this paper is our purpose.

We stress that the class of surfaces satisfying both Ahlfors David regular and rectifiable restrictions is very general and contains all classes of those classically considered in the literature, in particular the Lipschitz graphs.

Essential to our proofs is the effective use of the isotonic Clifford analysis, applying the simple connection between the two-sided biregular functions and the isotonic ones. In particular, the isotonic Cauchy transform tool will be used and some more sophisticated arguments given in [1] in relation with the existence of its continuous limit values on Ahlfors David regular and rectifiable surfaces.

We shall freely use the well-known properties of complex Clifford algebras which the reader can find in many sources such as for instance [4] but in many others as well.

We'll denote by $e_1, \ldots, e_n$ an orthonormal basis of the Euclidean space $\mathbb{R}^n$. Let $\mathbb{C}_n$ be the complex Clifford algebra constructed over $\mathbb{R}^n$.

The non-commutative multiplication in $\mathbb{C}_n$ is governed by the rules

$$e_j^2 = -1, \quad j = 1, 2, \ldots, n \quad \text{and} \quad e_j e_k + e_k e_j = 0, \quad 1 \le j \ne k \le n.$$

The Clifford algebra $\mathbb{C}_n$ is generated as a vector space by elements of the form

$$e_A = e_{j_1} \ldots e_{j_k},$$

where $A = \{j_1, \ldots, j_k\} \subset \{1, \ldots, n\}$ is such that $j_1 < \cdots < j_k$. For the empty set $\emptyset$, we put $e_\emptyset = 1$, the latter being the identity element. Any Clifford number $a \in \mathbb{C}_n$ may thus be written as

$$a = \sum_A a_A e_A, \quad a_A = \Re a_A + i \Im a_A \in \mathbb{C},$$

or still as $a = \Re a + i \Im a$, where $\Re a = \sum_A \Re a_A e_A$ and $\Im a = \sum_A \Im a_A e_A$ are the $\mathbb{R}_{0,n}$-valued real and imaginary parts of a. Here $\mathbb{R}_{0,n}$ denotes the universal Clifford algebra over $\mathbb{R}^n$.

The conjugation $a \to \bar{a}$ and the main involution $a \to \tilde{a}$ are respectively given by

$$\bar{a} = \sum_A \bar{a}_A \bar{e}_A, \quad \bar{e}_A = (-1)^{\frac{k(k+1)}{2}} e_A, \quad |A| = k,$$

$$\tilde{a} = \sum_A a_A \tilde{e}_A, \quad \tilde{e}_A = (-1)^k e_A, \qquad |A| = k.$$

It is easy to check that

$$\overline{ab} = \bar{b}\bar{a}, \quad \widetilde{ab} = \tilde{a}\tilde{b}, \quad a, b \in \mathbb{C}_n.$$

In this paper we continue the study of isotonic functions of two higher-dimensional variables, started in [1, 3], while the results on Bochner-Martinelli formulae are motivated by the original paper [16].

The isotonic Clifford analysis is a natural generalization of both holomorphic functions of several complex variables and two-sided biregular ones.

2. Isotonic functions theory

Generally speaking, we shall consider functions f on a domain Ω of $\mathbb{R}^{2n}$ with values in $\mathbb{C}_n$.

If we identify the $\mathbb{R}^n$-vector $(x_1, \ldots, x_n)$ with the real Clifford vector $\underline{x} = \sum_{j=1}^{n} e_j x_j$, then $\mathbb{R}^n$ may be considered as a subspace of $\mathbb{C}_n$. Next, we introduce the following higher-dimensional variables

$$\underline{x}_1 = \sum_{j=1}^{n} e_j x_j \quad \text{and} \quad \underline{x}_2 = \sum_{j=1}^{n} e_j x_{n+j}.$$

Definition 2.1. [13, 16] A function $f : \Omega \subseteq \mathbb{R}^{2n} \to \mathbb{C}_n$, is said to be isotonic in Ω if and only if f is continuously differentiable in Ω and moreover satisfies the equation

$$\partial_{\underline{x}_1} f + i\tilde{f}\partial_{\underline{x}_2} = 0,$$

with

$$\partial_{\underline{x}_1} = \sum_{j=1}^{n} e_j \partial_{x_j} \quad \text{and} \quad \partial_{\underline{x}_2} = \sum_{j=1}^{n} e_j \partial_{x_{n+j}}.$$

It is worth pointing out that if in particular f takes values in the space of scalars $\mathbb{C}$, then

$$(\partial_{x_j} + i\partial_{x_{n+j}})f = 0, \quad j = 1, \ldots, n,$$

which means that f is a holomorphic function with respect to the n complex variables $x_j + ix_{n+j}, j = 1, \ldots, n$.

On the other hand, if f, isotonic function, takes values in the real Clifford algebra $\mathbb{R}_{0,n}$, then

$$\begin{cases} \partial_{\underline{x}_1} f = 0, \\ \tilde{f}\partial_{\underline{x}_2} = 0. \end{cases}$$

or, equivalently, by the action of the main involution on the second equation we arrive to the overdetermined system:

$$\begin{cases} \partial_{\underline{x}_1} f = 0, \\ f\partial_{\underline{x}_2} = 0. \end{cases}$$

Thus, the definition of the two-sided biregular functions runs as follows.

Definition 2.2. A function $f : \Omega \subseteq \mathbb{R}^{2n} \to \mathbb{R}_{0,n}$ is said to be two-sided biregular in Ω iff it is of class C^1 and satisfies the above last system.

Due to

$$-\Delta_{2n} = \partial_{\underline{x}_1}^2 + \partial_{\underline{x}_2}^2,$$

where Δ_{2n} is the Laplacian in $\mathbb{R}^{2n}$, the two-sided biregular functions are harmonic.

We shall consider isotonic functions in the form $f = f_1 + if_2$, where f_1 and f_2 are $\mathbb{R}_{0,n}$-valued functions. Obviously any two-sided biregular mapping (f_1, f_2) in Ω defines an isotonic function $f = f_1 + if_2$. Furthermore, if one of the functions (say f_1) is two-sided biregular in Ω, then $f = f_1 + if_2$ is isotonic in Ω if f_2 is two-sided biregular in Ω.

The theory of two-sided biregular functions may be regarded as a generalization of holomorphic functions in domains of $\mathbb{C}^n \cong \mathbb{R}^{2n}$. Several properties of the holomorphic functions, such as Hartogs theorem and Bochner Martinelli formula, may be generalized for two-sided biregular functions, see [5, 6, 7, 15].

Let us finish the section with one fact expressing the analogy of the two-sided biregular functions with those holomorphic of several complex variables.

Likewise in complex analysis of several variables the following surprising statement for two-sided biregular functions is an easy consequence of the presence of a sufficiently overdeterminated setting. For this reason, we borrow the proof from those of [12].

Lemma 2.3. *Suppose that K is a compact subset of $\mathbb{R}^{2n}$, with $n \geq 2$, $\mathbb{R}^{2n} \setminus K$ is connected. Then every bounded two-sided biregular function f in $\mathbb{R}^{2n} \setminus K$ is constant, hence admits a two-sided biregular extension to $\mathbb{R}^{2n}$.*

Proof. Choose R so large that K lies in $\{\underline{x} \in \mathbb{R}^{2n} : |\underline{x}| < R\}$. Then for fixed $\underline{x}_2 \in \mathbb{R}^n$ with $|\underline{x}_2| > R$, the subspace $\{(\underline{x}_1, \underline{x}_2) : \underline{x}_1 \in \mathbb{R}^n\}$ does not meet K.

By the Liouville theorem in the Clifford analysis framework, see [4], Theorem 12.3.11, $f(\underline{x}_1, \underline{x}_2)$ is constant as a function of $\underline{x}_1$. Interchanging variables, one sees that f is constant in the variable $\underline{x}_2$ as well, provided the variable $\underline{x}_1$ is fixed with sufficiently large modulus. It follows that f is constant outside a large ball. Taking into account that f is harmonic in $\mathbb{R}^{2n} \setminus K$ the proof is completed by using the uniqueness theorem for such functions. □

3. Bochner-Martinelli formula for two-sided biregular functions

Let Ω be a bounded domain in $\mathbb{R}^{2n}$, $n \geq 2$, with a boundary Γ such that $\mathcal{H}^{2n-1}(\Gamma) < +\infty$, where $\mathcal{H}^{2n-1}$ denotes the $(2n-1)$-dimensional Hausdorff measure in $\mathbb{R}^{2n}$, see [11].

The same reasoning applied in [16] allows us to prove that for any $\mathbb{C}_n$-valued function f of class C^1 in $\overline{\Omega}$ the following Borel-Pompeiu type integral representation holds

$$
\begin{aligned}
&-\frac{1}{\omega_{2n}}\int_\Gamma\Bigg[\frac{(\underline{y}_1-\underline{x}_1)\big(\underline{\nu}_1(\underline{y})f(\underline{y})+i\tilde{f}(\underline{y})\underline{\nu}_2(\underline{y})\big)}{|\underline{y}-\underline{x}|^{2n}}\\
&\qquad+\frac{\big(f(\underline{y})\underline{\nu}_2(\underline{y})-i\underline{\nu}_1(\underline{y})\tilde{f}(\underline{y})\big)(\underline{y}_2-\underline{x}_2)}{|\underline{y}-\underline{x}|^{2n}}\Bigg]d\mathcal{H}^{2n-1}(\underline{y})\\
&+\frac{1}{\omega_{2n}}\int_\Omega\Bigg[\frac{(\underline{y}_1-\underline{x}_1)\big(\partial_{\underline{y}_1}f(\underline{y})+i\tilde{f}(\underline{y})\partial_{\underline{y}_2}\big)}{|\underline{y}-\underline{x}|^{2n}}\\
&\qquad+\frac{\big(f(\underline{y})\partial_{\underline{y}_2}-i\partial_{\underline{y}_1}\tilde{f}(\underline{y})\big)(\underline{y}_2-\underline{x}_2)}{|\underline{y}-\underline{x}|^{2n}}\Bigg]d\underline{y}=\begin{cases} f(\underline{x}), & \underline{x}\in\Omega,\\ 0, & \underline{x}\in\mathbb{R}^{2n}\setminus\overline{\Omega}\end{cases}
\end{aligned}
\tag{3.1}
$$

where

$$\underline{\nu}_1(\underline{y}) = \sum_{j=1}^{n} e_j \nu_j(\underline{y}), \quad \underline{\nu}_2(\underline{y}) = \sum_{j=1}^{n} e_j \nu_{n+j}(\underline{y}).$$

Hereby $\nu_s(\underline{y})$ for $s = 1, \dots, 2n$ denote the real components of the outward unit normal vector $\underline{\nu}(\underline{y}) = \sum_{s=1}^{2n} e_s \nu_s(\underline{y})$ at a point $\underline{y} \in \Gamma$, the latter being taking in Federer's sense, and ω_{2n} is the area of the unit sphere in $\mathbb{R}^{2n}$.

A Bochner-Martinelli formula for two-sided biregular functions will be derived from the formula (3.1).

First we introduce the following Cauchy kernels, defined in $\mathbb{R}^{2n}$ by

$$\underline{k}_1(\underline{x}) = \frac{1}{\omega_{2n}} \frac{\overline{\underline{x}_1}}{|\underline{x}|^{2n}} \quad \underline{x} \in \mathbb{R}^{2n} \setminus \{0\}, \quad \underline{k}_2(\underline{x}) = \frac{1}{\omega_{2n}} \frac{\overline{\underline{x}_2}}{|\underline{x}|^{2n}} \quad \underline{x} \in \mathbb{R}^{2n} \setminus \{0\}.$$

Clearly $\underline{k}_1$ and $\underline{k}_2$ are not two-sided biregular in $\mathbb{R}^{2n}$.

We now come to the Bochner-Martinelli formula for two-sided biregular functions.

Theorem 3.1. *Let $f : \Omega \subset \mathbb{R}^{2n} \to \mathbb{R}_{0,n}$ be a function of class C^1 in $\overline{\Omega}$. Then the formula*

$$\int_{\Gamma} [\underline{k}_1(\underline{y} - \underline{x})\underline{\nu}_1(\underline{y}) f(\underline{y}) + f(\underline{y})\underline{\nu}_2(\underline{y})\underline{k}_2(\underline{y} - \underline{x})] d\mathcal{H}^{2n-1}(\underline{y}) \tag{3.2}$$

$$- \int_{\Omega} [\underline{k}_1(\underline{y} - \underline{x})(\partial_{\underline{y}_1} f(\underline{y})) + (f(\underline{y})\partial_{\underline{y}_2})\underline{k}_2(\underline{y} - \underline{x})] d\underline{y} = \begin{cases} f(\underline{x}), & \underline{x} \in \Omega \\ 0, & \underline{x} \in \mathbb{R}^{2n} \setminus \overline{\Omega} \end{cases}$$

holds.

Proof. It is sufficient to take the real part of (3.1) □

Remark 3.2. By using different methods the formula (3.2) for a sufficiently smooth boundary was already proved by Brackx and Pincket in [5], Theorem 2.3.

3.1. The Bochner-Martinelli type integrals

From now on, Ω stands for a bounded domain in $\mathbb{R}^{2n}$, $n \geq 2$ with an Ahlfors David regular (AD-regular) boundary Γ, i.e., it satisfies

$$c^{-1} r^{2n-1} \leq \mathcal{H}^{2n-1}(\Gamma \cap \{|\underline{y} - \underline{x}| \leq r\}) \leq c\, r^{2n-1},$$

for all $\underline{x} \in \Gamma$ and all $0 < r \leq \operatorname{diam} \Gamma$, the constant c being independent of both $\underline{x}$ and r. The best general reference here is [8, 11].

We follow [9] in assuming that a boundary Γ is said to be rectifiable if it is the Lipschitz image of some bounded subset of $\mathbb{R}^{2n-1}$.

A rectifiable and AD-regular boundary Γ besides satisfying $\mathcal{H}^{2n-1}(\Gamma) < +\infty$, has still better geometric properties.

When necessary we shall use the temporary notation $\Omega^+ = \Omega$, $\Omega^- = \mathbb{R}^{2n} \setminus \overline{\Omega^+}$.

Let f be a $\mathbb{C}_n$-valued continuous function on Γ. The isotonic Cauchy transform of f will be denoted by $\mathbf{C}^{\rm isot} f$ and defined by

$$\mathbf{C}^{\rm isot} f(\underline{x}) = \int_\Gamma \Big(\underline{k}_1(\underline{y}-\underline{x})\big(\underline{\nu}_1(\underline{y})f(\underline{y}) + i\tilde{f}(\underline{y})\underline{\nu}_2(\underline{y})\big) + \big(f(\underline{y})\underline{\nu}_2(\underline{y}) - i\underline{\nu}_1(\underline{y})\tilde{f}(\underline{y})\big)\underline{k}_2(\underline{y}-\underline{x})\Big) d\mathcal{H}^{2n-1}(\underline{y}), \quad \underline{x} \in \mathbb{R}^{2n} \setminus \Gamma.$$

A trivial verification shows that $\mathbf{C}^{\rm isot} f$ is an isotonic function in $\mathbb{R}^{2n} \setminus \Gamma$, which vanishes at infinity.

On the other hand, the isotonic singular integral operator of f, denoted by $\mathbf{S}^{\rm isot} f$ is given by

$$\mathbf{S}^{\rm isot} f(\underline{z}) = 2\lim_{\epsilon\to 0} \int_{\Gamma\setminus\{|\underline{y}-\underline{z}|\le\epsilon\}} \Big(\underline{k}_1(\underline{y}-\underline{z})\big(\underline{\nu}_1(\underline{y})(f(\underline{y})-f(\underline{z})) + i(\tilde{f}(\underline{y})-\tilde{f}(\underline{z}))\underline{\nu}_2(\underline{y})\big) + \big((f(\underline{y})-f(\underline{z}))\underline{\nu}_2(\underline{y}) - i\underline{\nu}_1(\underline{y})(\tilde{f}(\underline{y})-\tilde{f}(\underline{z}))\big)\underline{k}_2(\underline{y}-\underline{z})\Big) d\mathcal{H}^{2n-1}(\underline{y}) + f(\underline{z}), \ \underline{z}\in\Gamma.$$

The following results will be needed in the paper. For the proof we refer the reader to [1, 3].

Theorem 3.3. *Let f be a $\mathbb{C}_n$-valued continuous function on $\overline{\Omega}$, which moreover is isotonic in Ω. Then*

$$\mathbf{C}^{\rm isot} f(\underline{x}) = \begin{cases} f(\underline{x}), & \underline{x} \in \Omega, \\ 0, & \underline{x} \in \mathbb{R}^{2n} \setminus \overline{\Omega}. \end{cases}$$

Theorem 3.4. *Let f be a Hölder continuous function on Γ. Then $\mathbf{C}^{\rm isot} f$ has Hölder continuous limit values on Γ and the following Sokhotski-Plemelj formulae hold:*

$$\mathbf{C}^{\rm isot}_{\pm} f(\underline{z}) := \lim_{\Omega^\pm \ni \underline{x} \to \underline{z}} \mathbf{C}^{\rm isot} f(\underline{x}) = \frac{1}{2}(\mathbf{S}^{\rm isot} f(\underline{z}) \pm f(\underline{z})), \quad \underline{z} \in \Gamma.$$

In what follows, we regard f as being an $\mathbb{R}_{0,n}$-valued function. Thus $\mathbf{C}^{\rm isot} f$ splits into its real and imaginary parts as follows,

$$\mathbf{C}^{\rm isot} f(\underline{x}) = \mathcal{M}_1 f(\underline{x}) + i\mathcal{M}_2 f(\underline{x}), \quad \underline{x} \notin \Gamma, \tag{3.3}$$

where

$$\mathcal{M}_1 f(\underline{x}) := \int_\Gamma [\underline{k}_1(\underline{y}-\underline{x})\underline{\nu}_1(\underline{y})f(\underline{y}) + f(\underline{y})\underline{\nu}_2(\underline{y})\underline{k}_2(\underline{y}-\underline{x})] d\mathcal{H}^{2n-1}(\underline{y}).$$

and

$$\mathcal{M}_2 f(\underline{x}) := \int_\Gamma [\underline{k}_1(\underline{y}-\underline{x})\tilde{f}(\underline{y})\underline{\nu}_2(\underline{y}) - \underline{\nu}_1(\underline{y})\tilde{f}(\underline{y})\underline{k}_2(\underline{y}-\underline{x})] d\mathcal{H}^{2n-1}(\underline{y})$$

The equality (3.3) shows the main idea of the application of the isotonic Clifford analysis to two-sided biregular functions of two higher-dimensional variables.

Probably, the boundary integral in (3.2) reminds the reader of the standard Bochner-Martinelli integral occurring in several complex variables. Based on this analogy, here we shall also refer $\mathcal{M}_1$ as the Bochner-Martinelli integral.

Note that $\mathcal{M}_1$ is precisely the real $\mathbb{R}_{0,n}$-coordinate of the isotonic Cauchy transform, i.e., $\mathcal{M}_1 = \Re\,\mathbf{C}^{\mathrm{isot}}$.

3.1.1. Basic remarks.

(I) Combining (3.3) with Theorem 3.4 we can deduce that for any $\mathbb{R}_{0,n}$-valued Hölder continuous function f on Γ, denoted $f \in C^{0,\alpha}(\Gamma, \mathbb{R}_{0,n})$, $0 < \alpha \leq 1$, the following limit values:

$$(\mathcal{M}_1{}^{\pm}[f])(\underline{z}) := \lim_{\Omega^{\pm} \ni \underline{x} \to \underline{z} \in \Gamma} (\mathcal{M}_1[f])(\underline{x}),$$
$$(\mathcal{M}_2{}^{\pm}[f])(\underline{z}) := \lim_{\Omega^{\pm} \ni \underline{x} \to \underline{z} \in \Gamma} (\mathcal{M}_2[f])(\underline{x}).$$

exist and become continuous functions on Γ.

(II) Moreover, the following Plemelj-Sokhotski formulae

$$(\mathcal{M}_1{}^{+}[f])(\underline{z}) = \frac{1}{2}[\mathcal{N}_1[f](\underline{z}) + f(\underline{z})], \tag{3.4}$$
$$(\mathcal{M}_1{}^{-}[f])(\underline{z}) = \frac{1}{2}[\mathcal{N}_1[f](\underline{z}) - f(\underline{z})], \tag{3.5}$$
$$(\mathcal{M}_2{}^{\pm}[f])(\underline{z}) = \frac{1}{2}\mathcal{N}_2[f](\underline{z}). \tag{3.6}$$

hold. Hereby the integrals

$$\begin{aligned}\mathcal{N}_1 f(\underline{z}) := {}& 2\int_\Gamma \underline{k}_1(\underline{y}-\underline{z})\underline{\nu}_1(\underline{y})(f(\underline{y}) - f(\underline{z}))d\mathcal{H}^{2n-1}(\underline{y}) \\ & + 2\int_\Gamma (f(\underline{y}) - f(\underline{z}))\underline{\nu}_2(\underline{y})\underline{k}_2(\underline{y}-\underline{z})d\mathcal{H}^{2n-1}(\underline{y}) + f(\underline{z}).\end{aligned}$$

and

$$\begin{aligned}\mathcal{N}_2 f(\underline{z}) := {}& 2\int_\Gamma \underline{k}_1(\underline{y}-\underline{z})(\tilde{f}(\underline{y}) - \tilde{f}(\underline{z}))\underline{\nu}_2(\underline{y})d\mathcal{H}^{2n-1}(\underline{y}) \\ & - 2\int_\Gamma \underline{\nu}_1(\underline{y})(\tilde{f}(\underline{y}) - \tilde{f}(\underline{z}))\underline{k}_2(\underline{y}-\underline{z})d\mathcal{H}^{2n-1}(\underline{y})\end{aligned}$$

are understood in the sense of the Cauchy principal value.

(III) As usual one may conclude immediately that the following jump relations hold

$$\begin{gathered}\mathcal{M}_1{}^{+}[f] - \mathcal{M}_1{}^{-}[f] = f, \quad \mathcal{M}_1{}^{+}[f] + \mathcal{M}_1{}^{-}[f] = \mathcal{N}_1[f], \\ \mathcal{M}_2^{+}[f] = \mathcal{M}_2^{-}[f].\end{gathered} \tag{3.7}$$

4. Main extension theorems

In this section we shall study the problem to determine necessary and sufficient conditions, in order to a continuous function defined on the boundary of a domain $\Omega \subset \mathbb{R}^{2n}$ may have a two-sided biregular extension on Ω.

Theorem 4.1. *Assume that* $\Omega \subset \mathbb{R}^{2n}$ *with AD-regular boundary* Γ *and that* $f \in C^{0,\alpha}(\Gamma, \mathbb{R}_{0,n})$, $0 < \alpha \leq 1$. *Then, the following conditions are equivalent:*

(i) f *has two-sided biregular extension on* Ω
(ii) $\mathbf{S}^{\text{isot}} f = f$ *on* Γ.

Proof. It follows from (i) that there exists a $\mathbb{R}_{0,n}$-valued function $\mathcal{F}$ on $\overline{\Omega}$ which is two-sided biregular on Ω, continuous in $\overline{\Omega}$, with $\mathcal{F}|_\Gamma = f$.

Theorem 3.3 now leads to

$$\mathbf{C}^{\text{isot}} f(\underline{x}) = \begin{cases} \mathcal{F}(\underline{x}), & \underline{x} \in \Omega, \\ 0, & \underline{x} \in \mathbb{R}^{2n} \setminus \overline{\Omega}. \end{cases}$$

Therefore, $\mathbf{C}^{\text{isot}}_{-} f = 0$ on Γ and, in consequence, $\mathbf{S}^{\text{isot}} f = f$.

Conversely, suppose that (ii) holds. We claim that the function $\mathbf{C}^{\text{isot}} f$ is a two-sided biregular extension of f to Ω. Indeed, from (ii) we conclude that $\mathbf{C}^{\text{isot}} f$ is an isotonic extension of f, then we are left with the task of proving that $\mathbf{C}^{\text{isot}} f$ is $\mathbb{R}_{0,n}$-valued.

In order to get this, we can use the Plemelj-Sokhotski formulae to conclude that the boundary limit values $\mathcal{M}_2^{\pm} f(\underline{z}) = 0$ for all $\underline{z} \in \Gamma$. Since $\mathcal{M}_2 f$ is real harmonic off Γ, we have by the classical Dirichlet problem that $\mathcal{M}_2 f \equiv 0$ in $\mathbb{R}^{2n}$. Then, $\mathbf{C}^{\text{isot}} f \equiv \mathcal{M}_1 f$ and the proof is complete. □

If only the continuity of f is assumed, the proof of (i)⇔(ii) more strongly depends on the assumption that $\mathbf{S}^{\text{isot}} f = f$, but now uniformly, since the isotonic Cauchy transform of a continuous function has not in general continuous boundary limit values even for C^1-smooth boundary.

Theorem 4.2. *Suppose that in* $\mathbb{R}^{2n}$ *we are given a domain* Ω *with AD-regular and rectifiable boundary* Γ, *and let* f *be a continuous function on* Γ. *Then,* f *has two-sided biregular extension to* Ω *if and only if* $\mathbf{S}^{\text{isot}} f = f$ *uniformly on* Γ.

Proof. The proof follows very closely that of Theorem 4.1, but it strongly depends on the uniform existence of $\mathbf{S}^{\text{isot}} f$ and the conclusion of Theorem 4 in [1], see also Theorem 6 in [2]. □

Acknowledgment

The authors wish to thank the referee for his/her valuable comments and suggestions that have considerably enhanced this paper.

References

[1] Abreu Blaya, R.; Bory Reyes, J.; Peña Peña, D. and Sommen, F. The isotonic Cauchy transform. *Adv. Appl. Clifford Algebras*, Vol. 17, (2007) no. 2, 145–152.

[2] Bory Reyes, J.; Abreu Blaya, R. Cauchy transform and rectifiability in Clifford analysis. Z. Anal. Anwendungen 24 (2005), no. 1, 167–178.

[3] Bory Reyes, J.; Peña Peña, D. and Sommen, F. A Davydov type theorem for the isotonic Cauchy transform. *J. Anal. Appl.* Vol. 5, (2007), no. 2, 109–121.

[4] Brackx, F.; Delanghe, R.; Sommen, F. Clifford analysis. Research Notes in Mathematics, 76. *Pitman (Advanced Publishing Program), Boston, MA*, 1982.

[5] Brackx, F.; Pincket, W. A Bochner-Martinelli formula for the biregular functions of Clifford analysis. Complex Variables Theory Appl. 4 (1984), no. 1, 39–48.

[6] Brackx, F.; Pincket, W. The biregular functions of Clifford analysis: some special topics. Clifford algebras and their applications in mathematical physics (Canterbury, 1985), 159–166, NATO Adv. Sci. Inst. Ser. C Math. Phys. Sci., 183, Reidel, Dordrecht, 1986.

[7] Brackx, F.; Pincket, W. Two Hartogs theorems for nullsolutions of overdetermined systems in Euclidean space. Complex Variables Theory Appl. 4 (1985), no. 3, 205–222.

[8] David, G.; Semmes, S. Analysis of and on uniformly rectifiable sets. Mathematical Surveys and Monographs, 38. *American Mathematical Society, Providence, RI*, 1993.

[9] Federer, H. Geometric measure theory. Die Grundlehren der mathematischen Wissenschaften, Band 153 *Springer-Verlag New York Inc., New York* 1969.

[10] Huang, Sha; Qiao, Yu Ying; Wen, Guo Chun. Real and complex Clifford analysis. Advances in Complex Analysis and its Applications, 5. Springer, New York, 2006.

[11] Mattila, P. Geometry of sets and measures in Euclidean spaces. Fractals and rectifiability. Cambridge Studies in Advanced Mathematics, 44. Cambridge University Press, Cambridge, 1995.

[12] Range, R. M. Complex analysis: a brief tour into higher dimensions. *Amer. Math. Monthly*, 110 (2003), no. 2, 89–108.

[13] Rocha Chávez, R.; Shapiro, M.; Sommen, F. Integral theorems for functions and differential forms in $\mathbb{C}^m$. Research Notes in Mathematics, 428. Chapman & Hall/CRC, Boca Raton, FL, 2002.

[14] Sommen, F. Plane waves, biregular functions and hypercomplex Fourier analysis. Proceedings of the 13th winter school on abstract analysis (Srn, 1985). Rend. Circ. Mat. Palermo (2) Suppl. No. 9 (1985), 205–219 (1986).

[15] Sommen, F. Martinelli-Bochner type formulae in complex Clifford analysis. Z. Anal. Anwendungen 6 (1987), no. 1, 75–82.

[16] Sommen, F.; Peña Peña, D. Martinelli-Bochner formula using Clifford analysis, *Archiv der Mathematik*, 88 (2007), no. 4, 358–363.

Ricardo Abreu Blaya
Facultad de Informática y Matemática
Universidad de Holguín
Holguín 80100, Cuba
e-mail: `rabreu@facinf.uho.edu.cu`

Juan Bory Reyes
Departamento de Matemática
Universidad de Oriente
Santiago de Cuba 90500, Cuba
e-mail: `jbory@rect.uo.edu.cu`

Quaternionic and Clifford Analysis
Trends in Mathematics, 11–36

The Hilbert Transform on the Unit Sphere in $\mathbb{R}^m$

F. Brackx and H. De Schepper

Abstract. As an intrinsically multidimensional function theory, Clifford analysis offers a framework which is particularly suited for the integrated treatment of higher-dimensional phenomena. In this paper a detailed account is given of results connected to the Hilbert transform on the unit sphere in Euclidean space and some of its related concepts, such as Hardy spaces and the Cauchy integral, in a Clifford analysis context.

Mathematics Subject Classification (2000). Primary 30G35; Secondary 44A15.
Keywords. Hilbert transform, Hardy space, Cauchy integral.

1. Introduction

In one-dimensional signal processing the Hilbert transform is an indispensable tool for global as well as local signal analysis, yielding information on various independent signal properties. The instantaneous amplitude, phase and frequency are estimated by means of so-called quadrature filters. Such filters are essentially based on the notion of *analytic signal*, which consists of the linear combination of a bandpass filter, selecting a small part of the spectral information, and its Hilbert transform, the latter basically being the result of a phase shift by $\frac{\pi}{2}$ on the original filter (see, e.g., [18]). More strictly, if $f(x) \in L_2(\mathbb{R})$ is a real-valued signal of finite energy, and $\mathcal{H}[f]$ denotes its Hilbert transform given by the Cauchy Principal Value

$$\mathcal{H}[f](x) = \frac{1}{\pi} \operatorname{Pv} \int_{-\infty}^{+\infty} \frac{f(y)}{x-y}\, dy$$

then the corresponding analytic signal is the function $\frac{1}{2}f + \frac{i}{2}\mathcal{H}[f]$, which belongs to the Hardy space $H^2(\mathbb{R})$ and arises as the L_2 non-tangential boundary value (NTBV) for $y \to 0+$ of the holomorphic Cauchy integral of f in the upper half of the complex plane. Though discovered by Hilbert, the concept of a *conjugated pair* $(f, \mathcal{H}[f])$, nowadays called a Hilbert pair, was developed mainly by Titchmarch and Hardy.

The multidimensional approach to the Hilbert transform usually is a tensorial one, considering the so-called Riesz transforms in each of the Cartesian variables separately. As opposed to these tensorial approaches Clifford analysis is particularly suited for a treatment of multidimensional phenomena encompassing all dimensions at once as an intrinsic feature. In its most simple but still useful setting, flat m-dimensional Euclidean space, Clifford analysis focusses on so-called *monogenic functions*, i.e., null solutions of the Clifford vector-valued Dirac operator $\underline{\partial} = \sum_{j=1}^{m} e_j \partial_{x_j}$ where $(e_1, \ldots, e_m)$ forms an orthogonal basis for the quadratic space $\mathbb{R}^m$ underlying the construction of the Clifford algebra $\mathbb{R}_{0,m}$. Monogenic functions have a special relationship with harmonic functions of several variables in that they are refining their properties. The reason is that, as does the Cauchy–Riemann operator in the complex plane, the rotation-invariant Dirac operator factorizes the m-dimensional Laplace operator. This has, a.o., allowed for a nice study of Hardy spaces of monogenic functions, see [7, 25, 8, 9, 1, 11]. In this context the Hilbert transform, as well as more general singular integral operators, have been studied in Euclidean space (see [14, 25, 32, 20, 10, 12]), on Lipschitz hypersurfaces (see [26, 22, 21, 23]) and also on smooth closed hypersurfaces, in particular the unit sphere (see [11, 3, 6]).

In a recent paper [5] an account was given of the Hilbert transform, within the Clifford analysis context, on the smooth boundary of a bounded domain in Euclidean space of dimension at least three. It goes without saying that the study of the triptych Hilbert transform – Hardy space – Dirichlet problem in the particular case of the unit sphere, which is the subject of the underlying paper, has much more concrete results to offer, in particular w.r.t. this last issue. However, on the unit sphere, some interesting features and insights from the general setting are inevitably lost, since the Hilbert transform becomes a self-adjoint operator. We have gathered the relevant results spread over the literature and have moulded them together with some new insights into a comprehensive text. Particular attention is paid to the similarities with the case of the unit circle in the complex plane. For a detailed study of the aforementioned triptych in the complex plane we refer the reader to the inspiring book [2].

2. Clifford analysis: the basics

In this section we present the basic definitions and results of Clifford analysis which are necessary for our purpose. For an in-depth study of this higher-dimensional function theory and its applications we refer to [4, 13, 14, 15, 16, 17, 27, 28, 29, 30, 31].

Let $\mathbb{R}^{0,m}$ be the real vector space $\mathbb{R}^m$, endowed with a non-degenerate quadratic form of signature $(0, m)$, let $(e_1, \ldots, e_m)$ be an orthonormal basis for $\mathbb{R}^{0,m}$, and let $\mathbb{R}_{0,m}$ be the universal Clifford algebra constructed over $\mathbb{R}^{0,m}$.

The non-commutative multiplication in $\mathbb{R}_{0,m}$ is governed by the rules

$$e_i e_j + e_j e_i = -2\delta_{i,j} \qquad i, j \in \{1, \ldots, m\}.$$

For a set $A = \{i_1, \ldots, i_h\} \subset \{1, \ldots, m\}$ with $1 \le i_1 < i_2 < \cdots < i_h \le m$, let $e_A = e_{i_1} e_{i_2} \ldots e_{i_h}$. Moreover, put $e_\emptyset = 1$, the latter being the identity element. Then $(e_A : A \subset \{1, \ldots, m\})$ is a basis for the Clifford algebra $\mathbb{R}_{0,m}$. Any $a \in \mathbb{R}_{0,m}$ may thus be written as $a = \sum_A a_A \, e_A$ with $a_A \in \mathbb{R}$ or still as $a = \sum_{k=0}^m [a]_k$ where $[a]_k = \sum_{|A|=k} a_A \, e_A$ is the so-called k-vector part of a ($k = 0, 1, \ldots, m$). If we denote the space of k-vectors by $\mathbb{R}^k_{0,m}$, then the Clifford algebra $\mathbb{R}_{0,m}$ decomposes as $\bigoplus_{k=0}^m \mathbb{R}^k_{0,m}$. We will identify an element $\underline{x} = (x_1, \ldots, x_m) \in \mathbb{R}^m$ with the one-vector (or vector) $\underline{x} = \sum_{j=1}^m x_j \, e_j$. The multiplication of any two vectors $\underline{x}$ and $\underline{y}$ is given by

$$\underline{x}\,\underline{y} = \underline{x} \circ \underline{y} + \underline{x} \wedge \underline{y}$$

with

$$\begin{aligned} \underline{x} \circ \underline{y} &= -\sum_{j=1}^m x_j y_j = \frac{1}{2}(\underline{x}\,\underline{y} + \underline{y}\underline{x}) = -\langle \underline{x}, \underline{y} \rangle \\ \underline{x} \wedge \underline{y} &= \sum_{i<j} e_{ij}(x_i y_j - x_j y_i) = \frac{1}{2}(\underline{x}\,\underline{y} - \underline{y}\underline{x}) \end{aligned}$$

being a scalar and a 2-vector (also called bivector), respectively. In particular one has that $\underline{x}^2 = -\langle \underline{x}, \underline{x} \rangle = -|\underline{x}|^2 = -\sum_{j=1}^m x_j^2$. Conjugation in $\mathbb{R}_{0,m}$ is defined as the anti-involution for which $\overline{e}_j = -e_j$, $j = 1, \ldots, m$. In particular for a vector $\underline{x}$ we have $\overline{\underline{x}} = -\underline{x}$.

The Dirac operator in $\mathbb{R}^m$ is the first-order vector-valued differential operator

$$\underline{\partial} = \sum_{j=1}^m e_j \partial_{x_j}$$

its fundamental solution being given by

$$E(\underline{x}) = \frac{1}{a_m} \frac{\overline{\underline{x}}}{|\underline{x}|^m}$$

where a_m denotes the area of the unit sphere in $\mathbb{R}^{m+1}$. We consider functions f defined in $\mathbb{R}^m$ and taking values in $\mathbb{R}_{0,m}$. Such a function may be written as $f(\underline{x}) = \sum_A f_A(\underline{x}) \, e_A$ and each time we assign a property such as continuity, differentiability, etc. to f it is meant that all components f_A share this property. We say that the function f is left (resp. right) monogenic in the open region Ω of $\mathbb{R}^m$ iff f is continuously differentiable in Ω and satisfies in Ω the equation $\underline{\partial} f = 0$, resp. $f \underline{\partial} = 0$. As $\overline{\underline{\partial} f} = \overline{f} \, \overline{\underline{\partial}} = -\overline{f} \underline{\partial}$, a function f is left monogenic in Ω iff $\overline{f}$ is right monogenic in Ω. As moreover the Dirac operator factorizes the Laplace operator Δ, $-\underline{\partial}^2 = \underline{\partial}\,\overline{\underline{\partial}} = \overline{\underline{\partial}}\,\underline{\partial} = \Delta$, a monogenic function in Ω is harmonic (and hence C_∞) in Ω, and so are its components.

3. The Hilbert transform on S^{m-1}

Let u be a C_∞-smooth function on the unit sphere S^{m-1} of $\mathbb{R}^m$. Its *Cauchy integral* is defined in the interior of the unit ball $B^+ = \overset{\circ}{B}(O;1)$ and in its exterior $B^- = \text{co}\,\overline{B}(O;1)$ by

$$\mathcal{C}[u](\underline{x}) = \int_{S^{m-1}} E(\underline{\zeta} - \underline{x})\, d\sigma_{\underline{\zeta}}\, u(\underline{\zeta}) = \frac{1}{a_m} \int_{S^{m-1}} \frac{\underline{x} - \underline{\zeta}}{|\underline{x} - \underline{\zeta}|^m}\, \nu(\underline{\zeta})\, u(\underline{\zeta})\, dS(\underline{\zeta})$$

where the Clifford vector-valued oriented surface element $d\sigma_{\underline{\zeta}}$ has been rewritten as $\nu(\underline{\zeta})\, dS(\underline{\zeta})$, with $\nu(\underline{\zeta})$ denoting the outward pointing unit normal vector at $\underline{\zeta} \in S^{m-1}$. However, as on S^{m-1} it holds that $\nu(\underline{\zeta}) = \underline{\zeta}$, this Cauchy integral takes the form

$$\mathcal{C}[u](\underline{x}) = \frac{1}{a_m} \int_{S^{m-1}} \frac{\underline{x} - \underline{\zeta}}{|\underline{x} - \underline{\zeta}|^m}\, \underline{\zeta}\, u(\underline{\zeta})\, dS(\underline{\zeta}) = \frac{1}{a_m} \int_{S^{m-1}} \frac{1 + \underline{x}\,\underline{\zeta}}{|1 + \underline{x}\,\underline{\zeta}|^m}\, u(\underline{\zeta})\, dS(\underline{\zeta}).$$

Introducing the *Cauchy kernel* for $\underline{x} \in B^+ \cup B^-$ and $\underline{\zeta} \in S^{m-1}$ by

$$C(\underline{\zeta}, \underline{x}) = \frac{1}{a_m}\, \overline{\nu}(\underline{\zeta})\, \frac{\overline{\underline{x}} - \overline{\underline{\zeta}}}{|\underline{x} - \underline{\zeta}|^m} = \frac{1}{a_m}\, \underline{\zeta}\, \frac{\overline{\underline{x}} - \overline{\underline{\zeta}}}{|\underline{x} - \underline{\zeta}|^m} = \frac{1}{a_m} \frac{1 + \underline{\zeta}\,\underline{x}}{|1 + \underline{\zeta}\,\underline{x}|^m} \tag{3.1}$$

the Cauchy integral may be rewritten in terms of the $L_2(S^{m-1})$ inner product as

$$\mathcal{C}[u](\underline{x}) = \langle C(\underline{\zeta}, \underline{x}), u(\underline{\zeta})\rangle = \int_{S^{m-1}} \overline{C(\underline{\zeta}, \underline{x})}\, u(\underline{\zeta})\, dS(\underline{\zeta}).$$

The Cauchy kernel $C(\underline{\zeta}, \underline{x})$ is right-monogenic in $\underline{x} \in B^+ \cup B^-$, yielding the left-monogenicity of the Cauchy integral $\mathcal{C}[u](\underline{x})$ in the same region. Moreover one has that $\lim_{\underline{x}\to\infty} \mathcal{C}[u](\underline{x}) = 0$. The Cauchy integral operator $\mathcal{C}$ is sometimes called the Cauchy–Bitsadze operator.

Following the general theory, the non-tangential boundary values [NTBVs] of the Cauchy integral are given by

$$\lim_{B^+ \ni \underline{x} \to \underline{\xi}} \mathcal{C}[u](\underline{x}) = \frac{1}{2} u(\underline{\xi}) + \frac{1}{2} H[u](\underline{\xi}) \tag{3.2}$$

$$\lim_{B^- \ni \underline{x} \to \underline{\xi}} \mathcal{C}[u](\underline{x}) = -\frac{1}{2} u(\underline{\xi}) + \frac{1}{2} H[u](\underline{\xi}) \tag{3.3}$$

where we have put for $\underline{\xi} \in S^{m-1}$:

$$\begin{aligned} H[u](\underline{\xi}) &= \frac{2}{a_m} \text{Pv} \int_{S^{m-1}} \frac{1 + \underline{\xi}\underline{\zeta}}{|1 + \underline{\xi}\,\underline{\zeta}|^m}\, u(\underline{\zeta})\, dS(\underline{\zeta}) \qquad (3.4)\\ &= \frac{2}{a_m} \int_{S^{m-1}} \frac{1 + \underline{\xi} \circ \underline{\zeta}}{|1 + \underline{\xi}\,\underline{\zeta}|^m}\, u(\underline{\zeta})\, dS(\underline{\zeta}) + \frac{2}{a_m} \text{Pv} \int_{S^{m-1}} \frac{\underline{\xi} \wedge \underline{\zeta}}{|\underline{\xi} - \underline{\zeta}|^m}\, u(\underline{\zeta})\, dS(\underline{\zeta}). \end{aligned}$$

This singular integral transform is mostly called the Hilbert transform. The first integral in (3.4) is, up to constants, the so-called direct value of the double-layer potential with density $u(\underline{\zeta})$ on S^{m-1}:

$$\widetilde{W}(\underline{\xi}) = -(m-2)\int_{S^{m-1}} \frac{1+\underline{\xi}\circ\underline{\zeta}}{|\underline{\xi}-\underline{\zeta}|^m}\, u(\underline{\zeta})\, dS(\underline{\zeta}), \quad \underline{\xi}\in S^{m-1}$$

which is a continuous function on S^{m-1} (see [24, p. 360]). The singularity in the Hilbert kernel

$$\frac{1+\underline{\xi}\,\underline{\zeta}}{|1+\underline{\xi}\,\underline{\zeta}|^m}$$

clearly is due to the bivector part, where the integral has to be taken as a Cauchy Principal Value. The *Plemelj–Sokhotzki formulae* (3.2)–(3.3) lead to the *Cauchy transforms* $\mathcal{C}^{\pm}$ defined on $\mathcal{C}_\infty(S^{m-1})$ by

$$\mathcal{C}^+[u] = \frac{1}{2}u + \frac{1}{2}H[u], \qquad \mathcal{C}^-[u] = -\frac{1}{2}u + \frac{1}{2}H[u].$$

It follows that

$$u = \mathcal{C}^+[u] - \mathcal{C}^-[u], \qquad H[u] = \mathcal{C}^+[u] + \mathcal{C}^-[u]$$

expressing the function $u \in C_\infty(S^{m-1})$ as the *jump* of its Cauchy integral over the boundary S^{m-1}.

In the next section the operators H and $\mathcal{C}^{\pm}$ will be extended to $L_2(S^{m-1})$.

4. The Hardy spaces $H_2^{\pm}(S^{m-1})$

We call $M_\infty(B^{\pm})$ the space of left-monogenic functions in $B^{\pm}$, also vanishing at infinity in the case of B^-, which are moreover $C_\infty(\overline{B^{\pm}})$. The Cauchy integral operator $\mathcal{C}$ maps $C_\infty(S^{m-1})$ into $M_\infty(B^{\pm})$, while the operators H and $\mathcal{C}^{\pm}$ map $\mathcal{C}_\infty(S^{m-1})$ into itself. We call $M_\infty^{\pm}(S^{m-1})$ the spaces of functions on S^{m-1} which are the NTBVs of the functions in $M_\infty(B^{\pm})$ respectively, and we define the Hardy spaces $H_2^{\pm}(S^{m-1})$ as the closure in $L_2(S^{m-1})$ of $M_\infty^{\pm}(S^{m-1})$. Note that the usual notation for $H_2^+(S^{m-1})$ is $H^2(S^{m-1})$, and that $H_2^-(S^{m-1})$ is mostly not considered. Our notation however reflects the symmetry in the properties of both Hardy spaces.

The operators $\mathcal{C}, H$ and $\mathcal{C}^{\pm}$ may be extended, through a density argument, to operators on $L_2(S^{m-1})$. Introducing the Hardy spaces $H_2(B^{\pm})$ of left-monogenic functions in $B^{\pm}$, also vanishing at infinity in the case of B^-, which have NTBVs in $L_2(S^{m-1})$, the following properties of those operators are obtained.

Theorem 4.1.

(i) *The Cauchy integral operator $\mathcal{C}$ maps $L_2(S^{m-1})$ into $H_2(B^{\pm})$ and the NTBVs of $\mathcal{C}[f], f \in L_2(S^{m-1})$, are given by*

$$\mathcal{C}^{\pm}[f] = \pm\frac{1}{2}f + \frac{1}{2}H[f].$$

(ii) *The Cauchy transforms $\mathcal{C}^{\pm}$ are bounded linear operators from $L_2(S^{m-1})$ into $H_2^{\pm}(S^{m-1})$.*

(iii) *The Hilbert transform* H *is a bounded linear operator from* $L_2(S^{m-1})$ *onto* $L_2(S^{m-1})$.
(iv) $H^2 = \mathbf{1}$ *or* $H^{-1} = H$ *on* $L_2(S^{m-1})$.
(v) $H_2^{\pm}(S^{m-1})$ *are eigenspaces of* H *with respective eigenvalues* ± 1.

The adjoint H^* of the Hilbert operator H is given in general by $H^* = \nu H \nu$. However in the case of the unit sphere it reduces to

$$H^*[f](\underline{\xi}) = \frac{2}{a_m} \int_{S^{m-1}} \underline{\xi} \, \frac{1 + \underline{\xi}\,\underline{\zeta}}{|1 + \underline{\xi}\,\underline{\zeta}|^m} \, \underline{\zeta} \, f(\underline{\zeta}) \, dS(\underline{\zeta}) = H[f](\underline{\xi})$$

since $\underline{\xi}(1 + \underline{\xi}\,\underline{\zeta})\underline{\zeta} = \underline{\xi}\,\underline{\zeta} + 1$. This means that for the specific case of the unit sphere the Hilbert operator is self-adjoint: $H^* = H$, which also implies that $H^* = H^{-1}$ and hence $HH^* = H^*H = \mathbf{1}$, i.e., the Hilbert operator is unitary. It thus follows that the Cauchy transforms $\mathcal{C}^{\pm}$ are self-adjoint and that the Kerzman–Stein operator $\mathcal{A} = H - H^*$, which, in the general setting, measures the "degree of non-selfadjointness" of the Hilbert operator defined on the smooth boundary of a bounded domain, here equals the null operator. It is a known result that the unit ball is the only bounded domain with smooth boundary for which the Hilbert transform is unitary (see, e.g., [34]); for the interplay between the unitary character of the Hilbert transform and the geometry of bounded and unbounded domains with more general boundary, we refer to the detailed study contained in [19].

By means of the operators H and $\mathcal{C}^{\pm}$ the Hardy spaces $H_2^{\pm}(S^{m-1})$ may now be characterized as follows.

Lemma 4.2. *A function* $g \in L_2(S^{m-1})$ *belongs to the Hardy space* $H_2^+(S^{m-1})$ *if and only if one of the following conditions is satisfied:*

(i) $\mathcal{C}^+[g] = g$; (ii) $\mathcal{C}^-[g] = 0$; (iii) $H[g] = g$.

A function $g \in H_2^+(S^{m-1})$ may be identified with its left-monogenic extension $\mathcal{C}[g] \in H_2(B^+)$; note that, due to Cauchy's Theorem, then $\mathcal{C}[g] = 0$ in B^-. The constant function 1 is a typical example of a function in $H_2^+(S^{m-1})$; in fact one has $\mathcal{C}[1] = 1$ in B^+, while $\mathcal{C}[1] = 0$ in B^-, yielding $H[1] = \mathcal{C}^+[1] = 1$ and $\mathcal{C}^-[1] = 0$.

Lemma 4.3. *A function* $h \in L_2(S^{m-1})$ *belongs to the Hardy space* $H_2^-(S^{m-1})$ *if and only if one of the following conditions is satisfied:*

(i) $\mathcal{C}^-[h] = -h$; (ii) $\mathcal{C}^+[h] = 0$; (iii) $H[h] = -h$.

A function $h \in H_2^-(S^{m-1})$ may thus be identified with its left-monogenic extension $\mathcal{C}[-h] \in H_2(B^-)$, while here $\mathcal{C}[h]$ vanishes in B^+.

The Cauchy transforms $\mathcal{C}^{\pm}$ are sometimes called the *Hardy projections*, since they indeed are projection operators of $L_2(S^{m-1})$, leading to the direct sum decomposition

$$L_2(S^{m-1}) = H_2^+(S^{m-1}) \oplus H_2^-(S^{m-1}) \tag{4.1}$$

which for a function $f \in L_2(S^{m-1})$ explicitly reads:

$$\begin{aligned} f &= \mathcal{C}^+[f] - \mathcal{C}^-[f] = \frac{1}{2}(\mathbf{1}+H)[f] + \frac{1}{2}(\mathbf{1}-H)[f] \\ H[f] &= \mathcal{C}^+[f] + \mathcal{C}^-[f] = \frac{1}{2}(\mathbf{1}+H)[f] - \frac{1}{2}(\mathbf{1}-H)[f]. \end{aligned}$$

For the C_∞-smooth boundary $\partial\Omega$ of a general bounded domain Ω one has that $\mathcal{C}^\pm$ are skew projections. For the specific case of the unit sphere, however, they are orthogonal projections.

Proposition 4.4. *The direct sum decomposition* (4.1) *is an orthogonal decomposition:* $H_2^+(S^{m-1})^\perp = H_2^-(S^{m-1})$, *or equivalently* $H_2^-(S^{m-1})^\perp = H_2^+(S^{m-1})$.

In order to prove this, we proceed as follows. First, as the Hardy space $H_2^+(S^{m-1})$ is a closed subspace of $L_2(S^{m-1})$, we may indeed write down the orthogonal direct sum decomposition $L_2(S^{m-1}) = H_2^+(S^{m-1}) \oplus H_2^+(S^{m-1})^\perp$. The orthogonal projections $\mathbb{P}$ and $\mathbb{P}^\perp$ on $H_2^+(S^{m-1})$ and $H_2^+(S^{m-1})^\perp$ respectively are called the *Szegö projections.* Also, the Hilbert space $H_2^+(S^{m-1})$ possesses a reproducing kernel $S(\underline{\zeta},\underline{x})$, $\underline{\zeta} \in S^{m-1}$, $\underline{x} \in B^+$, the so-called Szegö kernel, for which

$$\langle S(\underline{\zeta},\underline{x}), g(\underline{\zeta})\rangle = \mathcal{C}[g](\underline{x}), \quad \underline{x} \in B^+$$

for all $g \in H_2^+(S^{m-1})$. Strictly speaking the reproducing character is only obtained by identifying the function $g \in H_2^+(S^{m-1})$ with its left-monogenic extension $\mathcal{C}[g]$ to B^+. Note that the Szegö kernel $S(\underline{\zeta},\underline{x})$ is only defined for $\underline{x} \in B^+$. It is also the kernel function of the integral transform expressing the projection $\mathbb{P}$ of $L_2(S^{m-1})$ on $H_2^+(S^{m-1})$:

$$\langle S(\underline{\zeta},\underline{x}), f(\underline{\zeta})\rangle = \mathbb{P}[f](\underline{x}), \quad f \in L_2(S^{m-1}), \quad \underline{x} \in B^+.$$

It is well known, for a general domain Ω with C_∞-smooth boundary $\partial\Omega$, that the Szegö-kernel is the orthogonal (or Szegö-)projection of the Cauchy kernel $C(\underline{\zeta},\underline{x})$ on $H_2^+(\partial\Omega)$, see [5, Proposition 6.1]. However, in the case of the unit sphere, we have an even more intimate relationship. Indeed, from the general theory it is known that the Cauchy kernel $C(\underline{\zeta},\underline{x})$, (3.1), belongs to $H_2^+(S^{m-1})^\perp$ for all $\underline{x} \in B^-$ and hence does not admit, w.r.t. the variable $\underline{\zeta}$, a left-monogenic extension to B^+. So we restrict ourselves to $\underline{x} \in B^+$ for which it automatically holds that $\frac{\underline{x}}{|\underline{x}|^2} \in B^-$. We also know that, for $\underline{x} \in B^+$, the Cauchy kernel $C(\underline{\zeta},\underline{x})$ belongs to $H_2^-(S^{m-1})^\perp$ and thus has, w.r.t the variable $\underline{\zeta}$, no left-monogenic extension to B^- vanishing at infinity. Now we observe that

$$C(\underline{\zeta},\underline{x}) = \frac{1}{a_m}\frac{\left(\underline{\zeta} - \frac{\underline{x}}{|\underline{x}|^2}\right)\underline{x}}{\left|\left(\underline{\zeta} - \frac{\underline{x}}{|\underline{x}|^2}\right)\underline{x}\right|^m} = \frac{1}{a_m}\frac{\underline{\zeta} - \frac{\underline{x}}{|\underline{x}|^2}}{\left|\underline{\zeta} - \frac{\underline{x}}{|\underline{x}|^2}\right|^m}\frac{\underline{x}}{|\underline{x}|^m}.$$

As the function

$$C(\underline{y},\underline{x}) = \frac{1}{a_m} \frac{\underline{y} - \frac{\underline{x}}{|\underline{x}|^2}}{\left|\underline{y} - \frac{\underline{x}}{|\underline{x}|^2}\right|^m} \frac{\underline{x}}{|\underline{x}|^m} = \frac{1}{a_m} \frac{\underline{y}\,\underline{x}+1}{|\underline{y}\,\underline{x}+1|^m}$$

is left-monogenic in $\underline{y} \in B^+$ with

$$\lim_{B^+ \ni \underline{y} \to \underline{\xi}} C(\underline{y},\underline{x}) = C(\underline{\xi},\underline{x}), \quad \underline{\xi} \in S^{m-1}$$

it becomes clear that the Cauchy kernel for the unit sphere has a left-monogenic extension to B^+. It thus follows that, as long as $\underline{x} \in B^+$, the Cauchy kernel $C(\underline{\zeta},\underline{x})$ belongs to the Hardy space $H_2^+(S^{m-1})$, whence the Szegö-kernel in this case has to coincide with $C(\underline{\zeta},\underline{x})$, i.e.,

$$S(\underline{\zeta},\underline{x}) = C(\underline{\zeta},\underline{x}) = \frac{1}{a_m} \frac{1+\underline{\zeta}\,\underline{x}}{|1+\underline{\zeta}\,\underline{x}|^m}, \qquad \underline{\zeta} \in S^{m-1},\ \underline{x} \in B^+$$

with left-monogenic extension to B^+ given by

$$S(\underline{y},\underline{x}) = \frac{1}{a_m} \frac{1+\underline{y}\,\underline{x}}{|1+\underline{y}\,\underline{x}|^m}, \qquad \underline{y} \in B^+,\ \underline{x} \in B^+.$$

The Hermitian symmetry of this extended Szegö-kernel (see [5, Proposition 6.2]) is now readily obtained:

$$\overline{S(\underline{y},\underline{x})} = \frac{1+\underline{x}\,\underline{y}}{|1+\underline{x}\,\underline{y}|^m} = S(\underline{x},\underline{y})$$

since $|1+\underline{x}\,\underline{y}| = |1+\underline{y}\,\underline{x}|$, and moreover

$$S(\underline{x},\underline{x}) = \frac{1}{a_m} \frac{1}{(1-|\underline{x}|^2)^{m-1}}$$

indeed is positive for all $\underline{x} \in B^+$ (see [5, Proposition 6.3]). Finally observe that the extended Szegö-kernel $S(\underline{y},\underline{x})$, while being left-monogenic in $\underline{y} \in B^+$, at the same time is right-monogenic in $\underline{x} \in B^+$.

Now we can show that the Cauchy transforms (or Hardy projections) $\pm\mathcal{C}^\pm$ coincide with the orthogonal Szegö projections $\mathbb{P}$ and $\mathbb{P}^\perp$ on $H_2^+(S^{m-1})$ and $H_2^+(S^{m-1})^\perp$, respectively. Indeed, for a function $f \in L_2(S^{m-1})$, and still with $\underline{x} \in B^+$, we consecutively have

$$\begin{aligned} \mathcal{C}[f](\underline{x}) &= \langle C(\underline{\zeta},\underline{x}), f(\underline{\zeta})\rangle = \langle S(\underline{\zeta},\underline{x}), f(\underline{\zeta})\rangle \\ &= \langle S(\underline{\zeta},\underline{x}), \mathbb{P}[f](\underline{\zeta})\rangle = \langle C(\underline{\zeta},\underline{x}), \mathbb{P}[f](\underline{\zeta})\rangle = \mathcal{C}[\mathbb{P}[f]](\underline{x}) \end{aligned}$$

and so for $\underline{\xi} \in S^{m-1}$

$$\mathcal{C}^+[f](\underline{\xi}) = \mathcal{C}^+[\mathbb{P}[f]](\underline{\xi}) = \mathbb{P}[f](\underline{\xi})$$

and

$$\mathcal{C}^-[f](\underline{\xi}) = \mathcal{C}^+[f](\underline{\xi}) - f(\underline{\xi}) = -\mathbb{P}^\perp[f](\underline{\xi}).$$

The fact that the direct sum decomposition (4.1) of $L_2(S^{m-1})$ is orthogonal entails a number of consequences, listed below.

(i) In the case of a general bounded domain Ω with C_∞-smooth boundary $\partial\Omega$ we only know for a function $h \in H_2^+(S^{m-1})^\perp$ that it has no left-monogenic extension to Ω^+. In the case of the unit sphere however, such a function will belong to $H_2^-(S^{m-1})$ and thus has a left-monogenic extension to B^- vanishing at infinity, which is nothing else but $\mathcal{C}[-h]$. Moreover for each $\underline{x} \in B^+$ we have, by Cauchy's Theorem in B^-, that

$$\mathcal{C}[h](\underline{x}) = \frac{1}{a_m} \int_{S^{m-1}} \frac{\underline{x}-\underline{\zeta}}{|\underline{x}-\underline{\zeta}|^m}\, d\sigma_{\underline{\zeta}}\, h(\underline{\zeta}) = \mathcal{C}[h](\infty)$$

since $\frac{\underline{x}-\underline{\zeta}}{|\underline{x}-\underline{\zeta}|^m}$ is right-monogenic in B^-, while the extension of h is left-monogenic in B^-. As $\mathcal{C}[h](\infty) = 0$ it follows that $\mathcal{C}[h](\underline{x}) = 0$ for all $\underline{x} \in B^+$, confirming a property stated above concerning the Cauchy integral of functions in $H_2^-(S^{m-1})$.

(ii) For $\underline{x} \in B^-$ we know that the Cauchy kernel $C(\underline{\zeta}, \underline{x})$, $\underline{\zeta} \in S^{m-1}$, belongs to $H_2^+(S^{m-1})^\perp = H_2^-(S^{m-1})$ and thus must have a left-monogenic extension to B^- vanishing at infinity. If $\underline{x} \in B^-$ then $\frac{\underline{x}}{|\underline{x}|^2} \in B^+$, making the function

$$C(\underline{y}, \underline{x}) = \frac{1}{a_m} \frac{\underline{y} - \frac{\underline{x}}{|\underline{x}|^2}}{\left|\underline{y} - \frac{\underline{x}}{|\underline{x}|^2}\right|^m} \frac{\underline{x}}{|\underline{x}|^m} = \frac{1}{a_m} \frac{\underline{y}\,\underline{x}+1}{|\underline{y}\,\underline{x}+1|^m}$$

left-monogenic in $\underline{y} \in B^-$. Moreover

$$\lim_{B^- \ni \underline{y} \to \underline{\xi}} C(\underline{y}, \underline{x}) = \frac{1}{a_m} \frac{\underline{\xi}\,\underline{x}+1}{|\underline{\xi}\,\underline{x}+1|^m} = C(\underline{\xi}, \underline{x}), \quad \underline{\xi} \in S^{m-1}.$$

This means that the left-monogenic extension to B^-, vanishing at infinity, of the Cauchy kernel $C(\underline{\zeta}, \underline{x})$, $\underline{x} \in B^-$, is precisely that particular function $C(\underline{y}, \underline{x})$:

$$\mathcal{C}[-C(\underline{\zeta}, \underline{x})](\underline{y}) = C(\underline{y}, \underline{x}) = \frac{1}{a_m} \frac{\underline{y}\,\underline{x}+1}{|\underline{y}\,\underline{x}+1|^m}, \qquad \underline{x} \in B^-.$$

Moreover the Cauchy kernel $C(\underline{\zeta}, \underline{x})$ with $\underline{x} \in B^-$ is, up to a minus sign, the reproducing kernel of $H_2^-(S^{m-1}) = H_2^+(S^{m-1})^\perp$. Indeed, take $h \in H_2^-(S^{m-1})$ and identify it with its left-monogenic extension $\mathcal{C}[-h] \in H_2(B^-)$. Then for $\underline{x} \in B^-$ it holds that

$$\mathcal{C}[-h](\underline{x}) = \langle C(\underline{\zeta}, \underline{x}), -h(\underline{\zeta})\rangle = \langle -C(\underline{\zeta}, \underline{x}), h(\underline{\zeta})\rangle.$$

(iii) Also the unit normal vector function $\nu(\underline{\zeta}) = \underline{\zeta}$, $\underline{\zeta} \in S^{m-1}$, belongs to $H_2^+(S^{m-1})^\perp$ (or $H_2^-(S^{m-1})$), and must have a left-monogenic extension to B^- vanishing at infinity. Clearly this extension is given by

$$\mathcal{C}[-\underline{\zeta}](\underline{x}) = \frac{\underline{x}}{|\underline{x}|^m}, \qquad \underline{x} \in B^-$$

while $\mathcal{C}[-\underline{\zeta}](\underline{x}) = 0$ for $\underline{x} \in B^+$. It follows that $\mathcal{C}^+[-\underline{\zeta}] = 0$, $\mathcal{C}^-[-\underline{\zeta}] = \underline{\zeta}$ and $H[-\underline{\zeta}] = \underline{\zeta}$. This result may also be written as

$$\frac{1}{a_m} \int_{S^{m-1}} \frac{\underline{x} - \underline{\zeta}}{|\underline{x} - \underline{\zeta}|^m} \, dS(\underline{\zeta}) = \frac{\underline{x}}{|\underline{x}|^m}, \qquad \underline{x} \in B^-,$$

(iv) The Hardy space $H_2^+(S^{m-1})$ and its orthogonal complement

$$H_2^+(S^{m-1})^\perp = \underline{\zeta}\, H_2^+(S^{m-1}) = H_2^-(S^{m-1})$$

are eigenspaces of the unitary operator H with respective eigenvalues $+1$ and -1.

(v) For a general domain Ω with C_∞-smooth boundary $\partial\Omega$ the Garabedian kernel, i.e., the reproducing kernel for the Hilbert space $H_2^+(\partial\Omega)^\perp$, is given for $\underline{x} \in \Omega^+ = \overset{\circ}{\Omega}$ by

$$L(\underline{\zeta}, \underline{x}) = \nu(\underline{\zeta})\, S(\underline{\zeta}, \underline{x}) = \mathbb{P}^\perp \left[\frac{1}{a_m} \frac{\underline{\zeta} - \underline{x}}{|\underline{\zeta} - \underline{x}|^m} \right]$$

where $\frac{1}{a_m} \frac{\underline{\zeta}-\underline{x}}{|\underline{\zeta}-\underline{x}|^m} = E(\underline{x} - \underline{\zeta}) = \nu(\underline{\zeta})\, C(\underline{\zeta}, \underline{x})$ belongs to $H_2^-(\partial\Omega)$. For the specific case of the unit sphere we obtain for $\underline{x} \in B^+$ that

$$\begin{aligned} L(\underline{\zeta}, \underline{x}) &= \underline{\zeta}\, S(\underline{\zeta}, \underline{x}) \;=\; \underline{\zeta}\, C(\underline{\zeta}, \underline{x}) \\ &= \frac{1}{a_m} \underline{\zeta} \frac{1 + \underline{\zeta}\, \underline{x}}{|1 + \underline{\zeta}\, \underline{x}|^m} \;=\; \frac{1}{a_m} \frac{\underline{\zeta} - \underline{x}}{|\underline{\zeta} - \underline{x}|^m} \;=\; E(\underline{x} - \underline{\zeta}) \end{aligned}$$

since now $E(\underline{x} - \underline{\zeta})$ belongs to $H_2^+(S^{m-1})^\perp = H_2^-(S^{m-1})$. It follows that a left-monogenic extension to B^+ of the Garabedian kernel is impossible. In fact we observe that the Garabedian kernel does have an extension to B^+, namely

$$L(\underline{y}, \underline{x}) = \frac{1}{a_m} \frac{\underline{y} - \underline{x}}{|\underline{y} - \underline{x}|^m}$$

however showing a pointwise singularity at $\underline{y} = \underline{x}$. Moreover, this extension is both left- and right-monogenic in $\underline{x} \in B^+ \setminus \{\underline{y}\}$ and shows the anti-symmetry property $L(\underline{y}, \underline{x}) = -L(\underline{x}, \underline{y})$.

5. The Dirichlet problem for S^{m-1}

According to the orthogonal direct sum decomposition (4.1) of $L_2(S^{m-1})$, a function $f \in L_2(S^{m-1})$ may be written as

$$f = \mathbb{P}[f] + \mathbb{P}^\perp[f] = \mathbb{P}[f] - \underline{\zeta}\, \mathbb{P}[\underline{\zeta}\, f] = -\underline{\zeta}\, \mathbb{P}^\perp[\underline{\zeta}\, f] + \mathbb{P}^\perp[f]$$

from which it follows that

$$H[f] = H[\mathbb{P}[f]] + H[\mathbb{P}^\perp[f]] = \mathbb{P}[f] - \mathbb{P}^\perp[f] = -\underline{\zeta}\, \mathbb{P}^\perp[\underline{\zeta}\, f] - \mathbb{P}^\perp[f] = \mathbb{P}[f] + \underline{\zeta}\, \mathbb{P}[\underline{\zeta}\, f]$$

confirming that

$$\begin{array}{ll} \mathcal{C}^+[f] = \mathbb{P}[f] = -\underline{\zeta}\,\mathbb{P}^\perp[\underline{\zeta}\,f] & \in H_2^+(S^{m-1}) = H_2^-(S^{m-1})^\perp \\ \mathcal{C}^-[f] = \underline{\zeta}\,\mathbb{P}[\underline{\zeta}f] = -\mathbb{P}^\perp[f] & \in H_2^+(S^{m-1})^\perp = H_2^-(S^{m-1}) \end{array}$$

(see also Figure 1).

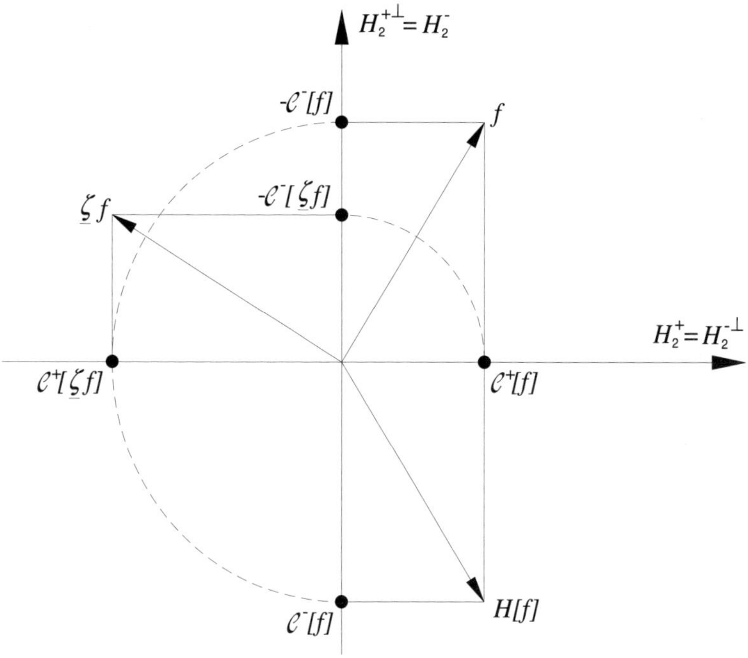

FIGURE 1.

Both $\mathbb{P}[f]$ and $\mathbb{P}[\underline{\zeta}f]$ have left-monogenic extensions to B^+, viz the functions $\mathcal{C}[\mathbb{P}[f]]$ and $\mathcal{C}[\mathbb{P}[\underline{\zeta}\,f]]$, respectively, which are both zero in B^-. Also $\mathbb{P}^\perp[f]$ and $\mathbb{P}^\perp[\underline{\zeta}\,f]]$ have left-monogenic extensions to B^- vanishing at infinity, viz the functions $\mathcal{C}[-\mathbb{P}^\perp[f]]$ and $\mathcal{C}[-\mathbb{P}^\perp[\underline{\zeta}\,f]]$ respectively, which moreover are zero in B^+. It follows that both f and $H[f]$ have harmonic extensions to B^+, viz the functions

$$\begin{aligned} f^+(\underline{x}) &= \mathcal{C}[\mathbb{P}[f]] - \underline{x}\,\mathcal{C}[\mathbb{P}[\underline{\zeta}\,f]] \\ H[f]^+(\underline{x}) &= \mathcal{C}[\mathbb{P}[f]] + \underline{x}\,\mathcal{C}[\mathbb{P}[\underline{\zeta}\,f]] \end{aligned}$$

which are zero in B^-. They also have harmonic extensions to B^- vanishing at infinity, viz the functions

$$\begin{aligned} f^-(\underline{x}) &= \underline{x}\,\mathcal{C}[\mathbb{P}^\perp[\underline{\zeta}\,f]] - \mathcal{C}[\mathbb{P}^\perp[f]] \\ H[f]^-(\underline{x}) &= \underline{x}\,\mathcal{C}[\mathbb{P}^\perp[\underline{\zeta}\,f]] + \mathcal{C}[\mathbb{P}^\perp[f]] \end{aligned}$$

which are zero in B^+.

By considering the appropriate linear combinations it is confirmed that:

(i) $\frac{1}{2}(f^+(\underline{x}) + H[f]^+(\underline{x})) = \mathcal{C}[\mathbb{P}[f]]$ is a left-monogenic extension to B^+ of the function $\mathcal{C}^+[f]$, which is zero in B^-;

(ii) $\frac{1}{2}(f^-(\underline{x}) - H[f]^-(\underline{x})) = \mathcal{C}[\mathbb{P}^\perp[f]]$ is a left-monogenic extension to B^- of the function $(-\mathcal{C}^-)[f]$, which is zero in B^+;

and it is found that

(iii) $\frac{1}{2}(-f^+(\underline{x}) + H[f]^+(\underline{x})) = \underline{x}\,\mathcal{C}[\mathbb{P}[\underline{\zeta}\, f]]$ is a harmonic extension to B^+ of the function $\mathcal{C}^-[f]$, which is zero in B^-;

(iv) $\frac{1}{2}(f^-(\underline{x}) + H[f]^-(\underline{x})) = \underline{x}\,\mathcal{C}[\mathbb{P}^\perp[\underline{\zeta}\, f]]$ is a harmonic extension to B^- of the function $\mathcal{C}^+[f]$, vanishing at infinity and being zero in B^+.

For those reasons the pairs $(f^+, H[f]^+)$ and $(f^-, -H[f]^-)$ may be called pairs of *harmonic conjugates* in B^+ and B^- respectively. This is nicely reflected in the fact that the Szegö and Garabedian kernels are the building blocks for the Poisson kernel and its conjugate in the unit ball. Indeed, as $f^+(\underline{x})$ is the harmonic extension of f to B^+, it may be obtained by the Poisson integral

$$f^+(\underline{x}) = \langle P(\underline{x}, \underline{\zeta}), f(\underline{\zeta})\rangle, \qquad \underline{x} \in B^+$$

where $P(\underline{x}, \underline{\zeta})$ denotes the Poisson kernel for the unit ball given by

$$P(\underline{x}, \underline{\zeta}) = \frac{1}{a_m} \frac{1 + \underline{x}^2}{|\underline{\zeta} - \underline{x}|^m}.$$

On the other hand we consecutively have

$$\begin{aligned} f^+(\underline{x}) = \mathcal{C}[\mathbb{P}[f]](\underline{x}) - \underline{x}\,\mathcal{C}[\mathbb{P}[\underline{\zeta}\, f]](\underline{x}) &= \langle C(\underline{\zeta}, \underline{x}), \mathbb{P}[f](\underline{\zeta})\rangle - \underline{x}\,\langle C(\underline{\zeta}, \underline{x}), \mathbb{P}[\underline{\zeta}\, f](\underline{\zeta})\rangle \\ &= \langle S(\underline{\zeta}, \underline{x}), f(\underline{\zeta})\rangle - \underline{x}\,\langle S(\underline{\zeta}, \underline{x}), \underline{\zeta}\, f(\underline{\zeta})\rangle \\ &= \langle S(\underline{\zeta}, \underline{x}), f(\underline{\zeta})\rangle + \underline{x}\,\langle \underline{\zeta}\, S(\underline{\zeta}, \underline{x}), f(\underline{\zeta})\rangle \;=\; \langle S(\underline{\zeta}, \underline{x}) - L(\underline{\zeta}, \underline{x})\,\underline{x}, f(\underline{\zeta})\rangle \end{aligned}$$

where the kernel function appearing is explicitly given by

$$S(\underline{\zeta}, \underline{x}) - L(\underline{\zeta}, \underline{x})\,\underline{x} = \frac{1}{a_m}\left(\frac{1 + \underline{\zeta}\,\underline{x}}{|1 + \underline{\zeta}\,\underline{x}|^m} - \frac{\underline{\zeta} - \underline{x}}{|\underline{\zeta} - \underline{x}|^m}\,\underline{x}\right) = \frac{1}{a_m} \frac{1 + \underline{x}^2}{|\underline{\zeta} - \underline{x}|^m}$$

in which the Poisson kernel for the unit ball is indeed recognized, or in other words:

$$P(\underline{x}, \underline{\zeta}) = S(\underline{\zeta}, \underline{x}) - L(\underline{\zeta}, \underline{x})\,\underline{x}, \qquad \underline{\zeta} \in S^{m-1},\ \underline{x} \in B^+.$$

In a similar way we obtain, still for $\underline{x} \in B^+$:

$$H[f]^+(\underline{x}) = \mathcal{C}[\mathbb{P}[f]](\underline{x}) + \underline{x}\,\mathcal{C}[\mathbb{P}[\underline{\zeta}\, f]](\underline{x}) = \langle S(\underline{\zeta}, \underline{x}) + L(\underline{\zeta}, \underline{x})\,\underline{x}, f(\underline{\zeta})\rangle$$

with

$$S(\underline{\zeta}, \underline{x}) + L(\underline{\zeta}, \underline{x})\,\underline{x} = \frac{1}{a_m} \frac{1 + 2\underline{\zeta}\,\underline{x} - \underline{x}^2}{|\underline{\zeta} - \underline{x}|^m} = \overline{Q(\underline{x}, \underline{\zeta})}$$

where the conjugate Poisson kernel is given by

$$Q(\underline{x}, \underline{\zeta}) = \frac{1}{a_m} \frac{1 - \underline{x}^2 + 2\underline{x}\,\underline{\zeta}}{|\underline{\zeta} - \underline{x}|^m}, \qquad \underline{\zeta} \in S^{m-1},\ \underline{x} \in B^+.$$

Summarizing we obtain, via the Szegö and Garabedian kernels, that for $\underline{x} \in B^+$

$$\begin{aligned} f^+(\underline{x}) &= \langle P(\underline{x}, \underline{\zeta}), f(\underline{\zeta}) \rangle = \mathcal{P}[f](\underline{x}) \\ H[f]^+(\underline{x}) &= \langle \overline{Q(\underline{x}, \underline{\zeta})}, f(\underline{\zeta}) \rangle = \mathcal{Q}[f](\underline{x}) \end{aligned}$$

confirming that $f^+ = \mathcal{P}[f]$ and $H[f]^+ = \mathcal{Q}[f]$ are harmonic extensions to B^+ of f and $H[f]$ respectively, for which moreover

$$\frac{1}{2} P(\underline{x}, \underline{\zeta}) + \frac{1}{2} \overline{Q(\underline{x}, \underline{\zeta})} = \frac{1}{a_m} \frac{1 + \underline{\zeta}\, \underline{x}}{|1 + \underline{\zeta}\, \underline{x}|^m} = C(\underline{\zeta}, \underline{x}), \quad \underline{\zeta} \in S^{m-1}, \ \underline{x} \in B^+$$

or, in terms of the Cauchy and the Poisson integrals:

$$\frac{1}{2} \mathcal{P}[f](\underline{x}) + \frac{1}{2} \mathcal{Q}[f](\underline{x}) = \mathcal{C}[f](\underline{x}), \qquad \underline{x} \in B^+$$

(see Figure 2).

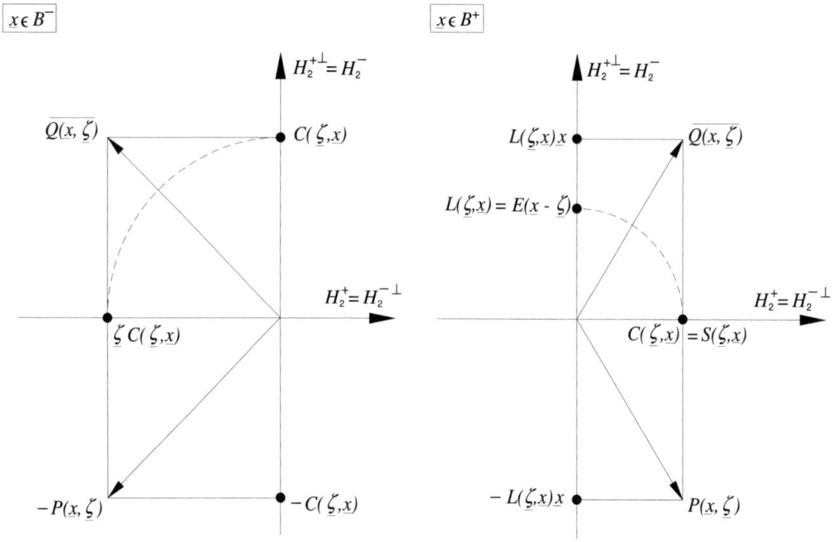

FIGURE 2.

In the same order of ideas we have for $\underline{x} \in B^-$ that

$$\begin{aligned} f^-(\underline{x}) &= \underline{x}\mathcal{C}[\mathbb{P}^\perp[\underline{\zeta}\, f]](\underline{x}) - \mathcal{C}[\mathbb{P}^\perp[f]](\underline{x}) \\ &= \underline{x} \langle C(\underline{\zeta}, \underline{x}), \mathbb{P}^\perp[\underline{\zeta}\, f](\underline{\zeta}) \rangle - \langle C(\underline{\zeta}, \underline{x}), \mathbb{P}^\perp[f](\underline{\zeta}) \rangle \\ &= \underline{x} \langle C(\underline{\zeta}, \underline{x}), \underline{\zeta}\, f(\underline{\zeta}) \rangle - \langle C(\underline{\zeta}, \underline{x}), f(\underline{\zeta}) \rangle = \langle \underline{\zeta}\, C(\underline{\zeta}, \underline{x})\, \underline{x} - C(\underline{\zeta}, \underline{x}), f(\underline{\zeta}) \rangle. \end{aligned}$$

The kernel function appearing here is found to be

$$\frac{1}{a_m} \left(\underline{\zeta} \frac{1 + \underline{\zeta}\, \underline{x}}{|1 + \underline{\zeta}\, \underline{x}|^m} \underline{x} - \frac{1 + \underline{\zeta}\, \underline{x}}{|1 + \underline{\zeta}\, \underline{x}|^m} \right) = -\frac{1}{a_m} \frac{1 + \underline{x}^2}{|1 + \underline{\zeta}\, \underline{x}|^m} = -P(\underline{x}, \underline{\zeta})$$

(see Figure 2) and hence

$$f^-(\underline{x}) = \langle -P(\underline{x},\underline{\zeta}), f(\underline{\zeta})\rangle, \qquad \underline{x} \in B^-.$$

Similarly we obtain, still for $\underline{x} \in B^-$, that

$$H[f]^-(\underline{x}) = \underline{x}\,\mathcal{C}[\mathbb{P}^\perp[\underline{\zeta} f]] + \mathcal{C}[\mathbb{P}^\perp[f]] = \langle \underline{\zeta}\, C(\underline{\zeta},\underline{x})\,\underline{x} + C(\underline{\zeta},\underline{x}), f(\underline{\zeta})\rangle$$

with

$$\underline{\zeta}\, C(\underline{\zeta},\underline{x})\,\underline{x} + C(\underline{\zeta},\underline{x}) = \frac{1}{a_m}\,\frac{1+2\underline{\zeta}\,\underline{x} - \underline{x}^2}{|1+\underline{\zeta}\underline{x}|^m} = \overline{Q(\underline{x},\underline{\zeta})}$$

and hence

$$H[f]^-(\underline{x}) = \langle \overline{Q(\underline{x},\underline{\zeta})}, f(\underline{\zeta})\rangle, \qquad \underline{x} \in B^-$$

(see also Figure 2).

Note that the above-mentioned concept of harmonic conjugate is built on the concept of the Hilbert transform H. Both concepts indeed are intimately related in the sense that either one of both implies the other notion. In fact we will give in Section 7 an alternative definition of harmonic conjugation which will lead to an alternative Hilbert transform on the unit sphere.

6. Fourier expansion

We will now characterize the Hardy spaces $H_2^\pm(S^{m-1})$ and the Hilbert transform H in terms of the Fourier expansion of a function $f \in L_2(S^{m-1})$. Initially one can expand such a function in spherical harmonics, but one of the strengths of Clifford analysis is that a refined series expansion in so-called spherical monogenics is possible.

A (solid) spherical monogenic of degree k is a (left-)monogenic homogeneous polynomial of degree k. An important result in Clifford analysis states that any (solid) spherical harmonic $S_k(\underline{x})$, i.e., any harmonic homogeneous polynomial of degree $k > 0$, may be decomposed as

$$S_k(\underline{x}) = P_k(\underline{x}) + \underline{x}\, R_{k-1}(\underline{x}) \tag{6.1}$$

where P_k and R_{k-1} both are spherical monogenics of degree k and $k-1$ respectively. In fact, given $S_k(\underline{x})$ $(k > 0)$, those components may be calculated via

$$R_{k-1}(\underline{x}) = -\frac{1}{m+2k-2}\,\partial S_k(\underline{x}).$$

When restricting the decomposition (6.1) to the unit sphere we obtain the corresponding decomposition of a (surface) spherical harmonic:

$$S_k(\underline{\zeta}) = P_k(\underline{\zeta}) + \underline{\zeta}\, R_{k-1}(\underline{\zeta}), \qquad \underline{\zeta} \in S^{m-1}. \tag{6.2}$$

Invoking the polar form of the Dirac operator

$$\underline{\partial} = \underline{\zeta}\,\partial_r - \frac{1}{r}\partial_{\underline{\zeta}} = \frac{1}{r}\,\underline{\zeta}\,(r\partial_r - \underline{\zeta}\,\partial_{\underline{\zeta}}) = \frac{1}{r}\,\underline{\zeta}\,(E + \Gamma)$$

where E and Γ are the scalar Euler operator and the bivector-valued spherical Dirac operator, respectively:

$$E = r\partial_r = \sum_{j=1}^{m} x_j\, \partial_{x_j}, \quad \Gamma = -\underline{\zeta}\, \partial_{\underline{\zeta}} = -\underline{\zeta} \wedge \partial_{\underline{\zeta}} = -\sum_{j<k} e_j e_k \,(x_j \partial_{x_k} - x_k\, \partial_{x_j})$$

the components of the decomposition (6.2) may be calculated as

$$\begin{aligned} P_k(\underline{\zeta}) &= \frac{1}{m+2k-2}\,(m+k-2-\Gamma)[S_k(\underline{\zeta})] \\ Q_{k-1}(\underline{\zeta}) &= \underline{\zeta}\, R_{k-1}(\underline{\zeta}) = \frac{1}{m+2k-2}\,(k+\Gamma)[S_k(\underline{\zeta})]. \end{aligned}$$

The first component $P_k(\underline{\zeta})$ is an eigenfunction of Γ with eigenvalue $-k$. It trivially has a left-monogenic extension $P_k(\underline{x})$ to B^+ obtained through multiplication by r^k; this extension thus clearly is a homogeneous polynomial of degree k. In terms of the Hardy spaces this observation is simply stated as $P_k(\underline{\zeta}) \in H_2^+(S^{m-1})$, and thus $P_k(\underline{\zeta}) = \mathbb{P}[S_k(\underline{\zeta})]$. The function $P_k(\underline{\zeta})$ is called an *inner spherical monogenic* of degree $k > 0$. The space of inner spherical monogenics of degree $k > 0$ is denoted by $\mathcal{M}_k^+$. The second component $Q_{k-1}(\underline{\zeta}) = \underline{\zeta}\, R_{k-1}(\underline{\zeta})$ is an eigenfunction of Γ with eigenvalue $m + k - 2$. It is obtained by multiplication of the inner spherical monogenic $R_{k-1}(\underline{\zeta}) \in H_2^+(S^{m-1})$ by the unit normal vector $\underline{\zeta}$, in other words: $Q_{k-1}(\underline{\zeta}) \in H_2^-(S^{m-1}) = H_2^+(S^{m-1})^\perp$, and thus it holds that $Q_{k-1}(\underline{\zeta}) = \mathbb{P}^\perp[S_k(\underline{\zeta})]$. Hence $Q_{k-1}(\underline{\zeta})$ does not admit a left-monogenic extension to B^+, but instead can be extended left-monogenically to B^- with limit zero at infinity. For that reason $Q_{k-1} = \underline{\zeta}\, R_{k-1}(\underline{\zeta})$ is called an *outer spherical monogenic* of degree $k-1$. Its left-monogenic extension to B^- is given by

$$\frac{1}{r^{m+k-2}}\,\underline{\zeta}\, R_{k-1}(\underline{\zeta}) = \frac{\underline{x}\, R_{k-1}(\underline{x})}{|\underline{x}|^{m+2k-2}}$$

which clearly is a homogeneous function of degree $-(m+k-2)$, which is left-monogenic in $\mathbb{R}^m \backslash \{0\}$ and vanishes at infinity, as it should. The space of outer spherical monogenics of degree $k-1$ is denoted by $\mathcal{M}_{k-1}^-$, where $\mathcal{M}_{k-1}^- = \underline{\zeta}\, \mathcal{M}_{k-1}^+$. This means that the space $\mathcal{H}_k$ of (surface) spherical harmonics of degree $k > 0$ may be decomposed as the direct sum $\mathcal{H}_k = \mathcal{M}_k^+ \oplus \mathcal{M}_{k-1}^- = \mathcal{M}_k^+ \oplus \underline{\zeta}\, \mathcal{M}_{k-1}^+$. Naturally for $k = 0$ one has that $\mathcal{H}_0 = \mathcal{M}_0^+ = \mathbb{R}$.

Now, given a function $f \in L_2(S^{m-1})$, it can be developed into its Fourier series of spherical harmonics

$$f(\underline{\zeta}) = \sum_{k=0}^{\infty} S_k[f](\underline{\zeta}), \qquad \underline{\zeta} \in S^{m-1} \tag{6.3}$$

where S_k denotes the projection operator from $L_2(S^{m-1})$ onto $\mathcal{H}_k$ given by

$$S_k[f](\underline{\zeta}) = \dim(\mathcal{H}_k)\,\frac{1}{a_m} \int_{S^{m-1}} P_{k,m}(\langle \underline{\zeta}, \underline{\xi} \rangle)\, f(\underline{\xi})\, dS(\underline{\xi})$$

with $P_{k,m}(t)$ the Legendre polynomials of degree k in dimension m. Subsequently f can be developed into spherical monogenics:

$$\begin{aligned} f(\underline{\zeta}) &= P_0[f] + \sum_{k=1}^{\infty} P_k[f](\underline{\zeta}) + Q_{k-1}[f](\underline{\zeta}) \\ &= S_0[f] + \sum_{k=1}^{\infty} \frac{(m+k-2-\Gamma)}{m+2k-2} S_k[f](\underline{\zeta}) + \frac{(k+\Gamma)}{m+2k-2} S_k[f](\underline{\zeta}). \end{aligned} \tag{6.4}$$

For all $f \in L_2(S^{m-1})$ it holds that $S_0[f](\underline{\zeta}) = P_0[f](\underline{\zeta}) = \text{constant}$, while P_k and Q_{k-1} stand for the projections of $L_2(S^{m-1})$ onto $\mathcal{M}_k^+$ and $\mathcal{M}_{k-1}^-$, respectively:

$$\begin{aligned} f \mapsto P_k[f](\underline{\zeta}) &= -\frac{1}{a_m} \underline{\zeta} \int_{S^{m-1}} \left\{ C_k^{\frac{m}{2}}(t)\, \underline{\zeta} - C_{k-1}^{\frac{m}{2}}(t)\, \underline{\omega} \right\} f(\underline{\omega})\, dS(\underline{\omega}) \\ f \mapsto Q_k[f](\underline{\zeta}) &= -\frac{1}{a_m} \underline{\zeta} \int_{S^{m-1}} \left\{ C_k^{\frac{m}{2}}(t)\, \underline{\zeta} - C_{k-1}^{\frac{m}{2}}(t)\, \underline{\omega} \right\} \underline{\omega}\, f(\underline{\omega})\, dS(\underline{\omega}) \end{aligned}$$

with $t = \langle \underline{\zeta}, \underline{\omega} \rangle$ and $C_k^{\frac{m}{2}}(t)$ the Gegenbauer polynomials of degree k in dimension m.

The Hardy spaces $H_2^{\pm}(S^{m-1})$ may thus be characterized by

$$\begin{aligned} \mathcal{M}^+ &= \sum_{k=0}^{\infty} \mathcal{M}_k^+ = H_2^+(S^{m-1}) \\ \mathcal{M}^- &= \sum_{k=1}^{\infty} \mathcal{M}_{k-1}^- = \bigoplus_{k=1}^{\infty} \underline{\zeta}\, \mathcal{M}_{k-1}^+ = \underline{\zeta}\, H_2^+(S^{m-1}) = H_2^+(S^{m-1})^{\perp} = H_2^-(S^{m-1}). \end{aligned}$$

Moreover the Hilbert transform of the function $f \in L_2(S^{m-1})$ is then given by

$$H[f](\underline{\zeta}) = P_0[f] + \sum_{k=1}^{\infty} P_k[f](\underline{\zeta}) - Q_{k-1}[f](\underline{\zeta}) \tag{6.5}$$

since for all k we have $P_k[f] \in H_2^+(S^{m-1})$ and $Q_{k-1}[f] \in H_2^-(S^{m-1})$. Note that this formula nicely illustrates the fact that $H^2 = \mathbf{1}$, as well as the particular examples $H[1] = 1$ and $H[\underline{\zeta}] = -\underline{\zeta}$. For the Szegö and Hardy projections of the function $f \in L_2(S^{m-1})$ with the expansion (6.4):

$$f(\underline{\zeta}) = P_0[f] + \sum_{k=1}^{\infty} P_k[f](\underline{\zeta}) + \underline{\zeta}\, R_{k-1}[f](\underline{\zeta}) \tag{6.6}$$

we thus obtain

$$\begin{aligned} \mathcal{C}^+[f] &= \mathbb{P}[f] = P_0[f] + \sum_{k=1}^{\infty} P_k[f](\underline{\zeta}) \\ -\mathcal{C}^-[f] &= \mathbb{P}^{\perp}[f] = \underline{\zeta} \sum_{k-1}^{\infty} R_{k-1}[f](\underline{\zeta}) \end{aligned}$$

and

$$\begin{aligned}\mathbb{P}[\underline{\zeta}\, f] &= -\sum_{k=1}^{\infty} R_{k-1}[f](\underline{\zeta}) \\ \mathbb{P}^{\perp}[\underline{\zeta}\, f] &= \underline{\zeta}\, P_0[f] + \sum_{k=1}^{\infty} \underline{\zeta}\, P_k[f](\underline{\zeta}).\end{aligned}$$

Each $P_k[f](\underline{\zeta})$ belongs to $H_2^+(S^{m-1})$, so its left-monogenic extension to B^+ results from the action of the Szegö kernel which is reproducing for $H_2^+(S^{m-1})$:

$$\langle S(\underline{\zeta}, \underline{x}), P_k[f](\underline{\zeta})\rangle = r^k\, P_k[f](\underline{\zeta}) = P_k[f](\underline{x}), \qquad \underline{x} \in B^+$$

or

$$\begin{aligned}P_k[f](\underline{x}) &= \frac{1}{a_m}\int_{S^{m-1}} \frac{1+\underline{x}\,\underline{\zeta}}{|1+\underline{x}\,\underline{\zeta}|^m}\, P_k[f](\underline{\zeta})\, dS(\underline{\zeta}), \qquad \underline{x} \in B^+ \\ &= \frac{1}{a_m}\int_{S^{m-1}} \frac{\underline{x}-\underline{\zeta}}{|\underline{x}-\underline{\zeta}|^m}\, d\sigma_{\underline{\zeta}}\, P_k[f](\underline{\zeta})\end{aligned}$$

which, naturally, is in accordance with Cauchy's Formula for the left-monogenic function $P_k[f](\underline{x})$. Each $\underline{\zeta}\, R_{k-1}(\underline{\zeta})$ belongs to $H_2^-(S^{m-1})$ which has the negative Cauchy kernel $-C(\underline{\zeta}, \underline{x})$, $\underline{x} \in B^-$, as its reproducing kernel. So

$$\langle -C(\underline{\zeta}, \underline{x}), \underline{\zeta}\, R_{k-1}(\underline{\zeta})\rangle = \frac{1}{r^{m+k-2}}\, \underline{\zeta}\, R_{k-1}(\underline{\zeta}), \qquad \underline{x} \in B^-$$

where at the right-hand side the left-monogenic extension of $\underline{\zeta}\, R_{k-1}(\underline{\zeta})$ to B^- appears. It follows that for $\underline{x} \in B^-$

$$\frac{1}{a_m}\int_{S^{m-1}} \frac{\underline{x}-\underline{\zeta}}{|\underline{x}-\underline{\zeta}|^m}\, R_{k-1}(\underline{\zeta})\, dS(\underline{\zeta}) = \frac{1}{r^{m+k-2}}\, \underline{\zeta}\, R_{k-1}(\underline{\zeta}) = \frac{\underline{x}\, R_{k-1}(\underline{x})}{r^{m+2k-2}}.$$

Note that in particular for $R_0(\underline{\zeta}) = 1$ we obtain

$$\frac{1}{a_m}\int_{S^{m-1}} \frac{\underline{x}-\underline{\zeta}}{|\underline{x}-\underline{\zeta}|^m}\, dS(\underline{\zeta}) = \frac{\underline{x}}{r^m}, \qquad \underline{x} \in B^-$$

confirming a result already mentioned in Section 4.

Remark 6.1. As mentioned above the function $f \in L_2(S^{m-1})$ with Fourier expansion (6.4) or alternatively (6.6) admits a harmonic extension to B^+; it is now seen that this extension is given by

$$\begin{aligned}f^+(\underline{x}) &= P_0[f] + \sum_{k=1}^{\infty} r^k P_k[f](\underline{\zeta}) + r^k\, \underline{\zeta}\, R_{k-1}[f](\underline{\zeta}) \\ &= P_0[f] + \sum_{k=1}^{\infty} P_k[f](\underline{x}) + \underline{x}\, R_{k-1}[f](\underline{x}).\end{aligned}$$

Moreover f also admits a harmonic extension to B^- vanishing at infinity which clearly is given by

$$\begin{aligned} f^-(\underline{x}) &= \frac{P_0[f]}{r^{m-2}} + \sum_{k=1}^{\infty} \frac{P_k[f](\underline{\zeta})}{r^{m+k-2}} + \frac{\underline{\zeta}}{r^{m+k-2}} R_{k-1}[f](\underline{\zeta}) \\ &= \frac{1}{r^{m-2}} P_0[f] + \sum_{k=1}^{\infty} \frac{1}{r^{m+2k-2}} P_k[f](\underline{x}) + \frac{\underline{x}}{r^{m+2k-2}} R_{k-1}[f](\underline{x}). \end{aligned}$$

Its Hilbert transform $H[f]$ with Fourier expansion (6.5) or alternatively

$$H[f](\underline{\zeta}) = P_0[f] + \sum_{k=1}^{\infty} P_k[f](\underline{\zeta}) - \underline{\zeta}\, R_{k-1}[f](\underline{\zeta}) \tag{6.7}$$

in its turn admits a harmonic extension to B^+ given by

$$\begin{aligned} H[f]^+(\underline{x}) &= P_0[f] + \sum_{k=1}^{\infty} r^k P_k[f](\underline{\zeta}) - r^k\, \underline{\zeta}\, R_{k-1}[f](\underline{\zeta}) \\ &= P_0[f] + \sum_{k=1}^{\infty} P_k[f](\underline{x}) - \underline{x}\, R_{k-1}[f](\underline{x}) \end{aligned}$$

as well as a harmonic extension to B^- vanishing at infinity given by

$$\begin{aligned} H[f]^-(\underline{x}) &= \frac{P_0[f]}{r^{m-2}} + \sum_{k=1}^{\infty} \frac{P_k[f](\underline{\zeta})}{r^{m+k-2}} - \frac{\underline{\zeta}}{r^{m+k-2}} R_{k-1}[f](\underline{\zeta}) \\ &= \frac{1}{r^{m-2}} P_0[f] + \sum_{k=1}^{\infty} \frac{1}{r^{m+2k-2}} P_k[f](\underline{x}) - \frac{\underline{x}}{r^{m+2k-2}} R_{k-1}[f](\underline{x}). \end{aligned}$$

It is now readily confirmed that $(f^+, H[f]^+)$ and $(f^-, -H[f]^-)$ are conjugate harmonic pairs in B^+ and B^- respectively, since

$$\begin{aligned} \frac{1}{2}\left(f^+(\underline{x}) + H[f]^+(\underline{x})\right) &= P_0[f] + \sum_{k=1}^{\infty} r^k P_k[f](\underline{\zeta}) \\ \frac{1}{2}\left(f^-(\underline{x}) - H[f]^-(\underline{x})\right) &= \sum_{k=1}^{\infty} \frac{1}{r^{m+k-2}}\, \underline{\zeta}\, R_{k-1}[f](\underline{\zeta}) \end{aligned}$$

are left-monogenic in B^+ and B^- respectively, the latter also vanishing at infinity. Note that

$$\mathbb{P}[f] = \frac{1}{2}\left(f + H[f]\right) = P_0[f] + \sum_{k=1}^{\infty} P_k[f]$$

while being left-monogenically extendable to B^+, only shows inner spherical monogenics in its Fourier expansion, or in other words: its outer spherical monogenic part is zero. Similarly

$$\mathbb{P}^{\perp}[f] = \frac{1}{2}(f - H[f]) = \sum_{k=1}^{\infty} Q_{k-1}[f]$$

while being left-monogenically extendable to B^-, only shows outer spherical monogenics in its Fourier expansion, or in other words: its inner spherical monogenic part is zero. For these reasons we may call $\mathbb{P}[f] \in H_2^+(S^{m-1})$ and $\mathbb{P}^{\perp}[f] \in H_2^-(S^{m-1})$ the analytic signal, respectively anti-analytic signal, associated to the original finite energy signal $f \in L_2(S^{m-1})$. As already mentioned in the introduction (Section 1) the concept of analytic signal is an important tool in signal analysis.

7. An alternative Hilbert transform

The Fourier series expansions (6.4) and (6.6) provide us with a transparent formalism to obtain the Hilbert transform $H[f]$ of a function $f \in L_2(S^{m-1})$. In the same order of ideas one could think of an alternative definition for a Hilbert transform by putting

$$H^{\text{alt}}[f](\underline{\zeta}) = a_0\, P_0[f](\underline{\zeta}) + \sum_{k=1}^{\infty} a_k P_k[f](\underline{\zeta}) + b_k Q_{k-1}[f](\underline{\zeta})$$

where the coefficients (a_0, a_k, b_k) have to be chosen in a meaningful way.

The approach followed in this section to devise a meaningful alternative Hilbert transform relies upon an alternative concept of harmonic conjugate, as was already announced at the end of Section 5. To this end, start with the decomposition (6.3) of a real-valued function $f \in L_2(S^{m-1})$ into spherical harmonics and observe that its unique harmonic extension to B^+ is given by

$$f^+(\underline{x}) = \mathcal{P}^+[f](\underline{x}) = \sum_{k=0}^{\infty} r^k S_k[f](\underline{\zeta}) = \sum_{k=0}^{\infty} S_k[f](\underline{x}).$$

The function

$$\mathcal{Q}^+[f](\underline{x}) = -\sum_{k=1}^{\infty} \frac{\Gamma(S_k[f])(\underline{x})}{m+k-2}$$

also is harmonic in B^+, and moreover $\mathcal{P}^+[f] + \mathcal{Q}^+[f]$ is left-monogenic in B^+ since

$$S_k[f] - \frac{1}{m+k-2}\Gamma(S_k[f]) = \frac{m+k-2-\Gamma}{m+k-2} S_k[f] = \frac{m+2k-2}{m+k-2} P_k[f].$$

The function $\mathcal{Q}^+[f](\underline{x})$ may thus be qualified as an alternative harmonic conjugate to $\mathcal{P}^+[f](\underline{x})$ in B^+; in the terminology of [6, 3] the function $\mathcal{Q}^+[f](\underline{x})$ is the so-called angular harmonic conjugate to $\mathcal{P}^+[f](\underline{x})$ in B^+. The $L_2(S^{m-1})$ NTBV of

$\mathcal{Q}^+[f](\underline{x})$ is then given by

$$\mathcal{Q}^+[f](\underline{\zeta}) = -\sum_{k=1}^{\infty} \frac{\Gamma(S_k[f](\underline{\zeta})}{m+k-2} = -\sum_{k=1}^{\infty} \frac{1}{m+k-2} \Gamma\left(P_k[f] + Q_{k-1}[f]\right)(\underline{\zeta})$$
$$= -\sum_{k=1}^{\infty} \frac{1}{m+k-2} \left((-k)P_k[f] + (m+k-2)Q_{k-1}[f]\right)(\underline{\zeta})$$
$$= \sum_{k=1}^{\infty} \frac{k}{m+k-2} P_k[f](\underline{\zeta}) - Q_{k-1}[f](\underline{\zeta}).$$

This leads to the following definition.

Definition 7.1. The interior Hilbert transform $H^+[f]$ of an L_2 function f on the unit sphere S^{m-1} is given in terms of the spherical monogenic decomposition of f by

$$H^+[f](\underline{\zeta}) = \sum_{k=1}^{\infty} \frac{k}{m+k-2} P_k[f](\underline{\zeta}) - Q_{k-1}[f](\underline{\zeta}). \tag{7.1}$$

The main properties of H^+ are summarized as follows.

Proposition 7.2.

(i) *The interior Hilbert transform $H^+ : L_2(S^{m-1}) \to L_2(S^{m-1})$ is a self-adjoint bounded linear operator.*

(ii) *The unique harmonic extensions to B^+ of $f \in L_2(S^{m-1})$ and of $H^+[f]$ are angular harmonic conjugates in B^+, adding up to a left-monogenic function in B^+.*

(iii) *The interior Hilbert transform of a constant function on S^{m-1} is zero.*

In a similar way we are able to construct an exterior Hilbert transform H^- on S^{m-1} based on the same concept of angular harmonic conjugate, however now in B^-. Starting point then is the unique harmonic function in B^-, vanishing at infinity, whose NTBV is the given function $f \in L_2(S^{m-1})$; this harmonic extension $\mathcal{P}^-[f](\underline{x})$ is easily obtained by the so-called *Kelvin inversion*, given by

$$\mathcal{K}[u(\underline{x})] = \mathcal{K}^{-1}[u(\underline{x})] = \frac{1}{|\underline{x}|^{m-2}} u(\frac{\underline{x}}{|\underline{x}|^2})$$

which transforms harmonic functions on B^+ into harmonic functions on B^- and vice versa. We readily obtain

$$\mathcal{P}^-[f](\underline{x}) = \sum_{k=0}^{\infty} \mathcal{K}\left[S_k[f]\right] = \sum_{k=0}^{\infty} r^{2-k-m} S_k[f](\underline{\zeta})$$
$$= \frac{1}{r^{m-2}} S_0[f](\underline{\zeta}) + \sum_{k=1}^{\infty} \frac{1}{r^{m+k-2}} S_k[f](\underline{\zeta})$$
$$= \frac{1}{r^{m-2}} P_0[f](\underline{\zeta}) + \sum_{k=1}^{\infty} \frac{1}{r^{m+k-2}} P_k[f](\underline{\zeta}) + \frac{1}{r^{m+k-2}} Q_{k-1}[f](\underline{\zeta}).$$

Its angular harmonic conjugate in B^- is then given by (see [6])

$$\mathcal{Q}^-[f](\underline{x}) = -\frac{1}{r^{m-2}} P_0[f](\underline{\zeta}) + \sum_{k=1}^{\infty} \frac{-1}{r^{m+k-2}} P_k[f](\underline{\zeta}) + \frac{m+k-2}{k} \frac{1}{r^{m+k-2}} Q_{k-1}[f](\underline{\zeta})$$

and indeed

$$\mathcal{P}^-[f](\underline{x}) + \mathcal{Q}^-[f](\underline{x}) = \sum_{k=1}^{\infty} \frac{m+2k-2}{k} \frac{1}{r^{m+k-2}} Q_{k-1}[f](\underline{\zeta}) = \sum_{k=1}^{\infty} \frac{m+k-2}{k} Q_{k-1}[f](\underline{x})$$

is left-monogenic in B^- and vanishes at infinity. For $r \to 1+$ the angular harmonic conjugate $\mathcal{Q}^-[f](\underline{x})$ tends to

$$\mathcal{Q}^-[f](\underline{\zeta}) = -P_0[f](\underline{\zeta}) + \sum_{k=1}^{\infty} -P_k[f](\underline{\zeta}) + \frac{m+k-2}{k} Q_{k-1}[f](\underline{\zeta})$$

which inspires the following definition.

Definition 7.3. The exterior Hilbert transform $H^-[f]$ of an L_2 function f on the unit sphere S^{m-1} is given in terms of the spherical monogenic decomposition of f by

$$H^-[f](\underline{\zeta}) = -P_0[f] + \sum_{k=1}^{\infty} -P_k[f](\underline{\zeta}) + \frac{m+k-2}{k} Q_{k-1}[f](\underline{\zeta}). \tag{7.2}$$

The main properties of H^- are summarized as follows.

Proposition 7.4.

(i) *The exterior Hilbert transform* $H^- : L_2(S^{m-1}) \to L_2(S^{m-1})$ *is a self-adjoint bounded linear operator.*

(ii) *The unique harmonic extensions to* B^-*, vanishing at infinity, of* $f \in L_2(S^{m-1})$ *and of* $H^-[f]$ *are angular harmonic conjugates in* B^-*, adding up to a left-monogenic function in* B^-*.*

(iii) *The exterior Hilbert transform of the constant function* 1 *on* S^{m-1} *is* -1.

However note that, unlike the Hilbert transform H from the previous section, the interior and exterior Hilbert transforms do *not* square to the identity operator.

Also note that the Fourier expansion of $F^{\pm} = f + H^{\pm}[f]$ only contains inner, respectively outer spherical monogenics:

$$\frac{1}{2}\left(f + H^+[f]\right) = \sum_{k=0}^{\infty} \frac{1}{2} \frac{m+2k-2}{m+k-2} P_k[f]$$

$$\frac{1}{2}\left(f + H^-[f]\right) = \sum_{k=1}^{\infty} \frac{1}{2} \frac{m+2k-2}{k} Q_{k-1}[f]$$

which implies that $F^\pm$ belong to $H_2^\pm(S^{m-1})$ and may be extended left-monogenically to $B^\pm$. In this way they may be considered as alternative analytic signals associated to the finite energy signal $f \in L_2(S^{m-1})$.

8. The complex plane

It is a known fact in Clifford analysis that simply putting the dimension m equal to 2 does, in general, not yield the corresponding result in the complex plane. In some cases the result from Clifford analysis even has no meaning for $m = 2$, in others it has to be reinterpreted in terms of the imaginary unit i, the Cauchy–Riemann operator, the notion of holomorphy, etc.

As is well known in the complex plane a (solid) spherical harmonic of degree k may be decomposed as

$$S_k(x, y) = P_k(z) + \overline{z}\, R_{k-1}(\overline{z}) \tag{8.1}$$

where $P_k(z)$ is a holomorphic homogeneous polynomial of degree k and $R_{k-1}(\overline{z})$ is an anti-holomorphic homogeneous polynomial of degree $k - 1$, explicitly given by $R_{k-1}(\overline{z}) = \frac{1}{2}\,(\partial_x + i\partial_y)[S_k]$. A spherical harmonic of degree k on the unit circle S^1 is then obtained by taking restriction in (8.1), yielding

$$S_k(\zeta) = P_k(\zeta) + \overline{\zeta}\, R_{k-1}(\overline{\zeta}), \qquad \zeta \in S^1,$$

or, more explicitly

$$S_k(\zeta) = \lambda_k\, \zeta^k + \mu_k\, \overline{\zeta}^k, \qquad \lambda_k, \mu_k \in \mathbb{C}.$$

For a function $f \in L_2(S^1)$ we then obtain

$$f(\zeta) = \lambda_0 + \sum_{k=1}^{\infty} \lambda_k\, \zeta^k + \mu_k\, \frac{1}{\zeta^k}$$

from which it easily follows that its harmonic extensions to B^+ and B^- are respectively given by

$$f^+(x, y) = \lambda_0 + \sum_{k=1}^{\infty} \lambda_k\, z^k + \mu_k\, \overline{z}^k, \quad f^-(x, y) = \lambda_0 + \sum_{k=1}^{\infty} \lambda_k\, \frac{1}{\overline{z}^k} + \mu_k\, \frac{1}{z^k}.$$

According to (6.5) its Hilbert transform $H[f]$ is then given by

$$H[f](\zeta) = \lambda_0 + \sum_{k=1}^{\infty} \lambda_k\, \zeta^k - \mu_k\, \frac{1}{\zeta^k}$$

which has harmonic extensions to B^+ and B^-, respectively given by

$$H[f]^+(x, y) = \lambda_0 + \sum_{k=1}^{\infty} \lambda_k\, z^k - \mu_k\, \overline{z}^k, \quad H[f]^-(x, y) = \lambda_0 + \sum_{k=1}^{\infty} \lambda_k\, \frac{1}{\overline{z}^k} - \mu_k\, \frac{1}{z^k}.$$

In particular we obtain

$$H[1] = 1; \quad H[\zeta^k] = \zeta^k, \quad H[\frac{1}{\zeta^k}] = -\frac{1}{\zeta^k}, \quad k = 1, 2, \ldots$$

confirming that ζ^k, $k = 0, 1, 2, \ldots$ belong to $H_2^+(S^1)$, while $\frac{1}{\zeta^k} = \overline{\zeta}^k$, $k = 1, 2, \ldots$ belong to $H_2^-(S^1) = H_2^+(S^1)^\perp$. As expected we observe that $(f^+, H[f]^+)$ and $(f^-, -H[f]^-)$ are harmonic conjugate pairs in B^+ and B^- respectively, since

$$\frac{1}{2}\left(f^+ + H[f]^+\right)(x, y) = \lambda_0 + \sum_{k=1}^{\infty} \lambda_k z^k,$$

$$\frac{1}{2}\left(f^- - H[f]^-\right)(x, y) = \sum_{k=1}^{\infty} \mu_k \frac{1}{z^k}$$

are holomorphic in respectively B^+ and B^-.

Similarly, according to (7.1) and (7.2) the interior and exterior Hilbert transforms of $f \in L_2(S^1)$ are given by

$$\begin{aligned} H^+[f](\zeta) &= \sum_{k=1}^{\infty} P_k[f](\zeta) - Q_{k-1}[f](\zeta) \\ &= \sum_{k=1}^{\infty} \lambda_k \zeta^k - \mu_k \frac{1}{\zeta^k} \end{aligned}$$

$$\begin{aligned} H^-[f](\zeta) &= -P_0[f] + \sum_{k=1}^{\infty} -P_k[f](\zeta) + Q_{k-1}[f](\zeta) \\ &= -\lambda_0 + \sum_{k=1}^{\infty} (-\lambda_k)\zeta^k + \mu_k \frac{1}{\zeta^k}. \end{aligned}$$

Observe that

$$H^+[1] = 0; \quad H^+[\zeta^k] = \zeta^k, \quad H^+[\frac{1}{\zeta^k}] = -\frac{1}{\zeta^k}, \quad k = 1, 2, \ldots$$

which means that, up to the imaginary unit i, the interior Hilbert transform H^+ on S^1 coincides with the Hilbert transform $\widetilde{H}$ defined in [33, p. 225]:

$$\widetilde{H}[1] = 0; \quad \widetilde{H}[\exp(ik\theta)] = \frac{1}{i}\,\mathrm{sgn}(k)\,\exp(ik\theta), \; k = 1, 2, \ldots.$$

Also note that

$$H^-[1] = -1; \quad H^-[\zeta^k] = -\zeta^k, \quad H^-[\frac{1}{\zeta^k}] = \frac{1}{\zeta^k}, \quad k = 1, 2, \ldots$$

and that $(H^+ + H^-)[f] = 0$ if and only if f shows a vanishing constant component λ_0.

9. Conclusions

In this paper we have presented an in-depth study of the Hilbert transform on the unit sphere S^{m-1} in Euclidean space $\mathbb{R}^m$. At the same time we have illustrated that there is an intimate relationship between the Hilbert transform on the one side and the concept of conjugate harmonic functions on the other. In Sections 3–6 we have treated the Hilbert transform defined as a part of the NTBVs of the Cauchy integral of an L_2 function on S^{m-1}. It was then shown that the Poisson transform of both the function and its Hilbert transform add up to a left-monogenic function in the unit ball, giving rise to a specific notion of harmonic conjugate. In Section 7 we have proceeded the other way around. Starting from an alternative concept of conjugate harmonicity, the so-called angular conjugate harmonicity, we have defined an inner and an outer Hilbert transform on S^{m-1} as the NTBVs of the angular harmonic conjugate to the Poisson transform of the given $L_2(S^{m-1})$ function. Both approaches illustrate that neither the Hilbert transform on the unit sphere, nor conjugate harmonicity in the unit ball or its exterior, are uniquely defined concepts, but the definition of either of both entails the other.

References

[1] S. Bernstein, L. Lanzani, Szegö projections for Hardy spaces of monogenic functions and applications, *IJMMS* **29(10)**, 2002, 613–624.

[2] S.R. Bell, *The Cauchy Transform, Potential Theory, and Conformal Mapping*, CRC Press (Boca Raton–Ann Arbor–London–Tokyo, 1992).

[3] F. Brackx, B. De Knock, H. De Schepper, D. Eelbode, On the interplay between the Hilbert transform and conjugate harmonic functions, *Math. Meth. Appl. Sci.* **29(12)**, 2006, 1435–1450.

[4] F. Brackx, R. Delanghe, F. Sommen, *Clifford Analysis*, Pitman Advanced Publishing Program (Boston–London–Melbourne, 1982).

[5] F. Brackx, H. De Schepper, The Hilbert Transform on a Smooth Closed Hypersurface, *CUBO, A Mathematical Journal* **10(2)**, 2008, 83–106.

[6] F. Brackx, H. De Schepper, D. Eelbode, A new Hilbert transform on the unit sphere in $\mathbb{R}^m$, *Complex Var. Elliptic Equ.* **51(5–6)**, 2006, 453–462.

[7] F. Brackx, N. Van Acker, H^p spaces of monogenic functions. In: A. Micali et al. (eds.), *Clifford Algebras and their Applications in Mathematical Physics*, Kluwer Academic Publishers (Dordrecht, 1992), 177–188.

[8] P. Calderbank, *Clifford analysis for Dirac operators on manifolds-with-boundary*, Max Planck–Institut für Mathematik (Bonn, 1996).

[9] J. Cnops, *An introduction to Dirac operators on manifolds*, Birkhäuser Verlag (Basel, 2002).

[10] R. Delanghe, Some remarks on the principal value kernel in $\mathbb{R}^m$, *Complex Var. Theory Appl.* **47**, 2002, 653–662.

[11] R. Delanghe, On the Hardy spaces of harmonic and monogenic functions in the unit ball of $\mathbb{R}^{m+1}$. In: *Acoustics, mechanics and the related topics of mathematical analysis*, World Scientific Publishing (River Edge, New Jersey, 2002), 137–142.

[12] R. Delanghe, On Some Properties of the Hilbert Transform in Euclidean Space, *Bull. Belg. Math. Soc. – Simon Stevin* **11**, 2004, pp. 163–180.

[13] R. Delanghe, F. Sommen, V. Souček, *Clifford Algebra and Spinor–Valued Functions*, Kluwer Academic Publishers (Dordrecht–Boston–London, 1992).

[14] J. Gilbert, M. Murray, *Clifford Algebras and Dirac Operators in Harmonic Analysis*, Cambridge University Press (Cambridge, 1991).

[15] K. Gürlebeck, W. Sprößig, *Quaternionic analysis and elliptic boundary value problems*, Birkhäuser Verlag (Basel, 1990).

[16] K. Gürlebeck, W. Sprößig, *Quaternionic and Clifford Calculus for Physicists and Engineers*, Wiley (Chichester, 1998).

[17] K. Gürlebeck, K. Habetha and W. Sprößig, *Funktionentheorie in der Ebene und im Raum*, Birkhäuser Verlag (Basel, 2006).

[18] S.L. Hahn, *Hilbert Transforms in Signal Processing*, Artech House (Boston–London, 1996).

[19] S. Hofmann, E. Marmolejo Olea, M. Mitrea, S. Pérez Esteva, M. Taylor, Hardy Spaces, Singular Integrals and the Geometry of Euclidean Domains of Locally Finite Perimeter, to appear.

[20] V.V. Kravchenko, M.V. Shapiro, *Integral Representations for Spatial Models of Mathematical Physics*, Pitman Research Notes in Mathematics Series 351, Longman Scientific and Technical (Harlow, 1996).

[21] C. Li, A. McIntosh, T. Qian, Clifford algebras, Fourier transforms and singular convolution operators on Lipschitz surfaces, *Rev. Math. Iberoamer.* **10**, 1994, 665–721.

[22] C. Li, A. McIntosh, S. Semmes, Convolution singular integrals on Lipschitz surfaces, *J. Amer. Math. Soc.* **5**, 1992, 455–481.

[23] A. McIntosh, Fourier theory, singular integrals, and harmonic functions on Lipschitz domains. In: J. Ryan (ed.), *Clifford Algebras in Analysis and Related Topics*, Studies in Advanced Mathematics, CRC Press (Boca Raton, 1996), 33–87.

[24] S.G. Mikhlin, *Mathematical Physics, an Advanced course*, North–Holland Publ. Co. (Amsterdam–London, 1970).

[25] M. Mitrea, *Clifford Wavelets, Singular Integrals and Hardy Spaces*, Lecture Notes in Mathematics 1575, Springer–Verlag (Berlin, 1994).

[26] M. Murray, The Cauchy integral, Calderon commutation, and conjugation of singular integrals in $\mathbb{R}^n$, *Trans. of the AMS* **298**, 1985, 497–518.

[27] T. Qian et al. (eds.), *Advances in analysis and geometry: new developments using Clifford algebras*, Birkhäuser Verlag (Basel–Boston–Berlin, 2004).

[28] J. Ryan (ed.), *Clifford Algebras in Analysis and Related Topics*, Studies in Advanced Mathematics, CRC Press (Boca Raton, 1996).

[29] J. Ryan, Basic Clifford Analysis, *CUBO, A Mathematical Journal* **2**, 2000, 226–256.

[30] J. Ryan, Clifford Analysis. In: R. Abłamowicz and G. Sobczyk (eds.), *Lectures on Clifford (Geometric) Algebras and Applications*, Birkhäuser (Boston–Basel–Berlin, 2004), 53–89.

[31] J. Ryan, D. Struppa (eds.), *Dirac operators in analysis*, Addison Wesley Longman Ltd. (Harlow, 1998).

[32] M.V. Shapiro, N.L. Vasilevski, Quaternionic ψ-holomorphic functions, singular integral operators and boundary value problems, Parts I and II, *Complex Var. Theory Appl.* **27**, 1995, 14–46 and 67–96.

[33] E.M. Stein, *Harmonic Analysis: Real-Variable Methods, Orthogonality, and Oscillatory Integrals*, Princeton University Press (Princeton, 1993).

[34] P. Van Lancker, The Kerzman-Stein Theorem on the Sphere, *Complex Var. Theory and Appl.* **45(1)**, 2001, 73-99.

F. Brackx and H. De Schepper
Clifford Research Group
Faculty of Engineering
Ghent University
Galglaan 2
9000 Gent, Belgium
e-mail: Freddy.Brackx@UGent.be
Hennie.DeSchepper@UGent.be

Quaternionic and Clifford Analysis
Trends in Mathematics, 37–53

Discrete Clifford Analysis: A Germ of Function Theory

F. Brackx, H. De Schepper, F. Sommen and L. Van de Voorde

Abstract. We develop a discrete version of Clifford analysis, i.e., a higher-dimensional discrete function theory in a Clifford algebra context. On the simplest of all graphs, the rectangular $\mathbb{Z}^m$ grid, the concept of a discrete monogenic function is introduced. To this end new Clifford bases are considered, involving so-called forward and backward basis vectors, controlling the support of the involved operators. Following a proper definition of a discrete Dirac operator and of some topological concepts, function theoretic results amongst which Stokes' theorem, Cauchy's theorem and a Cauchy integral formula are established.

Mathematics Subject Classification (2000). Primary 30G35.

Keywords. Discrete Clifford analysis, discrete function theory, discrete Cauchy integral formula.

1. Elements of continuous Clifford analysis

Clifford analysis (see, e.g., [3, 6, 13, 16, 17]) is a higher-dimensional function theory centred around the notion of monogenic functions, which are usually considered as higher-dimensional analogues of holomorphic functions in the complex plane. The underlying framework is introduced by endowing m-dimensional Euclidean space $\mathbb{R}^{0,m}$ with a non-degenerate quadratic form of signature $(0, m)$, and considering the corresponding orthonormal basis $(e_1, \ldots, e_m)$. Then $\mathbb{R}_{0,m}$ denotes the real Clifford algebra constructed over $\mathbb{R}^{0,m}$, see, e.g., [21]. The non-commutative multiplication in $\mathbb{R}_{0,m}$ is governed by $e_j e_k + e_k e_j = -2\delta_{jk}$, $j, k = 1, \ldots, m$. A basis for $\mathbb{R}_{0,m}$ is then obtained by considering for each set $A = \{j_1, \ldots, j_h\} \subset \{1, \ldots, m\}$ the element $e_A = e_{j_1} \ldots e_{j_h}$, with $1 \leq j_1 < j_2 < \cdots < j_h \leq m$. For the empty set $\emptyset$ one puts $e_\emptyset = 1$, the identity element. Any Clifford number a in $\mathbb{R}_{0,m}$ may thus be written as $a = \sum_A e_A a_A$, $a_A \in \mathbb{R}$. When allowing for complex constants, the same set of generators $(e_1, \ldots, e_m)$ also produces the complex Clifford algebra $\mathbb{C}_m$, as well as all real Clifford algebras $\mathbb{R}_{p,q}$ of any signature $(p + q = m)$.

The Euclidean space $\mathbb{R}^{0,m}$ is embedded in $\mathbb{R}_{0,m}$ by identifying $(x_1,\ldots,x_m)$ with the Clifford vector $\underline{x} = \sum_{j=1}^{m} e_j x_j$. The multiplication of two vectors $\underline{x}$ and $\underline{y}$ is given by $\underline{x}\,\underline{y} = \underline{x} \bullet \underline{y} + \underline{x} \wedge \underline{y}$ with

$$\begin{aligned} \underline{x} \bullet \underline{y} &= -\sum_{j=1}^{m} x_j y_j = \frac{1}{2}(\underline{x}\,\underline{y} + \underline{y}\underline{x}) \\ \underline{x} \wedge \underline{y} &= \sum_{i<j} e_{ij}(x_i y_j - x_j y_i) = \frac{1}{2}(\underline{x}\,\underline{y} - \underline{y}\underline{x}) \end{aligned}$$

being the scalar-valued dot product (equalling the Euclidean inner product up to a minus sign) and the bivector-valued wedge product, respectively. The square of a vector $\underline{x}$ is scalar valued and equals the norm squared up to a minus sign: $\underline{x}^2 = -\langle \underline{x}, \underline{x} \rangle = -|\underline{x}|^2$. Conjugation in $\mathbb{R}_{0,m}$ is defined as the anti-involution for which $\bar{e}_j = -e_j$, $j = 1,\ldots,m$. In particular for a vector $\underline{x}$ we have $\bar{\underline{x}} = -\underline{x}$.

The Fourier dual of the vector $\underline{x}$ is the vector-valued first-order differential operator $\partial_{\underline{x}} = \sum_{j=1}^{m} e_j \partial_{x_j}$, called Dirac operator. A function f defined and differentiable in an open region Ω of $\mathbb{R}^m$ and taking values in $\mathbb{R}_{0,m}$ is called left-monogenic in Ω if $\partial_{\underline{x}}[f] = 0$. Since the Dirac operator factorizes the Laplacian, $\Delta = -\partial_{\underline{x}}^2$, monogenicity may be regarded as a refinement of harmonicity. The fundamental group leaving the Dirac operator $\partial_{\underline{x}}$ invariant is the special orthogonal group $\mathrm{SO}(m)$, doubly covered by the $\mathrm{Spin}(m)$ group of the Clifford algebra $\mathbb{R}_{0,m}$, leading to the Dirac operator being called a rotation invariant operator. In the present context, we will refer to this setting as the continuous case, as opposed to the discrete framework treated in this paper.

Recently, several authors have shown interest in developing an appropriate framework for discrete counterparts of the basic notions and concepts of Clifford analysis, see a.o. [14, 15, 8, 9, 10, 11], this interest being triggered by the need for an adequate numerical treatment of problems from potential theory and of boundary value problems, see also [16, 17]. This paper aims at contributing to the further development of the corresponding discrete function theory and its fundamental results, centred around a suitably defined discrete Dirac operator. In view of the above-mentioned connection between continuous Clifford analysis and complex analysis in the plane, special attention should be paid to the important property of the discrete Dirac operator factorizing a discrete Laplacian. This also was the case in the study of holomorphic functions on $\mathbb{Z}^2$, see, e.g., [12, 18, 7] and, more recently [19, 20].

2. Definition of a discrete Dirac operator

As the basis for the development of our theory, we will consider the natural graph corresponding to the equidistant grid $\mathbb{Z}^m$; thus a Clifford vector $\underline{x}$ as introduced above will now only show integer co-ordinates. For the pointwise discretization of the partial derivatives $\frac{\partial}{\partial x_j}$ we then introduce the traditional one-sided forward and backward differences, respectively given by

$$\begin{aligned}\Delta_j^+[f](\underline{x}) &= f(\ldots, x_j + 1, \ldots) - f(\ldots, x_j, \ldots) \\ &= f(\underline{x} + e_j) - f(\underline{x}), \quad j = 1, \ldots, m \\ \Delta_j^-[f](\underline{x}) &= f(\ldots, x_j, \ldots) - f(\ldots, x_j - 1, \ldots) \\ &= f(\underline{x}) - f(\underline{x} - e_j), \quad j = 1, \ldots, m.\end{aligned}$$

With respect to the $\mathbb{Z}^m$ neighbourhood of $\underline{x}$, the usual definition of the discrete Laplacian explicitly reads

$$\Delta^*[f](\underline{x}) = \sum_{j=1}^{m} \left[\Delta_j^+[f](\underline{x}) - \Delta_j^-[f](\underline{x})\right] = \sum_{j=1}^{m} [f(\underline{x} + e_j) + f(\underline{x} - e_j)] - 2mf(\underline{x})$$

the notation Δ^* referring to this operator being called the "star Laplacian"; it involves the values of the considered function at the midpoints of the faces of the unit cube centred at $\underline{x}$. Note that it can also be written as

$$\Delta^*[f](\underline{x}) = \sum_{j=1}^{m} \Delta_j^+ \Delta_j^-[f](\underline{x}) = \sum_{j=1}^{m} \Delta_j^- \Delta_j^+[f](\underline{x}).$$

When passing to the Dirac operator, we will combine each difference, forward or backward, with corresponding forward or backward basis vectors e_j^+ and e_j^-, $j = 1, \ldots, m$, carrying an orientation as well. To this end, we need to embed the Clifford algebra $\mathbb{R}_{0,m}$ into a bigger one, with an underlying vector space of the double dimension, e.g., the complex Clifford algebra $\mathbb{C}_{2m}$, where we consider those $2m$ basis vectors, submitting to the following three assumptions.

Assumption 2.1. The forward and the backward basis vector in each particular cartesian direction add up to the traditional basis vector in that direction, i.e., $e_j^+ + e_j^- = e_j$, $j = 1, \ldots, m$.

Assumption 2.2. There are no preferential cartesian directions, or: all cartesian directions play the same role in the metric. This assumption may be seen as a rotational invariance.

Assumption 2.3. The positive and negative orientations of any cartesian direction play an equivalent role. This assumption may be seen as a reflection invariance.

Starting from these assumptions, direct calculations lead to the following multiplication rules, see, e.g., [11]:

- $e_j^+ e_k^+ + e_k^+ e_j^+ = e_j^- e_k^- + e_k^- e_j^- = -2g$, $j \neq k$
- $e_j^+ e_k^- + e_k^- e_j^+ = 2g$, $j \neq k$
- $(e_j^+)^2 = (e_j^-)^2 = -\lambda$, $j = 1, \ldots, m$
- $e_j^+ e_j^- + e_j^- e_j^+ = 2\lambda - 1$, $j = 1, \ldots, m$.

The definition of a discrete Dirac operator in this setting may now be given.

Definition 2.4. The discrete Dirac operator ∂ is the first-order, Clifford vector-valued difference operator given by $\partial = \partial^+ + \partial^-$ where the forward and backward discrete Dirac operators ∂^+ and ∂^- are respectively given by $\partial^+ = \sum_{j=1}^m e_j^+ \Delta_j^+$ and $\partial^- = \sum_{j=1}^m e_j^- \Delta_j^-$.

If, in addition, we require the support of ∂^2 to remain contained in at most the unit cube centred at $\underline{x}$, the isotropy of the forward and backward basis vectors needs to be imposed, i.e., we have to put $\lambda = (e_j^+)^2 = (e_j^-)^2 = 0$, whence it follows in addition that $2\lambda - 1 = -1$. One thus finally arrives at

- $e_j^+ e_k^+ + e_k^+ e_j^+ = e_j^- e_k^- + e_k^- e_j^- = -2g$, $j \neq k$
- $e_j^+ e_k^- + e_k^- e_j^+ = 2g$, $j \neq k$
- $(e_j^+)^2 = (e_j^-)^2 = 0$, $j = 1, \ldots, m$
- $e_j^+ e_j^- + e_j^- e_j^+ = -1$, $j = 1, \ldots, m$.

These relations completely determine the metric of the underlying $2m$-dimensional space in terms of one free scalar parameter g. Note that this is the direct consequence of the above assumptions that no cartesian direction or orientation is preferential above the others. This immediately reduces the number of degrees of freedom and eventually implies in a natural way the use of a constant metric, thus simplifying the framework considerably. Clearly, one may also consider other possibilities for the metric, such as the anisotropic or metrodynamical case, see, e.g., [4] for an extensive treatment in the continuous setting. However, such a framework is not under consideration at the moment, as it would make the derivation of the desired function theoretic results more cumbersome.

Returning to the present setting, if we require the metric to be non-degenerate, i.e., its determinant to be non-zero, we will need to impose in addition that $g \neq -\frac{1}{4}$ and $g \neq \frac{1}{4(m-1)}$. With the above multiplication rules, ∂^2 takes the form

$$\partial^2 = (4(m-1)g - 1)\Delta^* - 2g \sum_{j<k} \widetilde{\Delta}_{jk} \tag{2.1}$$

where, for $j < k$, we have denoted

$$\widetilde{\Delta}_{jk}[f] = f(\underline{x} + e_j + e_k) + f(\underline{x} + e_j - e_k) + f(\underline{x} - e_j + e_k) + f(\underline{x} - e_j - e_k) - 4f(\underline{x})$$

each $\widetilde{\Delta}_{jk}$ thus being interpretable as a "cross Laplacian" on the corresponding (e_j, e_k) plane, see also [11]. Note however that the grid points involved in these additional terms do not respect the neighbourhood of the vertex $\underline{x}$ in the originally chosen $\mathbb{Z}^m$ graph, but involve the vertices of the unit cube centred at $\underline{x}$.

Note that, in the special case where $g = 0$, the additional terms in (2.1) drop, whence we are left with a true factorization of the star Laplacian, i.e., $\partial^2 = -\Delta^*$, the support of the involved operators now respecting the $\mathbb{Z}^m$ neighbourhood of $\underline{x}$. As has been remarked in [10], there is a well-known model for these particular forward and backward vectors, namely the so-called Witt basis of the Clifford algebra

$\mathbb{C}_{2m}$, see, e.g., [1, 2, 5, 22, 23] for the continuous setting, nowadays called Hermitian Clifford analysis. The corresponding discrete setting was already mentioned in [20], however without any function theoretic aims.

Also in the general case where the metric scalar g does not equal zero, a feasible model for the forward and backward Clifford vectors can be given, in terms of so-called curvature vectors B_j, $j = 1, \ldots, m$. We put

$$e_j^+ = \frac{1}{2}(e_j + B_j) \quad \text{and} \quad e_j^- = \frac{1}{2}(e_j - B_j), \qquad j = 1, \ldots, m$$

ensuring that $e_j^+ + e_j^- = e_j$, $j = 1, \ldots, m$. In order to satisfy the above multiplication relations, we moreover have to require that

- $B_j^2 = +1$, $j = 1, \ldots, m$;
- $\{e_j, B_j\} = 2(e_j \bullet B_j) = 0$, $j = 1, \ldots, m$;
- $\{e_k, B_j\} = 2(e_k \bullet B_j) = 0$, $j, k = 1, \ldots, m$, $j \neq k$;
- $\{B_k, B_j\} = 2(B_k \bullet B_j) = -8g$, $j, k = 1, \ldots, m$, $j \neq k$.

Note that the space spanned by the curvature vectors and the original m-dimensional space with basis $(e_1, \ldots, e_m)$ are mutually orthogonal. The set $(B_1, \ldots, B_m)$ may be interpreted as an "umbrella" of vectors on the unit sphere S^{m-1} of $\mathbb{R}^m$, containing two by two the same fixed angle α, with $\cos(\alpha) = -4g$. However, in order for this to be possible, the metric scalar g needs to be restricted to the interval $]-\frac{1}{4}, \frac{1}{4}]$, while still the value $\frac{1}{4(m-1)}$ needs to be excluded. In particular, if $g = 0$ then $\alpha = \frac{\pi}{2}$, in agreement with the Witt case above. Still observe that, if (2.1) is to be interpreted as a genuinely similar result to the continuous factorization $\partial_{\underline{x}}^2 = -\Delta$, then we should in fact further restrict g to the range $[0, \frac{1}{4(m-1)}[$.

It is our aim to extend this first model in a forthcoming paper, introducing generalized curvature tensors which control the support of all involved operators, in particular of alternative Dirac operators.

3. Discrete monogenic functions

In order to define discrete monogenicity, we first need some notions of topology in the discrete setting. To this end, we consider a bounded set $B \subset \mathbb{Z}^m$ and its characteristic function

$$\chi_B(\underline{x}) = \begin{cases} 1 & \text{if } \underline{x} \in B \\ 0 & \text{if } \underline{x} \notin B \end{cases}$$

as well as the discrete operator

$$\breve{\partial} = \sum_{j=1}^{m} e_j^+ \Delta_j^- + \sum_{j=1}^{m} e_j^- \Delta_j^+.$$

The vector-valued function

$$\chi_B \breve{\partial} = \sum_{j=1}^{m} e_j^+ \Delta_j^- [\chi_B] + \sum_{j=1}^{m} e_j^- \Delta_j^+ [\chi_B]$$

is then called the oriented boundary of B. Observe that $\mathrm{supp}(\chi_B \breve{\partial})$ always contains points which do not belong to B. In fact, it consists of all vertices, the $\mathbb{Z}^m$ neighbourhood of which contains both points of B and points of $B^c \equiv \mathbb{Z}^m \setminus B$. In addition to this definition of the boundary, one may then also define the interior of B (respectively the exterior of B) to be the set of all points of B (respectively of B^c) which do not belong to $\mathrm{supp}(\chi_B \breve{\partial})$. The interior of a set B consists of all points of B, the $\mathbb{Z}^m$ neighbourhood of which contains only points of B, while the exterior of B is the interior of B^c. An example of a two-dimensional set B, its oriented boundary $\mathrm{supp}(\chi_B \breve{\partial})$ and its interior is given in the picture below. Note that the exact position of the point set within the considered grid is unimportant. The picture only illustrates how, starting from a given point set, its interior and its boundary may be visually determined.

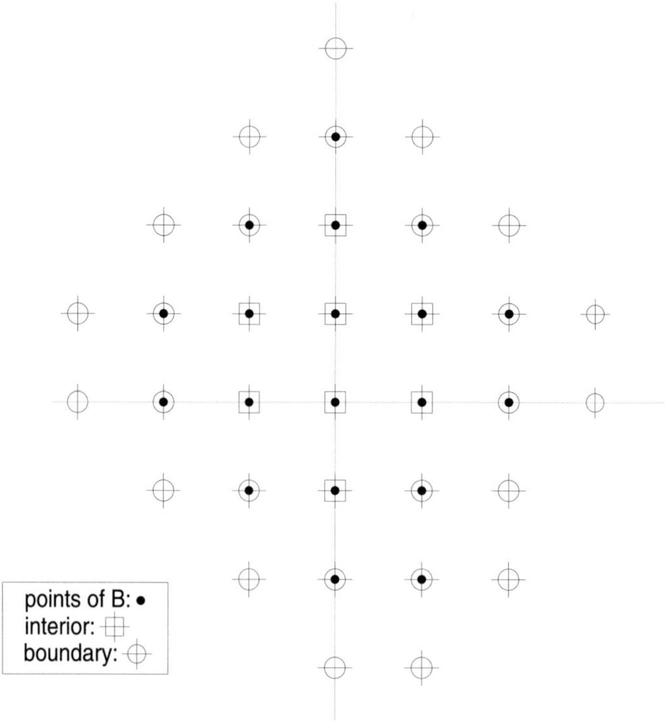

FIGURE 1. Illustration of topological notions

Each bounded set $B \subset \mathbb{Z}^m$ thus gives rise to a partition of $\mathbb{Z}^m$ into its interior, its exterior and the support of its oriented boundary.

The above concepts now allow for defining the notion of discrete monogenic function.

Definition 3.1. Let B be a bounded set in $\mathbb{Z}^m$ and let the Clifford algebra-valued function f be defined on $B \cup \mathrm{supp}(\chi_B \check{\partial})$. Then f is called *discrete (left) monogenic* in B if and only if $\partial[f](\underline{x}) = 0$ for all $\underline{x} \in B$.

Defined in this way, discrete monogenicity constitutes a proper generalization to higher dimension of discrete holomorphy, and in particular of the notion of so-called monodiffric functions of the second kind, investigated by Isaacs (see [18]), Ferrand (see [12]) and more recently also by Kiselman (see [19]). Moreover, it may be seen as a refinement of discrete harmonicity, since the right-hand side of (2.1) can be interpreted as a generalized discrete Laplacian, also called "mixed Laplacian", see [11], which even coincides with the star Laplacian when $g = 0$.

Example 3.2. To illustrate the definition of discrete monogenicity, consider the polynomial on $\mathbb{Z}^2$, given by

$$p^*(j,k) = e_1^+(2j+1) + e_2^+(-2k-1) + e_1^-(2j-1) + e_2^-(-2k+1), \qquad (j,k) \in \mathbb{Z}^2$$

It can be checked by direct calculation that p^* is monogenic in all points of $\mathbb{Z}^2$, viz $\partial[p^*](j,k) = 2e_1^- e_1^+ + (-2)e_2^- e_2^+ + 2e_1^+ e_1^- + (-2)e_2^+ e_2^- = 0$. However, there is more. The given function values are nothing but the restriction to $\mathbb{Z}^2$ of the polynomial p, given by

$$p(x,y) = e_1^+(2x+1) + e_2^+(-2y-1) + e_1^-(2x-1) + e_2^-(-2y+1), \qquad (x,y) \in \mathbb{R}^2$$

which is monogenic in the whole of $\mathbb{R}^2$: $\partial_{\underline{x}}[p](x,y) = 2e_1e_1^+ + (-2)e_2e_2^+ + 2e_1e_1^- + (-2)e_2e_2^- = 0$. In fact, we do not have to distinguish between p and p^*, as they determine each other uniquely. Moreover, the discrete monogenicity of p^* on $\mathbb{Z}^2$ is a consequence of the monogenicity of p, since the operator $\partial - \partial_{\underline{x}}$ turns a homogeneous polynomial of degree k into a polynomial of degree $k-2$, whence it turns any polynomial of the first degree into zero.

4. Some function theoretic results

We consider Clifford algebra-valued functions defined on $\mathbb{Z}^m$.

Then, first of all, a discrete version of Leibniz's rule is obtained by direct calculation. Observe that, as compared to its continuous counterpart, it contains an extra term, which fortunately will turn out to become small when considering finer grids.

Lemma 4.1 (Leibniz's rule). *Let f and g be Clifford algebra-valued functions defined on $\mathbb{Z}^m$. Then the following results hold.*

(i) $\Delta_j^\pm[fg] = \Delta_j^\pm[f]\, g + f\, \Delta_j^\pm[g] \pm \Delta_j^\pm[f]\, \Delta_j^\pm[g]$.

(ii) *If f is scalar valued, then*

$$[fg]\,\check{\partial} = g\,(f\,\check{\partial}) + f\,(g\,\check{\partial}) + \sum_{j=1}^{m} \left(\Delta_j^+[f]\,\Delta_j^+[g]\,e_j^- - \Delta_j^-[f]\,\Delta_j^-[g]\,e_j^+\right).$$

Proof. (i) For the case of the forward differences, calculation of the right-hand side gives

$$\begin{aligned}\Delta_j^+[f]\,g + f\,\Delta_j^+[g] &= [f(\underline{x}+e_j)g(\underline{x}) - f(\underline{x})g(\underline{x})] \\ &\quad + [f(\underline{x})g(\underline{x}+e_j) - f(\underline{x})g(\underline{x})] \\ \Delta_j^+[f]\,\Delta_j^+[g] &= f(\underline{x}+e_j)g(\underline{x}+e_j) - f(\underline{x})g(\underline{x}+e_j) \\ &\quad - f(\underline{x}+e_j)g(\underline{x}) + f(\underline{x})g(\underline{x})\end{aligned}$$

whence

$$\begin{aligned}\Delta_j^+[f]\,g + f\,\Delta_j^+[g] + \Delta_j^+[f]\,\Delta_j^+[g] &= f(\underline{x}+e_j)g(\underline{x}+e_j) - f(\underline{x})g(\underline{x}) \\ &= \Delta_j^+[fg].\end{aligned}$$

A similar calculation gives the result for the backward differences.

(ii) We subsequently obtain

$$\begin{aligned}[fg]\,\check{\partial} &= \sum_{j=1}^m \left[\Delta_j^-[fg]e_j^+ + \Delta_j^+[fg]e_j^-\right] \\ &= \sum_{j=1}^m \Big[\left(\Delta_j^-[f]\,g + f\,\Delta_j^-[g] - \Delta_j^-[f]\,\Delta_j^-[g]\right)e_j^+ \\ &\qquad + \left(\Delta_j^+[f]\,g + f\,\Delta_j^+[g] - \Delta_j^+[f]\,\Delta_j^+[g]\right)e_j^-\Big] \\ &= g\left[\sum_{j=1}^m\left(\Delta_j^-[f]e_j^+ + \Delta_j^+[f]\,e_j^-\right)\right] + f\left[\sum_{j=1}^m\left(\Delta_j^-[g]e_j^+ + \Delta_j^+[g]\,e_j^-\right)\right] \\ &\qquad + \sum_{j=1}^m\left(\Delta_j^+[f]\,\Delta_j^+[g]\,e_j^- - \Delta_j^-[f]\,\Delta_j^-[g]\,e_j^+\right)\end{aligned}$$

which exactly is the desired result, in view of the fact that f is scalar valued. □

Next, the integral of a discrete function f is quite naturally defined as

$$\int f = \sum_{\underline{x}\in\mathbb{Z}^m} f(\underline{x})$$

where, in order to ensure integrability, integrands are required to have compact (= bounded) supports. The following results are then directly obtained.

Lemma 4.2 (partial integration). *Let f and g be Clifford algebra-valued functions defined on $\mathbb{Z}^m$, where at least one of them has compact support, then*

$$\int f\,\Delta_j^\pm[g] = -\int \Delta_j^\mp[f]\,g.$$

Proof. Direct calculation yields

$$\begin{aligned}\int f\,\Delta_j^+[g] &= \sum_{\underline{x}\in\mathbb{Z}^m} f(\underline{x})\,g(\underline{x}+e_j) - \sum_{\underline{x}\in\mathbb{Z}^m} f(\underline{x})\,g(\underline{x})) \\ &= \sum_{\underline{y}\in\mathbb{Z}^m} f(\underline{y}-e_j)\,g(\underline{y}) - \sum_{\underline{x}\in\mathbb{Z}^m} f(\underline{x})\,g(\underline{x})) \\ &= \sum_{\underline{x}\in\mathbb{Z}^m} (f(\underline{x}-e_j) - f(\underline{x}))\,g(\underline{x}) \;=\; -\int \Delta_j^-[f]\,g\end{aligned}$$

with the substitution $\underline{y} = \underline{x} + e_j$ in the first summand. The other result is obtained similarly. □

Lemma 4.3 (Stokes' theorem). *Let f and g be Clifford algebra-valued functions defined on $\mathbb{Z}^m$, where at least one of them has compact support, then*

$$\int f\,(\partial g) = -\int (f\check{\partial})\,g \quad \textit{and} \quad \int f\,(\check{\partial} g) = -\int (f\partial)\,g.$$

Proof. Invoking partial integration we have

$$\begin{aligned}\int f\,(\partial g) &= \int f \sum_{j=1}^{m} \left(e_j^+ \Delta_j^+[g] + e_j^- \Delta_j^-[g] \right) \\ &= -\int \sum_{j=1}^{m} \left(\Delta_j^-[f] e_j^+ + \Delta_j^+[f] e_j^- \right) g \;=\; -\int (f\,\check{\partial})\,g\end{aligned}$$

and similarly for the second result. □

Pay attention to the fact that, in such kind of discrete integrals as in the above lemmata, the domains of integration on both sides of the formulae need not to be the same.

We now arrive at a first fundamental result.

Theorem 4.4 (Cauchy's theorem). *Let f be a Clifford algebra-valued function defined on $\mathbb{Z}^m$, which is discrete left monogenic in the bounded set B, then*

$$\int (\chi_B\,\check{\partial})\,f = 0.$$

Proof. On account of Stokes' theorem we have

$$\int (\chi_B\,\check{\partial})\,g = -\int \chi_B\,\partial[f] = 0$$

since the last integral only involves points of B, where f is left discrete monogenic. □

Corollary 4.5. *If B is a bounded set in $\mathbb{Z}^m$, then* $\displaystyle\int \chi_B\,\check{\partial} = 0.$

Proof. Take $f = 1$ in Cauchy's theorem. □

As an illustration of Corollary 4.5, again the same two-dimensional set B as in Figure 1 is considered in the picture below. As opposed to what was mentioned in Section 3, at this moment, the exact position of the grid points does become important. To that end we have indicated the co-ordinate axes in the picture and we recall that the grid length is 1; in this way the co-ordinates of all grid points are determined. For each point of $\text{supp}(\chi_B \breve{\partial})$ the value of $\chi_B \breve{\partial}$ is shown in the picture, whence it may easily be checked that the sum of all values indeed equals zero.

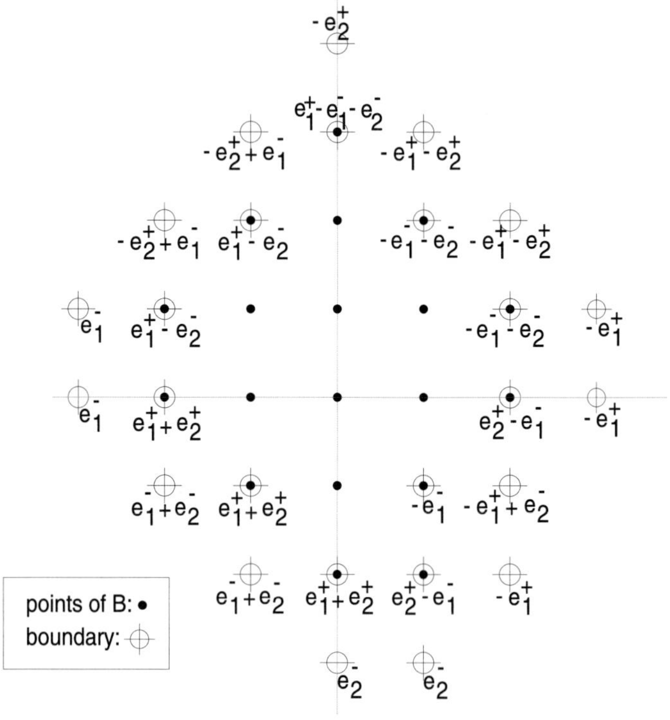

FIGURE 2. Illustration of Corollary 4.5

Furthermore, combination of Figure 2 with the polynomial values of Example 3.2 also provides an illustration of Cauchy's theorem.

Clearly, it is essential for the further development of this function theory to establish a Cauchy integral formula. To this end, let us assume that E is the fundamental solution of operator $\breve{\partial}$ defined by

$$E(\underline{x})\,\breve{\partial} = \delta(\underline{x}) = \left\{ \begin{array}{ll} 0, & \underline{x} \neq \underline{0} \\ 1 & \underline{x} = \underline{0} \end{array} \right\} = \prod_{j=1}^{m} \delta_{0\,x_j} \tag{4.1}$$

and

$$E(\underline{x}-\underline{y})\,\breve{\partial} = \delta(\underline{x}-\underline{y}) = \left\{ \begin{array}{ll} 0, & \underline{x} \neq \underline{y} \\ 1 & \underline{x} = \underline{y} \end{array} \right\} = \prod_{j=1}^{m} \delta_{x_j\, y_j}\,. \tag{4.2}$$

Next, consider once more a bounded set B, its characteristic function χ_B and put $f(\underline{x}) = \chi_B(\underline{x})E(\underline{x}-\underline{y})$. Then, by Leibniz's rule, we have

$$f(\underline{x})\breve{\partial} = E(\underline{x}-\underline{y})(\chi_B(\underline{x})\breve{\partial}) + \chi_B(\underline{x})\delta(\underline{x}-\underline{y}) + GT(\underline{x},\underline{y})$$

where we have put

$$GT(\underline{x},\underline{y}) = \sum_{j=1}^{m} \left(\Delta_j^+[\chi_B(\underline{x})]\Delta_j^+[E(\underline{x}-\underline{y})]\,e_j^- - \Delta_j^-[\chi_B(\underline{x})]\Delta_j^-[E(\underline{x}-\underline{y})]\,e_j^+\right). \tag{4.3}$$

Since f has compact support, we may apply Stokes' theorem, obtaining, for an arbitrary Clifford algebra-valued function g defined on $B \cup \operatorname{supp}(\chi_B\,\breve{\partial})$, that

$$\int E(\underline{x}-\underline{y})\,(\chi_B(\underline{x})\breve{\partial})\,g(\underline{x}) + \int \chi_B(\underline{x})\,\delta(\underline{x}-\underline{y})\,g(\underline{x}) + \int GT(\underline{x},\underline{y})\,g(\underline{x})$$
$$= -\int \chi_B(\underline{x})\,E(\underline{x}-\underline{y})\,\partial[g](\underline{x})$$

or still

$$\int E(\underline{x}-\underline{y})\,(\chi_B(\underline{x})\breve{\partial})\,g(\underline{x}) + \chi_B(\underline{y})\,g(\underline{y}) + \int GT(\underline{x},\underline{y})\,g(\underline{x})$$
$$= -\int \chi_B(\underline{x})\,E(\underline{x}-\underline{y})\,\partial[g](\underline{x}).$$

In fact, we have proved the following result.

Theorem 4.6 (Cauchy–Pompeiu formula). *Let B be a bounded set in $\mathbb{Z}^m$ and let g be a Clifford algebra-valued function defined on $B \cup \operatorname{supp}(\chi_B\,\breve{\partial})$, then for all points $\underline{y} \in B$ we have*

$$-g(\underline{y}) = \int \chi_B(\underline{x})\,E(\underline{x}-\underline{y})\partial g(\underline{x}) + \int (\chi_B\breve{\partial})\,E(\underline{x}-\underline{y})\,g(\underline{x}) + \int GT(\underline{x},\underline{y})\,g(\underline{x})$$

while for all points $\underline{y} \in B^c$:

$$0 = \int \chi_B(\underline{x})\,E(\underline{x}-\underline{y})\partial g(\underline{x}) + \int (\chi_B\breve{\partial})\,E(\underline{x}-\underline{y})\,g(\underline{x}) + \int GT(\underline{x},\underline{y})\,g(\underline{x})$$

where $GT(\underline{x},\underline{y})$ is given by (4.3).

Observe that the first and the second term at the right-hand side in the above formulae are 'traditional' terms, representing a volume integral over the bounded set B and a surface integral over the oriented boundary of B, respectively. On the contrary, the third term is an additional one, arising due to the grid (and more precisely: it originates from the additional term already arising in Leibniz's rule). We call this term the 'grid tension' term, which explains the notation $GT(\underline{x},\underline{y})$, introduced above.

Cauchy's integral formula now immediately follows.

Theorem 4.7 (Cauchy's integral formula). *Let B be a bounded set in $\mathbb{Z}^m$ and let the function f be discrete monogenic on B, then for all points $\underline{y} \in B$ we have*

$$-f(\underline{y}) = \int (\chi_B \check{\partial})\, E(\underline{x} - \underline{y})\, f(\underline{x}) + \int GT(\underline{x}, \underline{y})\, f(\underline{x})$$

while for all points $\underline{y} \in B^c$:

$$0 = \int (\chi_B \check{\partial})\, E(\underline{x} - \underline{y})\, f(\underline{x}) + \int GT(\underline{x}, \underline{y})\, f(\underline{x})$$

where $GT(\underline{x}, \underline{y})$ is given by (4.3).

Obviously, in the above results, an essential role is played by the so-called fundamental solution $E(\underline{x})$, defined by (4.1)–(4.2). In order to obtain $E(\underline{x})$ explicitly, we will pass to frequency space by means of the discrete-time Fourier transform, being defined for a discrete Clifford algebra-valued function $f(\underline{x})$ with compact support as follows (see also [14, 15]):

$$\mathcal{F}[f(\underline{x})](\underline{\xi}) = \int f(\underline{x}) \exp(-i\langle \underline{\xi}, \underline{x}\rangle) = \sum_{\underline{x} \in \mathbb{Z}^m} \exp(-i\langle \underline{\xi}, \underline{x}\rangle)\, f(\underline{x}), \qquad \underline{\xi} \in \mathbb{R}^m \tag{4.4}$$

and yielding a continuous and bounded multiperiodic function of $\underline{\xi}$ with period 2π in each of the m variables ξ_j. Elementary properties of this discrete-time Fourier transform are listed in the following lemma.

Lemma 4.8. *Let $f(\underline{x})$ be a Clifford algebra-valued function defined on $\mathbb{Z}^m$ with compact support and let its discrete-time Fourier transform be given by* (4.4), *then we have*

(i) $\mathcal{F}[f(\underline{x} \pm e_j)](\underline{\xi}) = \exp(\pm i\xi_j)\, \mathcal{F}[f(\underline{x})](\underline{\xi})$;

(ii) $\mathcal{F}[\Delta_j^{\pm} f(\underline{x})](\underline{\xi}) = \mp(1 - \exp(\pm i\xi_j))\, \mathcal{F}[f(\underline{x})](\underline{\xi})$;

(iii) $\mathcal{F}[f(\underline{x})\, \check{\partial}](\underline{\xi}) = \mathcal{F}[f(\underline{x})](\underline{\xi})\, G(\underline{\xi})$, *where*

$$G(\underline{\xi}) = \sum_{j=1}^{m} \left[(1 - \exp(-i\xi_j))\, e_j^+ + (\exp(i\xi_j) - 1)\, e_j^-\right] \tag{4.5}$$

(iv) $\mathcal{F}[\delta(\underline{x})](\underline{\xi}) = 1$.

Proof. (i) Direct calculation yields

$$\begin{aligned}\mathcal{F}[f(\underline{x} \pm e_j)](\underline{\xi}) &= \int f(\underline{x} \pm e_j) \exp(-i\langle \underline{\xi}, \underline{x}\rangle)\\ &= \int f(\underline{y}) \exp(-i\langle \underline{\xi}, \underline{y} \pm e_j\rangle) \;=\; \exp(\pm i\xi_j) \int f(\underline{y}) \exp(-i\langle \underline{\xi}, \underline{y}\rangle)\\ &= \exp(\pm i\xi_j)\, \mathcal{F}[f(\underline{x})](\underline{\xi})\end{aligned}$$

where we have used the substitution $\underline{y} = \underline{x} \pm e_j$.

(ii) The results immediately follow on account of (i), using the definitions of the forward and backward differences.

(iii) Starting from the definition of the operator $\breve{\partial}$, and invoking (ii), we obtain

$$\begin{aligned}\mathcal{F}[f(\underline{x})\,\breve{\partial}](\underline{\xi}) &= \mathcal{F}\left[\sum_{j=1}^{m}\left(\Delta_j^-[f](\underline{x})e_j^+ + \Delta_j^+[f](\underline{x})e_j^-\right)\right](\underline{\xi})\\ &= \sum_{j=1}^{m}\Big[(1-\exp(-\xi_j))\,\mathcal{F}[f(\underline{x})](\underline{\xi})\,e_j^+ + (\exp(\xi_j)-1)\,\mathcal{F}[f(\underline{x})](\underline{\xi})\,e_j^-\Big]\\ &= \mathcal{F}[f(\underline{x})](\underline{\xi})\sum_{j=1}^{m}\Big[(1-\exp(-\xi_j))\,e_j^+ + (\exp(\xi_j)-1)\,e_j^-\Big].\end{aligned}$$

(iv) We have

$$\mathcal{F}[\delta(\underline{x})](\underline{\xi}) = \int \delta(\underline{x})\exp(-i\langle\underline{\xi},\underline{x}\rangle) = \Big[\exp(-i\langle\underline{\xi},\underline{x}\rangle)\Big]_{\underline{x}=\underline{0}} = 1. \qquad \square$$

We will now use the above calculus rules to determine the fundamental solution E defined by (4.1)–(4.2) in frequency space, following similar lines as developed, e.g., in [14, 15] in the quaternionic case and in [19] in the complex case. Starting from the relation $E(\underline{x})\breve{\partial} = \delta(\underline{x})$, we directly obtain, combining (iii) and (iv), that

$$\hat{E}(\underline{\xi})\,G(\underline{\xi}) = 1$$

where $\hat{E}(\underline{\xi}) = \mathcal{F}[E(\underline{x})](\underline{\xi})$ and $G(\underline{\xi})$ is given by (4.5). Since $G(\underline{\xi})$ is vector valued, it is possible to solve this equation for $\hat{E}$; this yields

$$\hat{E}(\underline{\xi}) = \frac{G(\underline{\xi})}{(G(\underline{\xi}))^2}, \qquad \text{wherever } G(\underline{\xi}) \neq 0. \tag{4.6}$$

In the concrete model given above for the forward and backward Clifford bases, one has

$$G(\underline{\xi}) = \sum_{j=1}^{m}\Big[(1-\exp(-i\xi_j))(\tfrac{1}{2}e_j + \tfrac{1}{2}B_j) + (\exp(i\xi_j)-1)(\tfrac{1}{2}e_j - \tfrac{1}{2}B_j)\Big]$$

yielding

$$\begin{aligned}G(\underline{\xi}) &= \sum_{j=1}^{m}[(1-\cos\xi_j)\,B_j + i\,\sin\xi_j\,e_j]\\ (G(\underline{\xi}))^2 &= 4\sum_{j=1}^{m}\sin^2\frac{\xi_j}{2} - 32g\sum_{j<k}\sin^2\frac{\xi_j}{2}\sin^2\frac{\xi_k}{2}\end{aligned}$$

whence the fundamental solution $\hat{E}(\underline{\xi})$ in frequency space, (4.6), explicitly reads

$$\hat{E}(\underline{\xi}) = \frac{1}{4} \frac{\sum_{j=1}^{m} [(1-\cos\xi_j)\, B_j + i\, \sin\xi_j\, e_j]}{\sum_{j=1}^{m} \sin^2 \frac{\xi_j}{2} - 8g \sum_{j<k} \sin^2 \frac{\xi_j}{2} \sin^2 \frac{\xi_k}{2}}. \tag{4.7}$$

This explicit expression also allows us to investigate where the denominator of $\hat{E}(\underline{\xi})$ will be zero, i.e., where $G(\underline{\xi}) = 0$. Of course, in view of the inherent periodicity of the discrete-time Fourier transform, we may restrict ourselves to one basic interval $([0, 2\pi[)^m$. Let us start with some low-dimensional examples.

First, take $m = 2$. Here, we have $\underline{\xi} = (\xi_1, \xi_2)$, and the equation under investigation reads

$$\sin^2\left(\frac{\xi_1}{2}\right) + \sin^2\left(\frac{\xi_2}{2}\right) = 8g\, \sin^2\left(\frac{\xi_1}{2}\right)\, \sin^2\left(\frac{\xi_2}{2}\right)$$

which can be rewritten as

$$\left(1 - 4g\sin^2\left(\frac{\xi_2}{2}\right)\right)\, \sin^2\frac{\xi_1}{2} + \left(1 - 4g\sin^2\left(\frac{\xi_1}{2}\right)\right)\, \sin^2\frac{\xi_2}{2} = 0.$$

It is clear that this equation will never have any solution, apart from $\underline{\xi} = \underline{0}$, as long as $g < \frac{1}{4}$. In the picture below, for some values of g exceeding $\frac{1}{4}$, the corresponding curves of solutions are shown.

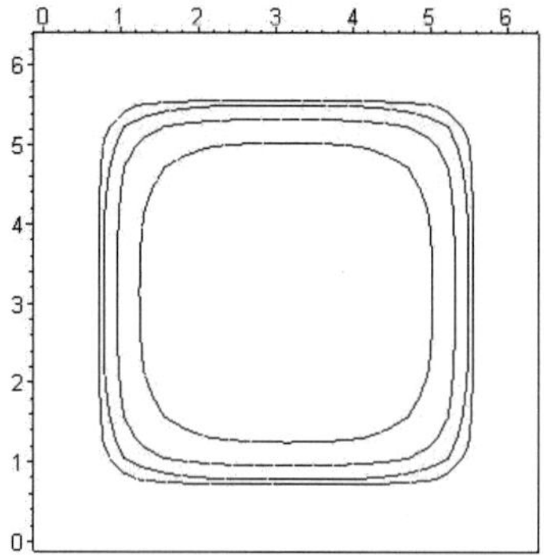

FIGURE 3. Curves in the plane where $G(\underline{\xi}) = 0$, for different values of $g > \frac{1}{4}$

Fortunately, the latter situation will never be encountered in practice, since we have already argued in Section 2 that the metric scalar g should be restricted to the interval $[0, \frac{1}{4(m-1)}[$, which here equals $[0, \frac{1}{4}[$.

Next, take $m = 3$ and $\underline{\xi} = (\xi_1, \xi_2, \xi_3)$. Here the equation reads

$$\begin{aligned}\sin^2\left(\frac{\xi_1}{2}\right) &+ \sin^2\left(\frac{\xi_2}{2}\right) + \sin^2\left(\frac{\xi_3}{2}\right)\\ = \; & 8g\left(\sin^2\left(\frac{\xi_1}{2}\right)\sin^2\left(\frac{\xi_2}{2}\right) + \sin^2\left(\frac{\xi_1}{2}\right)\sin^2\left(\frac{\xi_3}{2}\right) + \sin^2\left(\frac{\xi_2}{2}\right)\sin^2\left(\frac{\xi_3}{2}\right)\right)\end{aligned}$$

or still

$$\sum_{j=1}^{3}\left(1 - 4g\sum_{\substack{k=1\\k\neq j}}^{3}\sin^2\left(\frac{\xi_k}{2}\right)\right)\sin^2\left(\frac{\xi_j}{2}\right) = 0.$$

Again, we may directly see that the only solution of this equation will be $\underline{\xi} = \underline{0}$, provided that

$$1 - 4g\left(\sin^2\left(\frac{\xi_k}{2}\right) + \sin^2\left(\frac{\xi_\ell}{2}\right)\right) > 0, \qquad \text{for all } k, \ell = 1, 2, 3, \; k \neq \ell$$

which will be fulfilled whenever $g < \frac{1}{4(m-1)} = \frac{1}{8}$.

The general case now easily follows.

Proposition 4.9. *The expression* (4.7) *for the fundamental solution in frequency space, restricted to its basic period* $([0, 2\pi[)^m$, *holds for all* $\underline{\xi} \neq \underline{0}$, *provided that the metric scalar* g *is smaller then the critical value* $\frac{1}{4(m-1)}$.

Proof. It suffices to investigate the zeroes of the denominator of (4.7), i.e., the equation

$$\sum_{j=1}^{m}\sin^2\frac{\xi_j}{2} - 8g\sum_{j<k}\sin^2\frac{\xi_j}{2}\sin^2\frac{\xi_k}{2} = 0$$

which can be rewritten as

$$\sum_{j=1}^{m}\left(1 - 4g\sum_{\substack{k=1\\k\neq j}}^{m}\sin^2\left(\frac{\xi_k}{2}\right)\right)\sin^2\left(\frac{\xi_j}{2}\right) = 0.$$

This equation cannot have a solution different from $\underline{\xi} = \underline{0}$, whenever

$$\left(1 - 4g\sum_{\substack{k=1\\k\neq j}}^{m}\sin^2\left(\frac{\xi_k}{2}\right)\right) > 0, \qquad \text{for all } j = 1, \ldots, m.$$

Since for any $j = 1, \ldots, m$, the sum at the left-hand side of the above inequality consists of $m - 1$ terms, which all are smaller than 1, the statement follows. $\square$

5. Conclusions

We have laid the foundations of a function theory for a discrete Dirac operator, centred around the notion of discrete monogenic functions, a generalization of the discrete holomorphic functions grounded by Isaacs and Ferrand in the previous century. The present theory may also be interpreted as a refinement of discrete harmonic analysis, since the discrete Dirac operator factorizes a mixed discrete Laplacian or even the star Laplacian, according to the chosen model. Important results such as Cauchy's theorem and Cauchy's integral representation formula were established using the fundamental solution of an associated discrete Dirac operator, explicitly obtained in frequency space.

Acknowledgments

The financial support by the FWO Vlaanderen ("Krediet aan navorsers", no. 31506208) for research on the subject of "Discrete Clifford analysis" is gratefully acknowledged by Frank Sommen.

References

[1] F. Brackx, J. Bureš, H. De Schepper, D. Eelbode, F. Sommen, V. Souček, Fundaments of Hermitean Clifford Analysis. Part I: Complex structure, *Comp. Anal. Oper. Theory* **1(3)**, 2007, 341–365.

[2] F. Brackx, J. Bureš, H. De Schepper, D. Eelbode, F. Sommen, V. Souček, Fundaments of Hermitean Clifford Analysis. Part II: Splitting of h-monogenic equations, *Comp. Var. Elliptic Equ.* **52(10–11)**, 2007, 1063–1079.

[3] F. Brackx, R. Delanghe, F. Sommen, *Clifford Analysis*, Pitman Publishers, Boston-London-Melbourne, 1982.

[4] F. Brackx, N. De Schepper, F. Sommen, Metric dependent Clifford analysis with applications to wavelet analysis. In: D. Alpay, A. Luger and H. Woracek (eds.), *Wavelets, Multiscale Systems and Hypercomplex Analysis*, Book Series "Operator Theory: Advances and Applications", Birkhäuser, Basel, 2006, 17–67.

[5] F. Colombo, I. Sabadini, F. Sommen, D.C. Struppa, *Analysis of Dirac Systems and Computational Algebra*, Birkhäuser, Boston (2004).

[6] R. Delanghe, F. Sommen, V. Souček, *Clifford Algebra and Spinor-Valued Functions*, Kluwer Academic Publishers, Dordrecht, 1992.

[7] R. Duffin, Basic properties of discrete analytic functions, *Duke Math. J.* **23**, 1956, 335–363.

[8] N. Faustino, K. Gürlebeck, A. Hommel, U. Kaehler, Difference potentials for the Navier-Stokes equations in unbounded domains, *J. Diff. Equ. Appl.* **12(6)**, 2006, 577–595.

[9] N. Faustino, U. Kaehler, Fischer decomposition for difference Dirac operators, *Adv. Appl. Clifford Alg.* **17(1)**, 2007, 37–58.

[10] N. Faustino, U. Kaehler, On a correspondence principle between discrete differential forms, graph structure and multi-vector calculus on symmetric lattices, accepted for publication in *Adv. Appl. Clifford Alg.*

[11] N. Faustino, U. Kaehler, F. Sommen, Discrete Dirac operators in Clifford analysis, *Adv. Appl. Clifford Alg.* **17(3)**, 2007, 451–467.

[12] J. Ferrand, Fonctions préharmoniques et fonctions préholomorphes, *Bull. Sci. Math.* **68**, 1944, 152–180.

[13] J. Gilbert, M. Murray, *Clifford Algebras and Dirac Operators in Harmonic Analysis*, Cambridge University Press, Cambridge, 1991.

[14] K. Gürlebeck, A. Hommel, On finite difference potentials and their applications in a discrete function theory, *Math. Meth. Appl. Sci.* **25(16-18)**, 2002, 1563–1576.

[15] K. Gürlebeck, A. Hommel, On finite difference Dirac operators and their fundamental solutions, *Adv. Appl. Clifford Alg.* **11(S2)**, 2001, 89–106.

[16] K. Gürlebeck, W. Sprößig, *Quaternionic analysis and elliptic boundary value problems*, Intern. Series of Numerical Mathematics, **89**, Birkhäuser Verlag, Basel, 1990.

[17] K. Gürlebeck, W. Sprößig, *Quaternionic and Clifford calculus for engineers and physicists*, John Wiley & Sons, Chichester, 1997.

[18] R.P. Isaacs, *Monodiffric functions.* In: Construction and applications of conformal maps, Proceedings of a symposium, 257–266, National Bureau of Standards, *Appl. Math. Ser.* **no. 18**, U.S. Government Printing Office, Washington DC, 1952.

[19] C.O. Kiselman, Functions on discrete sets holomorphic in the sense of Isaacs, or monodiffric functions of the first kind, *Sci. China Ser. A* **48**, 2005, 86–96.

[20] S.P. Novikov, Schrödinger operators on graphs and symplectic geometry. In: E. Bierstone *et al.* (eds.), *The Arnoldfest*, **vol. 24**, Fields Inst. Comm., 1999, 397–413.

[21] I. Porteous, *Clifford algebras and the classical groups*, Cambridge Studies in Advanced Mathematics **50**, Cambridge University Press, Cambridge, 1995.

[22] R. Rocha–Chávez, M. Shapiro, F. Sommen, *Integral theorems for functions and differential forms in* $\mathbb{C}^m$, Research Notes in Mathematics **428**, Chapman & Hall/CRC, Boca Raton (2002).

[23] J. Ryan, Complexified Clifford analysis, *Comp. Var. Theory Appl.* **1(1)** (1982/83), 119–149.

F. Brackx, H. De Schepper, F. Sommen and L. Van de Voorde
Clifford Research Group
Faculty of Engineering
Ghent University
Galglaan 2
9000 Gent, Belgium

e-mail: `Freddy.Brackx@UGent.be`
`Hennie.DeSchepper@UGent.be`
`Frank.Sommen@UGent.be`
`Liesbet.Vandevoorde@UGent.be`

Quaternionic and Clifford Analysis
Trends in Mathematics, 55–68

On Factorization of Bicomplex Meromorphic Functions

K.S. Charak and D. Rochon

Abstract. In this paper the factorization theory of meromorphic functions of one complex variable is promoted to bicomplex meromorphic functions. Many results of one complex variable case are seen to hold in bicomplex case, and it is found that there are results for meromorphic functions of one complex variable which are not true for bicomplex meromorphic functions. In particular, we show that for any bicomplex transcendental meromorphic function F, there exists a bicomplex meromorphic function G such that GF is prime even if the set:

$$\{a \in \mathbb{T} : F(w) + a\phi(w) \text{ is not prime}\}$$

is empty or of cardinality $\aleph_1$ for any non-constant fractional linear bicomplex function ϕ. Moreover, as specific application, we obtain six additional possible forms of factorization of the complex cosine $\cos z$ in the bicomplex space.

Mathematics Subject Classification (2000). 30G, 30D30, 30G35, 32A, 32A30.

Keywords. Bicomplex Numbers, Factorization, Meromorphic Functions.

1. Introduction

The factorization theory of meromorphic functions of one complex variable is to study how a given meromorphic function can be factorized into other simpler meromorphic functions in the sense of composition. In number theory, every natural number can be factorized as a product of prime numbers. Therefore, prime numbers serve as building blocks of natural numbers and the theory of prime numbers is one of the main subarea of number theory. In our situation, we also have the so-called prime functions which play a similar role in the factorization theory of meromorphic functions as prime numbers do in number theory. More specifically, factorization theory of meromorphic functions essentially deals with the primeness,

The research of D.R. is partly supported by grants from CRSNG of Canada and FQRNT of Québec.

pseudo-primeness and unique factorizability of a meromorphic function. We start with the following concepts.

Definition 1.1. Let F be a meromorphic function. Then an expression

$$F(z) = f(g(z)) \tag{1.1}$$

where f is meromorphic and g is entire (g may be meromorphic when f is a rational function) is called a factorization of F with f and g as its left and right factors respectively. F is said to be non-factorizable or prime if for every representation of F of the form (1.1) we have that either f or g is linear. If every representation of F of the form (1.1) implies that f is rational or g is a polynomial (f is linear whenever g is transcendental, g is linear whenever f is transcendental), we say that F is pseudo-prime (left-prime, right-prime). If the factors are restricted to entire functions, the factorization is said to be in entire sense and we have the corresponding concepts of primeness in entire sense (called E-primeness), pseudo-primeness in entire sense (called E-pseudo-primeness) etc.

The first example of prime function is $F(z) = exp(z) + z$ given by Rosenbloom [23] who gave the definition of prime transcendental entire function by considering entire factors only, and asserted without proof that the function $F(z) = z + exp(z)$ is prime. In 1968, F. Gross [7] gave a complete proof of this assertion and extended the study of primeness to meromorphic functions and gave Definition 1.1. No systematic theory has actually been developed to handle the problems of factorization of transcendental meromorphic functions. However, recently, T.W. Ng [9, 10, 11, 12] proved some results which of course can solve some factorization problems in a systematic way. He introduced the methods from the Theory of Complex Analytic Sets and local holomorphic dynamics to solve some factorization problems. Classical function theory and the Nevanlinna Value Distribution theory are the main tools used in factorization theory of meromorphic functions. Most of the classes of functions which have been studied are concerned with the following one for several factors: (1) Growth of the function, (2) Distribution of zeros, (3) Periodicity, (4) Fixed-points, (5) Solutions of linear differential equations. For complete details on factorization theory of meromorphic functions one can refer to the books of C.T. Chuang and C.C. Yang [3], and F. Gross [5].

The main purpose of the present paper is to try to extend and promote the research on Factorization theory of meromorphic functions in one complex variable to two complex variables via Bicomplex Function Theory [14, 15, 16, 18]. In our study of factorization theory of bicomplex meromorphic functions, the idempotent representation of bicomplex meromorphic functions plays a vital role since the parallel definitions of factorization of meromorphic functions of one complex variable do not work. It is found that many results from one variable theory hold in bicomplex situation whereas some fail to hold, and this is the point of difference between the two situations and so makes sense to investigate. In particular, we show that for any bicomplex transcendental meromorphic function F, there exists a bicomplex meromorphic function G such that GF is prime even if the set:

$$\{a \in \mathbb{T} : F(w) + a\phi(w) \text{ is not prime}\}$$

is empty or of cardinality $\aleph_1$ for any non-constant fractional linear bicomplex function ϕ. Moreover, as specific application, we obtain six additional possible forms of factorization of the complex cosine $\cos z$ in the bicomplex space.

2. Preliminaries

2.1. Bicomplex numbers

Bicomplex numbers are defined as

$$\mathbb{T} := \{z_1 + z_2\mathbf{i_2} \mid z_1, z_2 \in \mathbb{C}(\mathbf{i_1})\} \tag{2.1}$$

where the imaginary units $\mathbf{i_1}, \mathbf{i_2}$ and $\mathbf{j}$ are governed by the rules: $\mathbf{i_1}^2 = \mathbf{i_2}^2 = -1$, $\mathbf{j}^2 = 1$ and

$$\begin{array}{rcccl} \mathbf{i_1}\mathbf{i_2} & = & \mathbf{i_2}\mathbf{i_1} & = & \mathbf{j}, \\ \mathbf{i_1}\mathbf{j} & = & \mathbf{j}\mathbf{i_1} & = & -\mathbf{i_2}, \\ \mathbf{i_2}\mathbf{j} & = & \mathbf{j}\mathbf{i_2} & = & -\mathbf{i_1}. \end{array} \tag{2.2}$$

Note that we define $\mathbb{C}(\mathbf{i}_k) := \{x + y\mathbf{i}_k \mid \mathbf{i}_k^2 = -1 \text{ and } x, y \in \mathbb{R}\}$ for $k = 1, 2$. Hence, it is easy to see that the multiplication of two bicomplex numbers is commutative. In fact, the bicomplex numbers

$$\mathbb{T} \cong \mathrm{Cl}_{\mathbb{C}}(1, 0) \cong \mathrm{Cl}_{\mathbb{C}}(0, 1)$$

are *unique* among the complex Clifford algebras in that they are commutative but not division algebra. It is also convenient to write the set of bicomplex numbers as

$$\mathbb{T} := \{w_0 + w_1\mathbf{i_1} + w_2\mathbf{i_2} + w_3\mathbf{j} \mid w_0, w_1, w_2, w_3 \in \mathbb{R}\}. \tag{2.3}$$

In particular, in equation (2.1), if we put $z_1 = x$ and $z_2 = y\mathbf{i_1}$ with $x, y \in \mathbb{R}$, then we obtain the following subalgebra of hyperbolic numbers, also called duplex numbers (see, e.g., [20, 26]):

$$\mathbb{D} := \{x + y\mathbf{j} \mid \mathbf{j}^2 = 1,\ x, y \in \mathbb{R}\} \cong \mathrm{Cl}_{\mathbb{R}}(0, 1).$$

Complex conjugation plays an important role both for algebraic and geometric properties of $\mathbb{C}$. For bicomplex numbers, there are three possible conjugations. Let $w \in \mathbb{T}$ and $z_1, z_2 \in \mathbb{C}(\mathbf{i_1})$ such that $w = z_1 + z_2\mathbf{i_2}$. Then we define the three conjugations as:

$$w^{\dagger_1} = (z_1 + z_2\mathbf{i_2})^{\dagger_1} := \overline{z}_1 + \overline{z}_2\mathbf{i_2}, \tag{2.4a}$$

$$w^{\dagger_2} = (z_1 + z_2\mathbf{i_2})^{\dagger_2} := z_1 - z_2\mathbf{i_2}, \tag{2.4b}$$

$$w^{\dagger_3} = (z_1 + z_2\mathbf{i_2})^{\dagger_3} := \overline{z}_1 - \overline{z}_2\mathbf{i_2}, \tag{2.4c}$$

where $\overline{z}_k$ is the standard complex conjugate of complex numbers $z_k \in \mathbb{C}(\mathbf{i_1})$. If we say that the bicomplex number $w = z_1 + z_2\mathbf{i_2} = w_0 + w_1\mathbf{i_1} + w_2\mathbf{i_2} + w_3\mathbf{j}$ has the "signature" $(++++)$, then the conjugations of type 1,2 or 3 of w have, respectively, the signatures $(+-+-)$, $(++--)$ and $(+--+)$. We can verify easily that the composition of the conjugates gives the four-dimensional abelian Klein group:

$$\begin{array}{|c||c|c|c|c|} \hline \circ & \dagger_0 & \dagger_1 & \dagger_2 & \dagger_3 \\ \hline\hline \dagger_0 & \dagger_0 & \dagger_1 & \dagger_2 & \dagger_3 \\ \hline \dagger_1 & \dagger_1 & \dagger_0 & \dagger_3 & \dagger_2 \\ \hline \dagger_2 & \dagger_2 & \dagger_3 & \dagger_0 & \dagger_1 \\ \hline \dagger_3 & \dagger_3 & \dagger_2 & \dagger_1 & \dagger_0 \\ \hline \end{array} \tag{2.5}$$

where $w^{\dagger_0} := w \ \forall w \in \mathbb{T}$.

The three kinds of conjugation all have some of the standard properties of conjugations, such as:

$$(s+t)^{\dagger_k} = s^{\dagger_k} + t^{\dagger_k}, \tag{2.6}$$

$$\left(s^{\dagger_k}\right)^{\dagger_k} = s, \tag{2.7}$$

$$(s \cdot t)^{\dagger_k} = s^{\dagger_k} \cdot t^{\dagger_k}, \tag{2.8}$$

for $s, t \in \mathbb{T}$ and $k = 0, 1, 2, 3$.

We know that the product of a standard complex number with its conjugate gives the square of the Euclidean metric in $\mathbb{R}^2$. The analogs of this, for bicomplex numbers, are the following. Let $z_1, z_2 \in \mathbb{C}(\mathbf{i_1})$ and $w = z_1 + z_2\mathbf{i_2} \in \mathbb{T}$, then we have that [20]:

$$|w|^2_{\mathbf{i_1}} := w \cdot w^{\dagger_2} = z_1^2 + z_2^2 \in \mathbb{C}(\mathbf{i_1}), \tag{2.9a}$$

$$|w|^2_{\mathbf{i_2}} := w \cdot w^{\dagger_1} = \left(|z_1|^2 - |z_2|^2\right) + 2\mathrm{Re}(z_1\overline{z}_2)\mathbf{i_2} \in \mathbb{C}(\mathbf{i_2}), \tag{2.9b}$$

$$|w|^2_{\mathbf{j}} := w \cdot w^{\dagger_3} = \left(|z_1|^2 + |z_2|^2\right) - 2\mathrm{Im}(z_1\overline{z}_2)\mathbf{j} \in \mathbb{D}, \tag{2.9c}$$

where the subscript of the square modulus refers to the subalgebra $\mathbb{C}(\mathbf{i_1}), \mathbb{C}(\mathbf{i_2})$ or $\mathbb{D}$ of $\mathbb{T}$ in which w is projected. Note that for $z_1, z_2 \in \mathbb{C}(\mathbf{i_1})$ and $w = z_1 + z_2\mathbf{i_2} \in \mathbb{T}$, we can define the usual (Euclidean in $\mathbb{R}^4$) norm of w as $|w| = \sqrt{|z_1|^2 + |z_2|^2} = \sqrt{\mathrm{Re}(|w|^2_{\mathbf{j}})}$.

It is easy to verify that $w \cdot \dfrac{w^{\dagger_2}}{|w|^2_{\mathbf{i_1}}} = 1$. Hence, the inverse of w is given by

$$w^{-1} = \frac{w^{\dagger_2}}{|w|^2_{\mathbf{i_1}}}. \tag{2.10}$$

From this, we find that the set $\mathcal{NC}$ of zero divisors of $\mathbb{T}$, called the *null-cone*, is given by $\{z_1 + z_2\mathbf{i_2} \mid z_1^2 + z_2^2 = 0\}$, which can be rewritten as

$$\mathcal{NC} = \{z(\mathbf{i_1} \pm \mathbf{i_2}) |\ z \in \mathbb{C}(\mathbf{i_1})\}. \tag{2.11}$$

2.2. Bicomplex holomorphic functions

It is also possible to define differentiability of a function at a point of $\mathbb{T}$:

Definition 2.1. Let U be an open set of $\mathbb{T}$ and $w_0 \in U$. Then, $f : U \subseteq \mathbb{T} \longrightarrow \mathbb{T}$ is said to be $\mathbb{T}$-differentiable at w_0 with derivative equal to $f'(w_0) \in \mathbb{T}$ if

$$\lim_{\substack{w \to w_0 \\ (w-w_0 \ inv.)}} \frac{f(w)-f(w_0)}{w-w_0} = f'(w_0).$$

We also say that the function f is $\mathbb{T}$-holomorphic on an open set U if and only if f is $\mathbb{T}$-differentiable at each point of U.

Using $w = z_1 + z_2\mathbf{i_2}$, a bicomplex number w can be seen as an element (z_1, z_2) of $\mathbb{C}^2$, so a function $f(z_1 + z_2\mathbf{i_2}) = f_1(z_1, z_2) + f_2(z_1, z_2)\mathbf{i_2}$ of $\mathbb{T}$ can be seen as a mapping $f(z_1, z_2) = (f_1(z_1, z_2), f_2(z_1, z_2))$ of $\mathbb{C}^2$. Here we have a characterization of such mappings:

Theorem 2.2. *Let U be an open set and $f : U \subseteq \mathbb{T} \longrightarrow \mathbb{T}$ such that $f \in C^1(U)$, and let $f(z_1 + z_2\mathbf{i_2}) = f_1(z_1, z_2) + f_2(z_1, z_2)\mathbf{i_2}$. Then f is $\mathbb{T}$-holomorphic on U if and only if*

$$f_1 \text{ and } f_2 \text{ are holomorphic in } z_1 \text{ and } z_2,$$

and

$$\frac{\partial f_1}{\partial z_1} = \frac{\partial f_2}{\partial z_2} \text{ and } \frac{\partial f_2}{\partial z_1} = -\frac{\partial f_1}{\partial z_2} \text{ on } U.$$

Moreover, $f' = \frac{\partial f_1}{\partial z_1} + \frac{\partial f_2}{\partial z_1}\mathbf{i_2}$ and $f'(w)$ is invertible if and only if $\det \mathcal{J}_f(w) \neq 0$.

This theorem can be obtained from results in [14] and [19]. Moreover, by the Hartogs theorem [25], it is possible to show that "$f \in C^1(U)$" can be dropped from the hypotheses. Hence, it is natural to define the corresponding class of mappings for $\mathbb{C}^2$:

Definition 2.3. The class of $\mathbb{T}$-holomorphic mappings on a open set $U \subseteq \mathbb{C}^2$ is defined as follows:

$$TH(U) := \{f{:}U \subseteq \mathbb{C}^2 \longrightarrow \mathbb{C}^2 | f \in \mathrm{H}(U) \text{ and } \frac{\partial f_1}{\partial z_1} = \frac{\partial f_2}{\partial z_2}, \frac{\partial f_2}{\partial z_1} = -\frac{\partial f_1}{\partial z_2} \text{ on } U\}.$$

It is the subclass of holomorphic mappings of $\mathbb{C}^2$ satisfying the complexified Cauchy-Riemann equations.

We remark that $f \in TH(U)$ in terms of $\mathbb{C}^2$ if and only if f is $\mathbb{T}$-differentiable on U. It is also important to know that every bicomplex number $z_1 + z_2\mathbf{i_2}$ has the following unique idempotent representation:

$$z_1 + z_2\mathbf{i_2} = (z_1 - z_2\mathbf{i_1})\mathbf{e_1} + (z_1 + z_2\mathbf{i_1})\mathbf{e_2}. \tag{2.12}$$

where $\mathbf{e_1} = \frac{1+\mathbf{j}}{2}$ and $\mathbf{e_2} = \frac{1-\mathbf{j}}{2}$. This representation is very useful because: addition, multiplication and division can be done term-by-term. It is also easy to verify the following characterization of the non-invertible elements.

Proposition 2.4. *An element* $w = z_1 + z_2\mathbf{i_2}$ *will be in the null-cone if and only if* $z_1 - z_2\mathbf{i_1} = 0$ *or* $z_1 + z_2\mathbf{i_1} = 0$.

The notion of holomorphicity can also be seen with this kind of notation. For this we need to define the projections $P_1, P_2 : \mathbb{T} \longrightarrow \mathbb{C}(\mathbf{i_1})$ as $P_1(z_1 + z_2\mathbf{i_2}) = z_1 - z_2\mathbf{i_1}$ and $P_2(z_1 + z_2\mathbf{i_2}) = z_1 + z_2\mathbf{i_1}$.

Definition 2.5. We say that $X \subseteq \mathbb{T}$ is a $\mathbb{T}$-cartesian set determined by X_1 and X_2 if $X = X_1 \times_e X_2 := \{z_1 + z_2\mathbf{i_2} \in \mathbb{T} : z_1 + z_2\mathbf{i_2} = w_1\mathbf{e_1} + w_2\mathbf{e_2}, (w_1, w_2) \in X_1 \times X_2\}$.

Now, it is possible to state the following striking theorems [14]:

Theorem 2.6. *Let* X_1 *and* X_2 *be open sets in* $\mathbb{C}(\mathbf{i_1})$. *If* $f_{e1} : X_1 \longrightarrow \mathbb{C}(\mathbf{i_1})$ *and* $f_{e2} : X_2 \longrightarrow \mathbb{C}(\mathbf{i_1})$ *are holomorphic functions of* $\mathbb{C}(\mathbf{i_1})$ *on* X_1 *and* X_2 *respectively, then the function* $f : X_1 \times_e X_2 \longrightarrow \mathbb{T}$ *defined as*

$$f(z_1 + z_2\mathbf{i_2}) = f_{e1}(z_1 - z_2\mathbf{i_1})\mathbf{e_1} + f_{e2}(z_1 + z_2\mathbf{i_1})\mathbf{e_2} \ \forall\ z_1 + z_2\mathbf{i_2} \in X_1 \times_e X_2$$

is $\mathbb{T}$*-holomorphic on the open set* $X_1 \times_e X_2$ *and*

$$f'(z_1 + z_2\mathbf{i_2}) = f'_{e1}(z_1 - z_2\mathbf{i_1})\mathbf{e_1} + f'_{e2}(z_1 + z_2\mathbf{i_1})\mathbf{e_2}$$

$\forall\ z_1 + z_2\mathbf{i_2} \in X_1 \times_e X_2$.

Theorem 2.7. *Let* X *be an open set in* $\mathbb{T}$, *and let* $f : X \longrightarrow \mathbb{T}$ *be a* $\mathbb{T}$*-holomorphic function on* X. *Then there exist holomorphic functions* $f_{e1} : X_1 \longrightarrow \mathbb{C}(\mathbf{i_1})$ *and* $f_{e2} : X_2 \longrightarrow \mathbb{C}(\mathbf{i_1})$ *with* $X_1 = P_1(X)$ *and* $X_2 = P_2(X)$, *such that:*

$$f(z_1 + z_2\mathbf{i_2}) = f_{e1}(z_1 - z_2\mathbf{i_1})\mathbf{e_1} + f_{e2}(z_1 + z_2\mathbf{i_1})\mathbf{e_2} \ \forall\ z_1 + z_2\mathbf{i_2} \in X.$$

3. Bicomplex meromorphic functions

3.1. Basic definitions

In the complex plane, it is well known (see [24]) that a function f is meromorphic in an open set U if and only if f is a quotient g/h of two functions which are holomorphic in U where h is not identically zero in any component of U. Based on this definition we define a bicomplex meromorphic function as follows.

Definition 3.1. A function f is said to be bicomplex meromorphic in an open set $X \subset \mathbb{T}$ if f is a quotient g/h of two functions which are bicomplex holomorphic in X where h is not identically in the null-cone in any component of X.

Theorem 3.2. *Let* $f : X \longrightarrow \mathbb{T}$ *be a bicomplex meromorphic function on the open set* $X \subset \mathbb{T}$. *Then there exist meromorphic functions* $f_{e1} : X_1 \longrightarrow \mathbb{C}(\mathbf{i_1})$ *and* $f_{e2} : X_2 \longrightarrow \mathbb{C}(\mathbf{i_1})$ *with* $X_1 = P_1(X)$ *and* $X_2 = P_2(X)$, *such that:*

$$f(z_1 + z_2\mathbf{i_2}) = f_{e1}(z_1 - z_2\mathbf{i_1})\mathbf{e_1} + f_{e2}(z_1 + z_2\mathbf{i_1})\mathbf{e_2} \ \forall\ z_1 + z_2\mathbf{i_2} \in X.$$

Proof. Let $f : X \longrightarrow \mathbb{T}$ be a bicomplex meromorphic function on X. Then f is a quotient g/h of two functions which are bicomplex holomorphic in X where h is not identically in the null-cone in any component of X. Therefore, from Theorem 2.7 and Proposition 2.4, there exist holomorphic functions $g_{e1}, h_{e1} : X_1 \longrightarrow \mathbb{C}(\mathbf{i_1})$ and $g_{e2}, h_{e2} : X_2 \longrightarrow \mathbb{C}(\mathbf{i_1})$ with $X_1 = P_1(X)$ and $X_2 = P_2(X)$, such that:

$$g(z_1 + z_2\mathbf{i_2}) = g_{e1}(z_1 - z_2\mathbf{i_1})\mathbf{e_1} + g_{e2}(z_1 + z_2\mathbf{i_1})\mathbf{e_2}$$

and

$$h(z_1 + z_2\mathbf{i_2}) = h_{e1}(z_1 - z_2\mathbf{i_1})\mathbf{e_1} + h_{e2}(z_1 + z_2\mathbf{i_1})\mathbf{e_2}$$

$\forall\ z_1 + z_2\mathbf{i_2} \in X$ where h_{ei} is not identically zero in any component of X_i for $i = 1, 2$. Hence,

$$\begin{aligned} f(z_1 + z_2\mathbf{i_2}) &= \frac{g_{e1}(z_1 - z_2\mathbf{i_1})\mathbf{e_1} + g_{e2}(z_1 + z_2\mathbf{i_1})\mathbf{e_2}}{h_{e1}(z_1 - z_2\mathbf{i_1})\mathbf{e_1} + h_{e2}(z_1 + z_2\mathbf{i_1})\mathbf{e_2}} \\ &= \frac{g_{e1}(z_1 - z_2\mathbf{i_1})}{h_{e1}(z_1 - z_2\mathbf{i_1})}\mathbf{e_1} + \frac{g_{e2}(z_1 + z_2\mathbf{i_1})}{h_{e2}(z_1 + z_2\mathbf{i_1})}\mathbf{e_2} \\ &= f_{e1}(z_1 - z_2\mathbf{i_1})\mathbf{e_1} + f_{e2}(z_1 + z_2\mathbf{i_1})\mathbf{e_2} \end{aligned}$$

where f_{ei} is meromorphic in X_i for i=1,2. □

Definition 3.3. Let $f : X \longrightarrow \mathbb{T}$ be a bicomplex meromorphic function on the open set $X \subset \mathbb{T}$. We will say that $w = (z_1 - z_2\mathbf{i_1})\mathbf{e_1} + (z_1 + z_2\mathbf{i_1})\mathbf{e_2} \in X$ is a (strong) pole for the bicomplex meromorphic function

$$F(w) = F_{e1}(z_1 - z_2\mathbf{i_1})\mathbf{e_1} + F_{e1}(z_1 + z_2\mathbf{i_1})\mathbf{e_2}$$

if $z_1 - z_2\mathbf{i_1} \in P_1(X)$ (and) or $z_1 + z_2\mathbf{i_1} \in P_2(X)$ are poles for $F_{e1} : P_1(X) \longrightarrow \mathbb{C}(\mathbf{i_1})$ and $F_{e2} : P_2(X) \longrightarrow \mathbb{C}(\mathbf{i_1})$ respectively.

Remark 3.4. Poles of bicomplex meromorphic functions are not isolated singularities.

It is also easy to obtain the following characterization of poles.

Proposition 3.5. *Let $f : X \longrightarrow \mathbb{T}$ be a bicomplex meromorphic function on the open set $X \subset \mathbb{T}$. If $w_0 \in X$ then w_0 is a pole of f if and only if*

$$\lim_{w \to w_0} |f(w)| = \infty.$$

Definition 3.6. The order of a bicomplex meromorphic function

$$F(w) = F_{e1}(z_1 - z_2\mathbf{i_1})\mathbf{e_1} + F_{e2}(z_1 + z_2\mathbf{i_1})\mathbf{e_2}$$

is defined as

$$\rho(F) = max\{\rho(F_{e1}), \rho(F_{e2})\}.$$

Finally, to avoid any confusion, we will say that a function $f : \mathbb{T} \longrightarrow \mathbb{T}$ is a **transcendental bicomplex meromorphic function** on $\mathbb{T}$ if $f_{ei} : \mathbb{C}(\mathbf{i_1}) \longrightarrow \mathbb{C}(\mathbf{i_1})$ is a transcendental meromorphic function for $i = 1, 2$.

3.2. Factorization of bicomplex meromorphic functions

In this subsection we introduce the bicomplex version of the factorization of meromorphic functions in the plane.

Definition 3.7. Let F be a bicomplex meromorphic function on $\mathbb{T}$. Then F is said to have f and g as left and right factors respectively if F_{ei} has f_{ei} and g_{ei} as left and right factors respectively for $i = 1, 2$, i.e., f_{ei} is meromorphic and g_{ei} is entire (g_{ei} may be meromorphic when f_{ei} is rational) for $i = 1, 2$.

Remark 3.8. If F has f and g as left and right factors respectively then we always have the following factorization: $F(w) = f(g(w))$.

Proof. Let $F_{ei} = f_{ei}(g_{ei}(z))$ on $\mathbb{C}(\mathbf{i_1})$ for $i = 1, 2$. Then

$$\begin{aligned} f(g(w)) &= f(g_{e1}(z_1 - z_2\mathbf{i_1})\mathbf{e_1} + g_{e2}(z_1 + z_2\mathbf{i_1})\mathbf{e_2}) \\ &= f_{e1}(g_{e1}(z_1 - z_2\mathbf{i_1}))\mathbf{e_1} + f_{e2}(g_{e2}(z_1 + z_2\mathbf{i_1}))\mathbf{e_2}) \\ &= F_{e1}(z_1 - z_2\mathbf{i_1})\mathbf{e_1} + F_{e2}(z_1 + z_2\mathbf{i_1})\mathbf{e_2} \\ &= F(w). \end{aligned}$$

□

Theorem 3.9. *Let $F(w)$ be a bicomplex meromorphic function on $\mathbb{T}$. If $F(w) = f(g(w))$ where f is bicomplex meromorphic and g is bicomplex entire (g may be bicomplex meromorphic when f is bicomplex rational) then F has f and g as left and right factors respectively.*

Proof. From Theorem 3.2,

$$F(z_1 + z_2\mathbf{i_2}) = F_{e1}(z_1 - z_2\mathbf{i_1})\mathbf{e_1} + F_{e2}(z_1 + z_2\mathbf{i_1})\mathbf{e_2} \text{ on } \mathbb{T}$$

where F_{ei} is meromorphic on $\mathbb{C}(\mathbf{i_1})$ for $i = 1, 2$. Moreover, $\forall w \in \mathbb{T}$

$$\begin{aligned} F(w) &= f(g(w)) \\ &= f(g_{e1}(z_1 - z_2\mathbf{i_1})\mathbf{e_1} + g_{e2}(z_1 + z_2\mathbf{i_1})\mathbf{e_2}) \\ &= f_{e1}(g_{e1}(z_1 - z_2\mathbf{i_1}))\mathbf{e_1} + f_{e2}(g_{e2}(z_1 + z_2\mathbf{i_1}))\mathbf{e_2}. \end{aligned}$$

Hence, $F_{ei} = f_{ei}(g_{ei}(z))$ on $\mathbb{C}(\mathbf{i_1})$ where f_{ei} is meromorphic and g_{ei} is entire (g_{ei} may be meromorphic when f_{ei} is rational) for $i = 1, 2$. □

Proposition 3.10. *The converse of Theorem* 3.9 *is false.*

Proof. Supposed that F_{ei} has f_{ei} and g_{ei} as left and right factors respectively. In that case, the functions $F = f(g(w))$. However, in the situation where you have a rational function (with poles) for f_{e1} with a meromorphic function (with poles) for g_{e1} and an entire function for f_{e2} and g_{e2} then the complete function g will be bicomplex meromorphic (with poles) where f is not bicomplex rational. □

It is now possible to define the concept of prime (pseudo-prime) function in terms of the idempotent representation.

Definition 3.11. A bicomplex meromorphic function

$$F(z_1 + z_2\mathbf{i_2}) = F_{e1}(z_1 - z_2\mathbf{i_1})\mathbf{e_1} + F_{e2}(z_1 + z_2\mathbf{i_1})\mathbf{e_2} \text{ on } \mathbb{T}$$

is said prime (pseudo-prime), if the meromorphic functions F_{e1} and F_{e2} are prime (pseudo-prime).

Remark 3.12. All bicomplex polynomials are pseudo-prime, and bicomplex polynomials of prime degree are prime.

Theorem 3.13. *If every factorization of a bicomplex meromorphic function $F(w) = f(g(w))$ into left and right factors implies that f or g is bicomplex linear (bicomplex polynomial or f is rational) then F is prime (pseudo-prime).*

Proof. First, we note that a bicomplex meromorphic function $h(z_1 + z_2\mathbf{i_2}) = h_{e1}(z_1 - z_2\mathbf{i_1})\mathbf{e_1} + h_{e2}(z_1 + z_2\mathbf{i_1})\mathbf{e_2}$ is bicomplex linear (bicomplex polynomial) if and only if h_{ei} is linear (polynomial) for $i = 1, 2$. Now, since every factorization of $F(w)$ of the form $f(g(w))$ implies that either f or g is bicomplex linear (bicomplex polynomial or f is bicomplex rational), then f_{ei} or g_{ei} is linear (polynomial or f_{ei} is rational) for $i = 1, 2$. This further implies that F_{ei} is prime (pseudo-prime) for $i = 1, 2$. □

Proposition 3.14. *The converse of Theorem* 3.13 *is false.*

Proof. Supposed that every factorization of $F_{ei}(w) = f_{ei}(g_{ei}(w))$ into left and right factors implies that f_{ei} or g_{ei} is polynomial or f_{ei} is rational for $i = 1, 2$. In that case, the function F is supposed to be pseudo-prime but in the situation where you have a polynomial for f_{e1} and a rational function for f_{e2} then the complete function f in the bicomplex space will be neither a bicomplexe polynomial nor a bicomplex rational function. □

In 1973, Gross, Osgoods and Yang posed the following problem (see [6]): Given any transcendental entire function f, does there exist a meromorphic function g such that fg is prime? In [13], Noda gave an affirmative answer to the above problem and in [8] Qiao and Yongxing extended this to meromorphic functions. The next theorem will show that the same result is also true in the bicomplex case.

Theorem 3.15. *Let F be any bicomplex transcendental meromorphic function, then there exists a bicomplex meromorphic function G such that GF is prime.*

Proof. Let $F(w) = F_{e1}(z_1 - z_2\mathbf{i_1})\mathbf{e_1} + F_{e1}(z_1 + z_2\mathbf{i_1})\mathbf{e_2}$ be any bicomplex transcendental meromorphic function. Therefore, F_{ei} is a transcendental meromorphic function for $i = 1, 2$. Hence, there exists a meromorphic function G_{ei} such that $F_{ei}G_{ei}$ is prime for $i = 1, 2$ (see Qiao and Yongxing [8]). Now, from Definition 3.11, FG is prime when G is defined as follows:

$$G(w) := G_{e1}(z_1 - z_2\mathbf{i_1})\mathbf{e_1} + G_{e1}(z_1 + z_2\mathbf{i_1})\mathbf{e_2}.$$ □

Theorem 3.16. *A bicomplex transcendental entire function of finite order such that $F(\mathbb{T}) \subset \mathbb{T}^{-1}$ is pseudo-prime.*

Proof. Let F be a bicomplex transcendental entire function such that $F(\mathbb{T}) \subset \mathbb{T}^{-1}$. Since $\rho(F)$ is finite, $\rho(F_{ei})$ is also finite for $i = 1, 2$. Moreover, since $F(\mathbb{T})$ is always invertible, it follows from Proposition 2.4 that the entire function F_{ei} has no zeros for $i = 1, 2$. Thus by a result of Gross (see [5], p. 215, Theorem 1) it follows that each F_{ei} is pseudo-prime. Hence by Definition 3.11, F is pseudo-prime. □

Example. $\exp(z_1 + z_2\mathbf{i_2})$ is pseudo-prime.

In [5] Fred Gross conjectured that if f and g are non-linear entire functions, at least one of them transcendental, then the composite function $f \circ g$ has infinitely many fix-points. Its factorization version is: if P is a polynomial and if α is a non-constant entire function, then the function $F(z) = P(z)\exp(\alpha(z)) + z$ is prime. Bergweiler [2] proved this long pending conjecture in its general form as: if P and Q are polynomials and α is an entire function such that Q and α are non-constant, and P, Q and α do not have a non-linear common right factor, then $P(z)\exp(\alpha(z)) + Q(z)$ is prime, and conversely also. The bicomplex analogue of Bergweiler's result also holds with stronger hypotheses.

Definition 3.17. Let $F(w) = F_{e1}(z_1 - z_2\mathbf{i_1})\mathbf{e_1} + F_{e1}(z_1 + z_2\mathbf{i_1})\mathbf{e_2} : D \longrightarrow \mathbb{T}$ be any bicomplex function. The function F is said to be strongly non-constant (non-linear) on D if F_{ei} is non-constant (non-linear) on $P_i(D)$ for $i = 1, 2$.

Theorem 3.18. *Let P and Q be bicomplex polynomials and α be a bicomplex entire function. Suppose that Q and α are strongly non-constant for $i = 1, 2$ and let $F(w) = P(w)\exp(\alpha(w)) + Q(w)$. Then F is prime if and only if P, Q, and α do not have strongly non-linear bicomplex common right factor.*

Proof. Let

$$
\begin{aligned}
P(z_1 + z_2\mathbf{i_2}) &= P_{e1}(z_1 - z_2\mathbf{i_1})\mathbf{e_1} + P_{e2}(z_1 + z_2\mathbf{i_1})\mathbf{e_2} \\
Q(z_1 + z_2\mathbf{i_2}) &= Q_{e1}(z_1 - z_2\mathbf{i_1})\mathbf{e_1} + Q_{e2}(z_1 + z_2\mathbf{i_1})\mathbf{e_2} \\
\alpha(z_1 + z_2\mathbf{i_2}) &= \alpha_{e1}(z_1 - z_2\mathbf{i_1})\mathbf{e_1} + \alpha_{e2}(z_1 + z_2\mathbf{i_1})\mathbf{e_2}.
\end{aligned}
$$

Then it follows that

$$g(z_1 + z_2\mathbf{i_2}) = g_{e1}(z_1 - z_2\mathbf{i_1})\mathbf{e_1} + g_{e2}(z_1 + z_2\mathbf{i_1})$$

is a strongly non-linear bicomplex common right factor of P, Q, and α if and only if g_{ei} is a non-linear common right factor of P_{ei}, Q_{ei}, and α_{ei} for $i = 1, 2$.

Now since

$$f \circ g = (f_{e1} \circ g_{e1})\mathbf{e_1} + (f_{e2} \circ g_{e2})\mathbf{e_2}$$

and

$$f + g = (f_{e1} + g_{e1})\mathbf{e_1} + (f_{e2} + g_{e2})\mathbf{e_2},$$

we have

$$\begin{aligned}F(w) &= P(w)\exp(\alpha(w)) + Q(w)\\ &= (P_{e1}(w_1)\mathbf{e_1} + P_{e2}(w_2)\mathbf{e_2})(e^{\alpha_{e1}(w_1)}\mathbf{e_1} + e^{\alpha_{e2}(w_2)}\mathbf{e_2})\\ &\quad + (Q_{e1}(w_1)\mathbf{e_1} + Q_{e2}(w_2)\mathbf{e_2})\\ &= (P_{e1}(w_1)e^{\alpha_{e1}(w_1)} + Q_{e1}(w_1))\mathbf{e_1} + (P_{e2}(w_2)e^{\alpha_{e2}(w_2)} + Q_{e2}(w_2))\mathbf{e_2}\end{aligned}$$

where $w = z_1 + z_2\mathbf{i_2}$, $w_1 = z_1 - z_2\mathbf{i_1}$ and $w_2 = z_1 + z_2\mathbf{i_1}$. Writing $F(w) = F_{e1}(w_1)\mathbf{e_1} + F_{e2}(w_2)\mathbf{e_2}$ we find from above that

$$F_{ei}(w_i) = P_{ei}(w_i)e^{\alpha_{ei}(w_i)} + Q_{ei}(w_i)$$

which by Bergweiler's result (see [2]) is prime if and only if P_{ei}, Q_{ei}, and α_{ei} do not have a non-linear common right factor for $i = 1, 2$. Therefore, since an arbitrary bicomplex function $f(w)$ is strongly non-constant if and only if $P_1(f(w))$ and $P_2(f(w))$ is non-constant, we obtain from our hypotheses and Theorem 3.13 that $F(w) = P(w)\exp(\alpha(w)) + Q(w)$ is prime if and only if P, Q, and α do not have strongly non-linear bicomplex common right factor. □

Remark 3.19. As for one complex variable, the Theorem 3.18 implies that if f and g are strongly non-linear bicomplex entire functions, at least one of them transcendental, then the composite function $f \circ g$ has infinitely many fix-points.

In [1], Baoqin and Guodong proved the following: if f is any transcendental meromorphic function, then for any non-constant fractional linear function ϕ, the set

$$\{a \in \mathbb{C} : f(z) + a\phi(z) \text{ is not prime}\}$$

is at most countable. We will now show that bicomplex version of this result is false.

Theorem 3.20. *Let f be any bicomplex transcendental meromorphic function, then for any non-constant fractional linear bicomplex function ϕ, the set*

$$\{a \in \mathbb{T} : f(w) + a\phi(w) \textit{ is not prime}\}$$

is empty or of cardinality $\aleph_1$.

Proof. For $w = z_1 + z_2\mathbf{i_2}$, $w_1 = z_1 - z_2\mathbf{i_1}$, and $w_2 = z_1 + z_2\mathbf{i_1}$ writing

$$f(w) = f_{e1}(w_1)\mathbf{e_1} + f_{e2}(w_2)\mathbf{e_2}$$

$$\phi(w) = \phi_{e1}(w_1)\mathbf{e_1} + \phi_{e2}(w_2)\mathbf{e_2},$$

where f_{ei} is transcendental meromorphic and ϕ_{ei} is fractional linear for $i = 1, 2$ where ϕ_{ei} is non-constant for $i = 1$ or $i = 2$. Without loss of generality, let ϕ_{e1} be non-constant. Then, the set

$$\{\alpha \in \mathbb{C}(\mathbf{i}_1) : f_{e1}(z) + \alpha\phi_{e1}(z) \text{ is not prime}\}$$

is at most countable.

Now by taking $a = a_{e1}\mathbf{e_1} + a_{e2}\mathbf{e_2}$ and since

$$f(w) + a\phi(w) = (f_{e1}(w_1) + a_{e1}\phi_{e1}(w_1))\mathbf{e_1} + (f_{e2}(w_2) + a_{e2}\phi_{e2}(w_2))\mathbf{e_2},$$

it follows from Definition 3.11 that:

$$\begin{aligned} f(w) + a\phi(w) \text{ is not prime } &\Leftrightarrow \quad f_{e1}(z) + a_{e1}\phi_{e1}(z) \text{ is not prime} \\ &\quad\ \ \textbf{or} \quad f_{e2}(z) + a_{e2}\phi_{e2}(z) \text{ is not prime.} \end{aligned}$$

Since $|\{\alpha \in \mathbb{C}(\mathbf{i_1}) : f_{e2}(z) + \alpha\phi_{e2}(z) \text{ is prime or not}\}| = |\mathbb{C}(\mathbf{i_1})| = \aleph_1$ then

$$\{a \in \mathbb{T} : f(w) + a\phi(w) \text{ is not prime}\}$$

is also of cardinality $\aleph_1$ except if the set is empty. □

Finally, by using idempotent representations of bicomplex numbers and bicomplex functions, and by using Definition 3.11 and Theorem 4.11 in [3] we have the following result.

Theorem 3.21. *Every bicomplex meromorphic solution of an nth-order ordinary bicomplex differential equation with bicomplex rational functions as coefficients is pseudo-prime.*

Example. $\cos(z_1 + z_2\mathbf{i_2})$ is pseudo-prime since it satisfies the ordinary bicomplex linear differential equation $y''(w) - y(w) = 0$.

Theorem 3.22. *Let* $F(z_1 + z_2\mathbf{i_2}) = \cos(z_1 + z_2\mathbf{i_2})$, *the possible forms of the factorization of* $F = f \circ g$ *are as follows:*

$$f(\zeta_1 + \zeta_2\mathbf{i_2}) = f_{e1}(\zeta_1 - \zeta_2\mathbf{i_1})\mathbf{e_1} + f_{e2}(\zeta_1 + \zeta_2\mathbf{i_1})\mathbf{e_2}$$

and

$$g(z_1 + z_2\mathbf{i_2}) = g_{e1}(z_1 - z_2\mathbf{i_1})\mathbf{e_1} + g_{e2}(z_1 + z_2\mathbf{i_1})\mathbf{e_2}$$

where the couple $(f_{ei}(z), g_{ei}(z))$ *is chosen from the following possible forms of factorization of* $\cos z$ *in the complex plane:*

(i) $f_{ei}(z) = \cos\sqrt{z}$, $g_{ei}(z) = z^2$;
(ii) $f_{ei}(z) = T_n(z)$, $g_{ei}(z) = \cos\frac{z}{n}$, *where* $T_n(z)$ *denotes the nth Chebyshev polynomial* $(n \geq 2)$;
(iii) $f_{ei}(z) = \frac{1}{2}(z^{-n} + z^n)$, $g_{ei}(z) = e^{\frac{iz}{n}}$, *where n denotes a non-negative integer*

for $i = 1, 2$. *Moreover, if the couple* $(f_{ei}(z), g_{ei}(z))$ *is of the form* (i) *or* (ii) *for each* $i = 1$ *and* 2, *then f and g are, in particular, entire holomorphic mappings of two complex variables.*

Proof. The proof is a direct consequence of the Theorems 2.2, 2.6 and 3.9 used with the three possible forms of factorization of $\cos z$ in the complex plane (see [3]). □

Corollary 3.23. *The complex cosine* $\cos z$ *has three possible forms of factorization in the complex plane and six additional possible forms of factorization in the bicomplex space.*

Proof. From Theorem 3.22 we have nine possible forms of factorization of the bicomplex cosine: $\cos(z_1 + z_2\mathbf{i_2})$. If we put $z_2 = 0$, we obtain automatically nine possible forms of factorization of the cosine in the complex plane. However, when $f(\zeta_1+\zeta_2\mathbf{i_2}) = f_{e1}(\zeta_1-\zeta_2\mathbf{i_1})\mathbf{e_1}+f_{e1}(\zeta_1+\zeta_2\mathbf{i_1})\mathbf{e_2}$ and $g(z_1+z_2\mathbf{i_2}) = g_{e1}(z_1-z_2\mathbf{i_1})\mathbf{e_1}+g_{e1}(z_1+z_2\mathbf{i_1})\mathbf{e_2}$ with $z_2 = 0$, we have that $g(z_1) = g_{e1}(z_1)\mathbf{e_1}+g_{e1}(z_1)\mathbf{e_2} = g_{e1}(z_1)$ and $(f \circ g)(z_1) = (f_{e1} \circ g_{e1})(z_1)$. In that case, we come back to the three classical complex forms of factorization of $\cos z_1$. Hence, the complex cosine has exactly six new possible forms of factorization in the bicomplex space. □

References

[1] L. Baoqin and S. Guodong, *On Goldbach problem concerning factorization of meromorphic functions*, Acta Math. Sinica **5** (4) (1997), 337–344.

[2] W. Bergweiler, *On factorization of certain entire functions*, J. Math.Anal. Appl. **193** (1995), 1003–1007.

[3] C.T. Chuang and C.C. Yang, *Fix-points and Factorization of meromorphic functions*, World Scientific, 1990.

[4] N. Fleury, M. Rausch de Traubenberg and R.M. Yamaleev, *Commutative extended complex numbers and connected trigonometry*, J. Math. Ann. and Appl. **180** (1993), 431–457.

[5] F. Gross, *Factorization of Meromorphic Function*, U.S. Government Printing Office, Washington, D. C., 1972.

[6] F. Gross, C.C. Yang and C. Osgoods, *Primeable entire function*, Nagoya Math J **51** (1973), 123–130.

[7] F. Gross, *On factorization of meromorphic functions*, Trans. Amer. Math. Soc. **131** (1968), 215–222.

[8] Q. Jin and G. Yongxing, *On Factorization of Meromorphic Function*, Acta Math. Sinica **13** (4) (1997), 509–512.

[9] T.W. Ng, *Imprimitive parameterization of analytic curves and factorization of entire functions*, J. London Math. Soc. **64** (2001), 385–394.

[10] T.W. Ng and C.C. Yang, *On the common right factors of meromorphic functions*, Bull. Austral. Math. Soc. **55** (1997), 395–403.

[11] T.W. Ng and C.C. Yang, *Certain criteria on the existence of transcendental entire common right factor*, Analysis **17** (1997), 387–393.

[12] T.W. Ng and C.C. Yang, *On the composition of prime transcendental function and a prime polynomial*, Pacific J. Math. **193** (2000), 131–141.

[13] Y. Noda, *On factorization of entire function*, Kodai Math J **4** (1981), 480–494.

[14] G.B. Price, *An introduction to multicomplex spaces and functions*, Marcel Dekker Inc., New York, 1991.

[15] D. Rochon, *A bicomplex Riemann zeta function*, Tokyo J. Math. **27** (2004), 357–369.

[16] D. Rochon, *A Bloch Constant for Hyperholomorphic Functions*, Complex Variables **44** (2001), 85–101.

[17] D. Rochon, *A generalized Mandelbrot set for bicomplex numbers*, Fractal **8** (2000), 355–368.

[18] D. Rochon, *On a relation of bicomplex pseudoanalytic function theory to the complexified stationary Schrödinger equation*, Complex Variables, **53** (6) (2008), 501–521.

[19] D. Rochon, *Sur une généralisation des nombres complexes: les tétranombres*, M. Sc. Université de Montréal, 1997.

[20] D. Rochon and M. Shapiro, *On algebraic properties of bicomplex and hyperbolic numbers*, Anal. Univ. Oradea, fasc. math., **11** (2004) , 71–110.

[21] D. Rochon and S. Tremblay, *Bicomplex Quantum Mechanics: I. The Generalized Schrödinger Equation*, Adv. App. Cliff. Alg. **12** (2) (2004), 231–248.

[22] D. Rochon and S. Tremblay, *Bicomplex Quantum Mechanics: II. The Hilbert Space*, Adv. App. Cliff. Alg. **16** (2) (2006), 135–157.

[23] P.C. Rosenbloom, *The fix-points of entire functions*, Medd. Lunds Univ. Math. Sem. M.Riesz (1952), 187–192.

[24] W. Rudin, *Real and Complex Analysis* 3rd *ed.*, New York, McGraw-Hill, 1976.

[25] B.V. Shabat, *Introduction to Complex Analysis part II: Functions of Several Variables*, American Mathematical Society, 1992.

[26] G. Sobczyk, *The hyperbolic number plane*, Coll. Maths. Jour. **26** (4) (1995), 268–280.

K.S. Charak
Department of Mathematics
University of Jammu
Jammu-180 006
India
e-mail: `kscharak7@rediffmail.com`

D. Rochon
Département de mathématiques et d'informatique
Université du Québec à Trois-Rivières
C.P. 500, Trois-Rivières
Québec, G9A 5H7
Canada
e-mail: `Dominic.Rochon@UQTR.CA`

Quaternionic and Clifford Analysis
Trends in Mathematics, 69–99

An Overview on Functional Calculus in Different Settings

Fabrizio Colombo, Graziano Gentili, Irene Sabadini and Daniele C. Struppa

Abstract. In this paper we give an overview of some different versions of functional calculus in a noncommutative setting. In particular, we will focus on a recent functional calculus based on the notion of slice-hyperholomorphy. This notion, in suitable versions, will allow us to study the case of linear quaternionic operators, as well as the case of n-tuples of linear (real or complex) operators.

Mathematics Subject Classification (2000). Primary 47A10, 30G35; Secondary 47A60.

Keywords. Functional calculus, spectral theory, bounded and unbounded operators, linear quaternionic operators, n-tuples of linear operators.

1. Introduction

Traditionally, the term functional calculus refers to what should more appropriately be indicated as "holomorphic functional calculus"; given a holomorphic function f of one complex variable, and an operator T, functional calculus attempts to define an operator $f(T)$, thus extending the function f from the complex plane to the space of operators. More generally, of course, we can think of f as a function is some suitable linear space of functions $\mathcal{F}$, of T as an operator in a suitable linear space of operators, and of "functional calculus" as a way to define $f(T)$. The classical theory of functional calculus is very well developed, and its importance for operator theory, as well as its applications to physics, are well established, see, e.g., [11].

In order to set the stage for what will follow, let X be a complex Banach space, and let T be a linear bounded operator on X. Suppose furthermore that f is a holomorphic function of one complex variable, so that it can be written as a power series $f(z) = \sum_{n\geq 0} a_n z^n$, with a_n, $z \in \mathbb{C}$ converging in an open disc that contains the set $|z| \leq \|T\|$. Under these natural conditions, one can easily define the value of f on T by simply writing $f(T) = \sum_{n\geq 0} a_n T^n$. Maybe the first and most fundamental example of such process is the definition of the operator e^T, which is defined simply as

$$e^T = \sum_{n\geq 0} \frac{T^n}{n!}.$$

As it is well known, however, this approach (while quite natural) is not fully satisfactory. In order to generalize such an approach, one first needs to realize that $f(T)$ can only be expected to be defined on the spectrum of T. Under this condition, we can now consider a function f, holomorphic on the spectrum of T, and we can use the Cauchy integral formula to define $f(T)$; this will be shown in detail in the next section. This approach is clearly of great importance as it can now be used to define a functional calculus when the operator T is unbounded (in which case the series $\sum_{n\geq 0} a_n T^n$ does not converge).

The situation quickly becomes more complicated if one wants to extend this approach to the case of several operators, and tries to define $f(T)$ when f is a holomorphic function of n complex variables and $T = (T_1, \dots, T_n)$ is an n-tuple of operators. For the case of bounded operators, one can attempt to follow the two different approaches inspired by the discussion above. On one hand, one can define $f(T)$ on a dense subset of $\mathcal{F}$ and then extend this definition by continuity. On the other hand, one can consider a reproducing integral formula for functions in $\mathcal{F}$ and try to extend it to get a function of operators.

This second procedure can be realized by using a Fourier type transform or a Cauchy type integral formula. In the literature, there are several different approaches to the topic according to the set $\mathcal{F}$ of functions and to the approach adopted. The choice of $\mathcal{F}$ may impose restrictions on the commutativity of operators considered or to the type of operators. For example, one can generalize the discussion in one complex variable to the case of several complex variables as in the earlier works of J.L. Taylor [23], [24], but this approach can be done only for n-tuples of commuting operators. Another possibility is to consider the Weyl calculus, see [25] and [2], for noncommuting, self-adjoint operators. One can also choose to use the noncommutative setting of Clifford algebra-valued functions. This approach is not new in the literature and among the works in this direction, we mention here the works of Jefferies, Kisil, MacIntosh and their coworkers [16], [17], [18], [19], [20], the book [15] and the references therein. Note that, despite the noncommutative setting which is useful in the case of several operators, one may still have restriction on the n-tuples of operators and on their spectrum. For example, the functional calculus in [20] works for commuting operators while the Riesz-Clifford calculus in [18] works for bounded self-adjoint operators. Of course, for the purpose of applications to physics, and for sake of generality, it would be of great interest to be able to lift these restrictions. This goal can be achieved by considering the n-tuple $T = (T_1, \dots, T_n)$ of operators as a unique operator to be substituted in place of a suitable hypercomplex variable. We will show how

this can be accomplished in our own approach based on the new theory of slice hyperholomorphy introduced in the papers [10] and [14]. For quaternionic operators we have developed our theory in the papers [4], [5], [6], while for n-tuples of noncommuting bounded and unbounded operators see [9].
The plan of the paper is as follows. In Section 2 we recall the classical Riesz-Dunford functional calculus, which is the inspiration (and the test-case) for all later theories. The following sections are devoted to our own recent work in this area, and to a variety of different approaches. Specifically, in Section 3 we recall the notion of slice-regular quaternionic functions, [14], and we use it to define and develop a new quaternionic functional calculus. Section 4 is devoted to an important variation of those ideas. Namely we recall the notion of slice-monogenic functions, [10], and we use it to introduce and develop a functional calculus for n-tuples of noncommuting operators. In Section 5 we describe one of the first generalizations of Riesz-Dunford functional calculus, namely the use of monogenic functions, to introduce monogenic functional calculus (just like Section 2, this section does not contain any of our results, but it is included to offer the background and justification for our work). The last section is devoted to a different way to use holomorphicity to define a functional calculus. As it is well known, the first and so far most successful notion of quaternionic holomorphicity is due to Fueter, and goes under the name of Cauchy-Fueter regularity. Indeed, it can be said that until the development of slice-regularity, this definition was the dominating one. As our last section will demonstrate, it is indeed possible to build a functional calculus based on this notion of regularity, but the kind of difficulties which arise are probably one of the best arguments for the development of an alternative definition of regularity, namely the slice-regularity we introduce in Section 3.

2. The Riesz-Dunford functional calculus

We begin by quickly revising the Riesz-Dunford functional calculus for one operator, which is based on the theory of holomorphic functions in one complex variable. For all the details and the proofs of the results in this section see [11]. Let us consider the polynomial

$$P(z) = a_0 + a_1 z + \cdots + a_n z^n$$

where $a_j, z \in \mathbb{C}$. We observe that if we replace z by a bounded linear operator T we get the linear operator

$$P(T) = a_0 \mathcal{I} + a_1 T + \cdots + a_n T^n,$$

where $\mathcal{I}$ denotes the identity operator. This approach can be extended to the case in which $P(z)$ is a power series converging on $|z| \leq \|T\|$. In general, functions of operators can be defined using the Cauchy integral formula:

Theorem 2.1. *Let $f : \Omega \to \mathbb{C}$ be a holomorphic function over an open set $\Omega \subseteq \mathbb{C}$ and let $z \in \Omega$. Let $\partial\Omega$ be a closed Jordan curve surrounding z. Then*

$$f(z) = \frac{1}{2\pi i} \int_{\partial\Omega} \frac{f(\lambda)}{\lambda - z}\, d\lambda.$$

The Cauchy kernel $(\lambda - z)^{-1}$ plays here an important role. Indeed, in terms of operators, it can be interpreted as $(\lambda\mathcal{I} - T)^{-1}$. So, in analogy to what happens in the theory of holomorphic functions, we write

$$f(T) = \frac{1}{2\pi i} \int_{\Gamma} (\lambda\mathcal{I} - T)^{-1} f(\lambda)\, d\lambda,$$

where Γ is a rectifiable Jordan curve that surrounds the "singularities" of $(\lambda\mathcal{I} - T)^{-1}$. Such singularities are the spectrum of T. Note that in the complex case the commutativity plays an important role. Indeed, if we consider the expansion

$$(\lambda\mathcal{I} - T)^{-1} = \sum_{n\geq 0} \lambda^{-1-n} T^n$$

we observe that, where the series converges, i.e., for $\|T\| < |\lambda|$, we can write

$$\sum_{n\geq 0} \lambda^{-1-n} T^n = \sum_{n\geq 0} T^n \lambda^{-1-n}.$$

Thus, we can compute an integral in the complex variable λ, isolating the powers of T:

$$\frac{1}{2\pi i} \int_{\Gamma} (\lambda\mathcal{I} - T)^{-1} f(\lambda)\, d\lambda, = \frac{1}{2\pi i} \sum_{n\geq 0} T^n \int_{\Gamma} \lambda^{-1-n} f(\lambda)\, d\lambda$$

and we set the value of the integral equal to $f(T)$.

Remark 2.2. We point out the fundamental fact that if T and λ do not commute and we want to write the integral

$$\frac{1}{2\pi i} \sum_{n\geq 0} T^n \int_{\Gamma} \lambda^{-1-n} f(\lambda)\, d\lambda$$

we must find the sum of the series

$$\sum_{n\geq 0} T^n \lambda^{-1-n}, \quad \text{for} \quad T\lambda \neq \lambda T.$$

Supposing that T and λ do not commute the sum of this series is not $(\lambda\mathcal{I} - T)^{-1}$. This is the crucial fact that will lead us to define, for slice hyperholomorphic function, a new notion of resolvent called S-resolvent operator and consequently a new notion of spectrum called S-spectrum.

We will recall below some of the main results for the functional calculus in the complex case.

2.1. Bounded operators

In the sequel, we will denote by X a complex Banach space and by T a bounded linear operator in X. We give the following definitions. The *resolvent set* $\rho(T)$ of T is the set of complex numbers λ, for which $(\lambda\mathcal{I} - T)^{-1}$ exists as a bounded operator with domain X. The *spectrum* $\sigma(T)$ of T is the complement of $\rho(T)$ and the number

$$r(T) = \sup\{|\lambda| \ : \ \lambda \in \sigma(T)\}$$

is called the *spectral radius* of T. Finally, the function $R(\lambda, T) = (\lambda\mathcal{I} - T)^{-1}$, defined on $\rho(T)$, is called the *resolvent* of T.

We have the following properties:

Proposition 2.3. *Let T be a bounded linear operator in X.*

1. *The resolvent set $\rho(T)$ is open.*
2. *The closed set $\sigma(T)$ is bounded and nonempty.*
3. *The function $R(\lambda, T)$ is analytic in $\rho(T)$.*
4. $r(T) = \lim_{n\to\infty} \sqrt[n]{\|T^n\|}$.
5. *(The resolvent equation) For every pair λ, $\mu \in \rho(T)$ we have*

$$R(\lambda, T) - R(\mu, T) = (\mu - \lambda)R(\lambda, T)R(\mu, T).$$

Definition 2.4. Let T be a bounded linear operator in X. By $\mathcal{F}(T)$ we denote the family of functions f which are analytic on some neighborhood of $\sigma(T)$.

Definition 2.5. Let $f \in \mathcal{F}(T)$, and let U be an open set whose boundary ∂U consists of a finite number of rectifiable Jordan curves, oriented in the positive sense. Suppose that $\sigma(T) \subseteq U$ and that $U \cup \partial U$ is contained in the domain of analyticity of f. The operator $f(T)$ is defined by

$$f(T) = \frac{1}{2\pi i} \int_{\partial U} R(\lambda, T) f(\lambda)\, d\lambda. \tag{2.1}$$

It is a classical result the fact the integral (2.1) depends only on f and does not depend on the open set U. The map associating to any function $f \in \mathcal{F}(T)$ the function $f(T)$ is the classical Riesz-Dunford functional calculus whose main properties are summarized below:

Theorem 2.6. *Let f, $g \in \mathcal{F}(T)$, α_1, $\alpha_2 \in \mathbb{C}$. Then*

- $\alpha_1 f + \alpha_2 g \in \mathcal{F}(T)$ *and* $(\alpha_1 f + \alpha_2 g)(T) = \alpha_1 f(T) + \alpha_2 g(T)$.
- $f \cdot g \in \mathcal{F}(T)$ *and* $f(T)g(T) = (f \cdot g)(T)$.
- *If* $f(\lambda) = \sum_{n\geq 0} \alpha_k \lambda^k$ *converges in a neighborhood of* $\sigma(T)$, *then* $f(T) = \sum_{n\geq 0} \alpha_k T^k$.

Theorem 2.7. *(The Spectral Mapping Theorem) If $f \in \mathcal{F}(T)$, then $f(\sigma(T)) = \sigma(f(T))$.*

Theorem 2.8. *Let $f \in \mathcal{F}(T)$, $g \in \mathcal{F}(f(T))$, and $F(\lambda) = g(f(\lambda))$. Then $F \in \mathcal{F}(T)$ and $F(T) = g(f(T))$.*

Theorem 2.9. (*Perturbation*) *Let T be a linear bounded operator, $f \in \mathcal{F}(T)$ and $\varepsilon > 0$. Then there is a $\delta > 0$ such that if T_1 is a bounded operator and $\|T_1 - T\| < \delta$, then $f \in \mathcal{F}(T_1)$ and $\|f(T_1) - f(T)\| < \varepsilon$.*

2.2. Unbounded closed operators

In the case of unbounded operators the spectrum can be defined as in the previous subsection. However, in this case, it may be a bounded set, an unbounded set, the empty set, or even the whole plane. We will avoid this last case, as we will assume $\rho(T) \neq \emptyset$. The following calculus is based on the calculus of bounded operators.

Definition 2.10. By $\mathcal{F}_\infty(T)$ we denote the family of functions f which are analytic on some neighborhood of $\sigma(T)$ and at ∞.

The neighborhood need not be connected and it can depend on f. Let $\alpha \in \rho(T)$ and define

$$A = (T - \alpha\mathcal{I})^{-1} = -R(\alpha, T).$$

The operator A defines a one-to-one mapping from X onto the domain $D(T)$ of T. We can now define a functional calculus for T in terms of the bounded operator A. Denote by $\mathcal{K}$ the complex sphere, with its usual topology, and define the homeomorphism

$$\mu = \Phi(\lambda) = (\lambda - \alpha)^{-1}, \qquad \Phi(\infty) = 0, \qquad \Phi(\alpha) = \infty.$$

Theorem 2.11. *Let $\alpha \in \rho(T)$. Then $\Phi(\sigma(T) \cup \{\infty\}) = \sigma(A)$ and the relation*

$$\phi(\mu) = f(\Phi^{-1}(\mu))$$

determines a one-to-one correspondence between $f \in \mathcal{F}_\infty(T)$ and $\phi \in \mathcal{F}(A)$.

Definition 2.12. Let $f \in \mathcal{F}_\infty(T)$. We define

$$f(T) = \phi(A),$$

where $\phi \in \mathcal{F}(A)$ and $\phi(\mu) = f(\Phi^{-1}(\mu))$.

Theorem 2.13. *Let $f \in \mathcal{F}_\infty(T)$. Then $f(T)$ is independent of the choice of $\alpha \in \rho(T)$. Let $V \supset \sigma(T)$ be an open set whose boundary ∂V consists of a finite number of rectifiable Jordan arcs. Let f be analytic on $V \cup \partial V$ and suppose that ∂V has positive orientation with respect to the set V. Then*

$$f(T) = f(\infty)\mathcal{I} + \frac{1}{2\pi i}\int_{\partial V} R(\lambda, T)\, f(\lambda)\, d\lambda. \tag{2.2}$$

Theorem 2.14. *Let f, $g \in \mathcal{F}_\infty(T)$, α_1, $\alpha_2 \in \mathbb{C}$. Then*

- *$\alpha_1 f + \alpha_2 g \in \mathcal{F}(T)$ and $(\alpha_1 f + \alpha_2 g)(T) = \alpha_1 f(T) + \alpha_2 g(T)$.*
- *$f \cdot g \in \mathcal{F}(T)$ and $f(T)g(T) = (f \cdot g)(T)$.*
- *$\sigma(f(T)) = f(\sigma(T) \cup \{\infty\})$.*

3. Slice quaternionic functions and the quaternionic functional calculus

3.1. Slice regular functions of one quaternionic variable

In the sequel, we will denote by $\mathbb{H}$ the algebra of real quaternions. Given a quaternion $q = x_0 + ix_1 + jx_2 + kx_3 \in \mathbb{H}$ let us denote its real part x_0 by $\mathrm{Re}[q]$ and its imaginary part $ix_1 + jx_2 + kx_3$ by $\mathrm{Im}[q]$. By $\mathbb{S}$ we will denote the sphere of purely imaginary unit quaternions, i.e.,

$$\mathbb{S} = \{q = ix_1 + jx_2 + kx_3 \mid x_1^2 + x_2^2 + x_3^2 = 1\}.$$

Definition 3.1. Let $U \subseteq \mathbb{H}$ be an open set and let $f : U \to \mathbb{H}$ be a real differentiable function. Let $I \in \mathbb{S}$ and let f_I be the restriction of f to the complex plane $L_I := \mathbb{R} + I\mathbb{R}$ passing through 1 and I. We say that f is a left slice regular function, in short regular function, if for every $I \in \mathbb{S}$

$$\frac{1}{2}\left(\frac{\partial}{\partial x} + I\frac{\partial}{\partial y}\right) f_I(x + Iy) = 0.$$

A crucial fact in the theory of regular functions is that polynomials and, more in general, power series in the variable q are regular. Conversely, every regular function can be represented as a power series as shown in the next result (see [14]:

Theorem 3.2. *If $B = B(0, R) \subseteq \mathbb{H}$ is the open ball centered in the origin with radius $R > 0$ and $f : B \to \mathbb{H}$ is a left regular function, then f has a series expansion of the form*

$$f(q) = \sum_{n=0}^{+\infty} q^n \frac{1}{n!}\frac{\partial^n f}{\partial x^n}(0)$$

converging on B.

Note that, even though the definition of regular function involves the direction of the unit quaternion I, the coefficients of the series expansion do not depend on the choice of I. Despite these evident advantages, the theory of regular functions still presents some disappointing facts which appear also with the classical regular functions in the sense of Cauchy-Fueter. Indeed, in general, the product or the composition of two regular functions is not regular.

Remark 3.3. A statement analogous to Theorem 3.2 holds for regular functions in an open ball centered in $p_0 \in \mathbb{R}$. For regular functions in an open ball $B(q_0, R)$ centered in a non real quaternion $q_0 \in \mathbb{H} \setminus \mathbb{R}$, the situation is significantly more complicated. First of all, to obtain an interesting family of regular functions one needs to require that $B(q_0, R) \cap \mathbb{R} \neq \emptyset$. Moreover, the problem of the series expansion centered at q_0 has to be answered in a different fashion since the binomial $(q - q_0)^n$ is not a regular function. These peculiarities, which appear explicitly for example in [21], [12], are presently under scrutiny.

A main result in the theory of regular functions is the analogue of the Cauchy integral formula. In order to state the result we need some notation. Given a quaternion q, we set

$$I_q = \begin{cases} \dfrac{\mathrm{Im}[q]}{|\mathrm{Im}[q]|} & \text{if } \mathrm{Im}[q] \neq 0 \\ \text{any element of } \mathbb{S} \text{ otherwise.} \end{cases}$$

We have the following, see [14]:

Theorem 3.4. *Let $f : B(0,R) \to \mathbb{H}$ be a regular function and let $q \in B(0,R)$. Then*

$$f(q) = \frac{1}{2\pi} \int_{\partial\Delta_q(0,r)} (\zeta - q)^{-1} \, d\zeta_{I_q} \, f(\zeta)$$

where $d\zeta_{I_q} = -I_q d\zeta$ and $r > 0$ is such that

$$\overline{\Delta_q(0,r)} := \{x + I_q y \mid x^2 + y^2 \leq r^2\}$$

contains q and is contained in $B(0,R)$.

It is important to point out that the function $R(q) = (q - p_0)^{-1}$, is regular if and only if $p_0 \in \mathbb{R}$. This implies that the Cauchy formula in Theorem 3.4 has a non regular kernel. By noting that, in the complex case, the Cauchy kernel $(\zeta - z)^{-1}$ is the sum of the series $\sum_n z^n \zeta^{-n-1}$ we are naturally lead to the problem of finding the sum of the series $\sum_n q^n \zeta^{-n-1}$ and give the following definition:

Definition 3.5. Let q, $s \in \mathbb{H}$ such that $sq \neq qs$. We will call noncommutative Cauchy kernel series (shortly Cauchy kernel series) the expansion

$$S^{-1}(s,q) := \sum_{n \geq 0} q^n s^{-1-n},$$

for $|q| < |s|$.

Remark 3.6. The sum of the Cauchy kernel series is not, in general, $(s - q)^{-1}$. Indeed, let q, $s \in \mathbb{H}$ and consider the equality

$$(s-q)^{-1} = [(1 - qs^{-1})s]^{-1} = s^{-1}(1 - qs^{-1})^{-1} = \sum_{n \geq 0} s^{-1}(qs^{-1})^n \tag{3.1}$$

which holds in the domain of convergence of the power series, i.e., for $|q| < |s|$. If we regard s as a fixed (non real) quaternion and q as a quaternionic variable, then when $q \in L_{I_s}$ we have that q and s commute and thus, in this particular case, we obtain

$$(s-q)^{-1}_{|L_{I_s}} = \sum_{n \geq 0} q^n s^{-1-n} \tag{3.2}$$

for all $q \in L_{I_s}$ such that $|q| < |s|$.

To find the sum of the Cauchy kernel series we will use the following result:

Theorem 3.7. *Let $S^{-1}(s,q)$ be the Cauchy kernel series for $sq \neq qs$. The inverse $S(s,q)$ of $S^{-1}(s,q)$ is solution to the equation*

$$S^2 + Sq - sS = 0. \tag{3.3}$$

The non trivial solution of (3.3) *is given by*

$$S(s,q) = (s+q-2\ \mathrm{Re}[s])^{-1}(sq-|s|^2)-q = -(q-\bar{s})^{-1}(q^2-2q\mathrm{Re}[s]+|s|^2), \tag{3.4}$$

and the solution coincides with the function $R(s,q) := s-q$ if and only if $sq = qs$.

Based on this result, we set the following

Definition 3.8. The function

$$S^{-1}(s,q) = -(q^2 - 2q\mathrm{Re}[s] + |s|^2)^{-1}(q-\bar{s})$$

which is defined for $q^2 - 2q\mathrm{Re}[s] + |s|^2 \neq 0$, is called noncommutative Cauchy kernel.

Remark 3.9. Note that if $s \in \mathbb{H}$ is a non real quaternion, then (see Lemma 3.8 in [5]) there exists no degree-one quaternionic polynomial $Q(q)$ such that

$$q^2 - 2q\mathrm{Re}[s] + |s|^2 = (q-\bar{s})Q(q). \tag{3.5}$$

Moreover, see Theorem 3.9 in [5], the function

$$S^{-1}(s,q) = -(q^2 - 2q\mathrm{Re}[s] + |s|^2)^{-1}(q-\bar{s})$$

cannot be extended continuously to any point of the set

$$\{(s,q) \in \mathbb{H}\times\mathbb{H} \ :\ q^2 - 2q\mathrm{Re}[s] + |s|^2 = 0\ \}.$$

3.2. Functional calculus for bounded operators

To introduce our functional calculus we will need, beside the basics on the theory of regular functions introduced in the previous section, also some preliminaries on the theory of linear operators on a right quaternionic vector space. Thus, we begin this section with a quick discussion on quaternionic linear operators.

Definition 3.10. Let V be a right vector space on $\mathbb{H}$. A map $T : V \to V$ is said to be right linear if

$$T(u+v) = T(u) + T(v)$$
$$T(us) = T(u)s$$

for all $s \in \mathbb{H}$ and all $u, v \in V$.

Definition 3.11. Let V be a bilateral quaternionic Banach space. We will denote by $\mathcal{B}(V)$ the bilateral vector space of all right linear bounded operators on V.

The set of right linear maps is not a quaternionic left or right vector space. If V is both left and right vector space, then the set $\mathrm{End}(V)$ of right linear maps on V is both a left and a right vector space on $\mathbb{H}$. In that case we can define $(aT)(v) := aT(v)$ and $(Ta)(v) := T(av)$. The composition of operators can be defined in the usual way. In particular, we have the identity operator $\mathcal{I}(u) = u$, for all $u \in V$ and setting $T^0 = \mathcal{I}$ we can define powers of a given operator $T \in \mathrm{End}(V)$:

$T^n = T(T^{n-1})$ for any $n \in \mathbb{N}$. An operator T is said to be invertible if there exists S such that $TS = ST = \mathcal{I}$ and we will write $S = T^{-1}$.

From now on we will only consider bilateral vector spaces V. The vector space $\mathrm{End}(V)$ is not an $\mathbb{H}$-algebra with respect to the composition of operators, in fact the property $s(TS) = (sT)S = T(sS)$ is not fulfilled for any $T, S \in \mathrm{End}(V)$ and any $s \in \mathbb{H}$. In this case we have $T(sS)(u) = T(sS(u))$ while $(sT)S(u)) = sT(S(u))$. Note on the other hand that $\mathrm{End}(V)$ is trivially an algebra over $\mathbb{R}$.

Mimicking Definition 3.5, we now give the notion of S-resolvent operator series:

Definition 3.12. (The S-resolvent operator series). Let $T \in \mathcal{B}(V)$ and let $s \in \mathbb{H}$. We define the S-resolvent operator series as

$$S^{-1}(s,T) := \sum_{n\geq 0} T^n s^{-1-n} \tag{3.6}$$

for $\|T\| < |s|$.

We have the following result (see [5]):

Theorem 3.13. *Let $T \in \mathcal{B}(V)$ and let $s \in \mathbb{H}$. The sum of the series* (3.6) *is given by*

$$S^{-1}(s,T) = -(T^2 - 2\mathrm{Re}[s]T + |s|^2\mathcal{I})^{-1}(T - \bar{s}\mathcal{I}), \tag{3.7}$$

for $\|T\| < |s|$.

Note that the operator $S^{-1}(s,T)$ is not the inverse of $s\mathcal{I} - T$. Indeed, as we proved in [5], the operator

$$\sum_{n\geq 0} (s^{-1}T)^n s^{-1}\mathcal{I}$$

is the right and left algebraic inverse of $s\mathcal{I} - T$. Moreover, the series converges in the operator norm for $\|T\| < |s|$. If $Ts\mathcal{I} = sT$, the operator $S^{-1}(s,T)$ equals $(s\mathcal{I} - T)^{-1}$ when the series (3.6) converges.

Definition 3.14. (The S-resolvent operator) Let $T \in \mathcal{B}(V)$ and let $s \in \mathbb{H}$. We define the S-resolvent operator as

$$S^{-1}(s,T) := -(T^2 - 2\mathrm{Re}[s]T + |s|^2\mathcal{I})^{-1}(T - \bar{s}\mathcal{I}). \tag{3.8}$$

Remark 3.15. With an abuse of notation we will denote the S-resolvent series and the S-resolvent operator with the same symbol $S^{-1}(s,T)$.

By direct computations one can show (see [5]) that the following result holds:

Theorem 3.16. *Let $T \in \mathcal{B}(V)$ and let $s \in \rho_S(T)$. Then the S-resolvent operator defined in* (3.8) *satisfies the* (S*-resolvent*) *equation*

$$S^{-1}(s,T)s - TS^{-1}(s,T) = \mathcal{I}.$$

In the theory of quaternionic linear operators we have more than one notion of spectrum depending on the side of the multiplication. Here we introduce also the new notion of S-spectrum related to the S-resolvent operator (3.8):

Definition 3.17. (The spectra of quaternionic operators) Let $T : V \to V$ be a linear quaternionic operator on the Banach space V.

- We define the S-spectrum $\sigma_S(T)$ of T related to the S-resolvent operator (3.8) as:
$$\sigma_S(T) = \{s \in \mathbb{H} \ : \ T^2 - 2\operatorname{Re}[s]T + |s|^2\mathcal{I} \quad \text{is not invertible}\}.$$
- We denote by $\sigma_L(T)$ the left spectrum of T related to the resolvent $(s\mathcal{I}-T)^{-1}$ that is
$$\sigma_L(T) = \{s \in \mathbb{H} \ : \ s\mathcal{I} - T \quad \text{is not invertible}\}.$$

Remark 3.18. We may attempt to define the right spectrum $\sigma_R(T)$ of T as
$$\sigma_R(T) = \{s \in \mathbb{H} \ : \ \mathcal{I} \cdot s - T \quad \text{is not invertible}\},$$
where the notation $\mathcal{I} \cdot s$ means that the multiplication by s is on the right, i.e., $\mathcal{I} \cdot s(v) = \mathcal{I}(v)s$. However, the operator $\mathcal{I} \cdot s - T$ is not linear, so we will never refer to this notion.

We collect in the following theorem the main properties of the quaternionic spectra:

Theorem 3.19. *Let $T \in \mathcal{B}(V)$.*

1. *The spectra $\sigma_S(T)$ and $\sigma_L(T)$ are contained in the set $\{s \in \mathbb{H} : |s| \le \|T\|\}$.*
2. *(Compactness of S-spectrum) The S-spectrum $\sigma_S(T)$ is a compact nonempty set.*
3. *(Structure of the S-spectrum) Let $p = p_0 + p_1 I \in p_0 + p_1\mathbb{S} \subset \mathbb{H} \setminus \mathbb{R}$ belong to the S-spectrum $\sigma_S(T)$ of T. Then all the elements of the sphere $p_0 + p_1\mathbb{S}$ belong to $\sigma_S(T)$.*

We now have to specify the class of open sets containing the S-spectrum of an operator T which are suitable to define our functional calculus:

Definition 3.20. Let $T \in \mathcal{B}(V)$. Let $U \subset \mathbb{H}$ be an open set containing $\sigma_S(T)$ and such that

(i) $\partial(U \cap L_I)$ is union of a finite number of rectifiable Jordan curves for every $I \in \mathbb{S}$,

(ii) $\sigma_S(T)$ is contained in a finite union of open balls $B_i \subset U$ with center in real points and annular domains $A_j \subset U$ with center in real points whose boundaries do not intersect $\sigma_S(T)$.

A function f is said to be locally s-regular on $\sigma_S(T)$ if there exists an open set $U \subset \mathbb{H}$, as above, on which f is s-regular.
We will denote by $\mathcal{R}_{\sigma_S(T)}$ the set of locally s-regular functions on $\sigma_S(T)$.

Note that the class $\mathcal{R}_{\sigma_S(T)}$ contains functions which admits local power series expansions. If we restrict these functions to a specific plane L_I, the coefficients of the expansion do not depend on the choice of $I \in \mathbb{S}$ (see Theorem 3.2). A termwise integration allows to prove the following:

Theorem 3.21. *Let $T \in \mathcal{B}(V)$ and $f \in \mathcal{R}_{\sigma_S(T)}$. Let $U \subset \mathbb{H}$ be an open set as in Definition 3.20 and let $U_I = U \cap L_I$ for $I \in \mathbb{S}$. Then the integral*

$$\frac{1}{2\pi} \int_{\partial U_I} S^{-1}(s,T)\; ds_I\; f(s) \tag{3.9}$$

does not depend on the choice of the imaginary unit I and on the open set U.

This result allows us to give the following:

Definition 3.22. Let $T \in \mathcal{B}(V)$ and $f \in \mathcal{R}_{\sigma_S(T)}$. Let $U \subset \mathbb{H}$ be an open set as in Definition 3.20, and set $U_I = U \cap L_I$ for $I \in \mathbb{S}$. We define

$$f(T) = \frac{1}{2\pi} \int_{\partial U_I} S^{-1}(s,T)\; ds_I\; f(s). \tag{3.10}$$

Note that, in the special case of monomials, we have

Theorem 3.23. *Let $s,\ a \in \mathbb{H}$, $m \in \mathbb{N}$. Consider the monomial $s^m a$ and $T \in \mathcal{B}(V)$. Let $U \subset \mathbb{H}$ be an open set as in Definition 3.20, and set $U_I = U \cap L_I$ for $I \in \mathbb{S}$. Then*

$$T^m a = \frac{1}{2\pi} \int_{\partial U_I} S^{-1}(s,T)\; ds_I\; s^m\; a. \tag{3.11}$$

3.3. Functional calculus for unbounded operators

We now describe how, based on our result on bounded operators, we can extend our functional calculus to the case of unbounded operators:

Definition 3.24. Let V be a quaternionic Banach space. We consider the linear closed densely defined operator $T : \mathcal{D}(T) \subset V \to V$ where $\mathcal{D}(T)$ denotes the domain of T. Let us assume that

1) $\mathcal{D}(T)$ is dense in V,
2) $T - \overline{s}\mathcal{I}$ is densely defined in V,
3) $\mathcal{D}(T^2) \subset \mathcal{D}(T)$ is dense in V,
4) $T^2 - 2T\mathrm{Re}[s] + |s|^2\mathcal{I}$ is one-to-one with range V.

The S-resolvent operator is defined by

$$S^{-1}(s,T) = -(T^2 - 2T\mathrm{Re}[s] + |s|^2\mathcal{I})^{-1}(T - \overline{s}\mathcal{I}). \tag{3.12}$$

Definition 3.25. Let $T : \mathcal{D}(T) \subset V \to V$ be a linear closed densely defined operator as in Definition 3.24. We define the S-resolvent set of T to be the set

$$\rho_S(T) = \{s \in \mathbb{H} \text{ such that } S^{-1}(s,T) \text{ exists and it is in } \mathcal{B}(V)\}. \tag{3.13}$$

We define the S- spectrum of T as the set

$$\sigma_S(T) = \mathbb{H} \setminus \rho_S(T). \tag{3.14}$$

Theorem 3.26. (*Structure of the spectrum*) *Let $T : V \to V$ be a closed operator such that $\sigma_S(T) \neq \emptyset$. If $p = p_0 + p_1 I \in \mathbb{H} \setminus \mathbb{R}$ belong to the S-spectrum $\sigma_S(T)$ of T, then all the elements of the sphere $p_0 + p_1\mathbb{S}$ belong to $\sigma_S(T)$. The S-spectrum $\sigma_S(T)$ is a union of real points and 2-spheres.*

Let V be a bilateral quaternionic Banach space and let $T : \mathcal{D}(T) \subset V \to V$, $T = T_0 + iT_1 + jT_2 + kT_3$ be a linear operator. If at least one of its components T_j is an unbounded operator then its resolvent is not defined at infinity. It is therefore natural to consider closed operators T for which the resolvent $S^{-1}(s,T)$ is not defined at infinity and to define the extended spectrum as

$$\overline{\sigma}_S(T) := \sigma_S(T) \cup \{\infty\}.$$

Let us consider $\overline{\mathbb{H}} = \mathbb{H} \cup \{\infty\}$ endowed with the natural topology: a set is open if and only if it is union of open discs $D(q,r)$ with center at points in $q \in \mathbb{H}$ and radius r, for some r, and/or union of sets the form $\{q \in \mathbb{H} \mid |q| > r\} \cup \{\infty\} = D'(\infty, r) \cup \{\infty\}$, for some r.

Definition 3.27. We say that f is a regular function at ∞ if $f(q)$ is a regular function in a set $D'(\infty, r)$ and $\lim_{q\to\infty} f(q)$ exists and it is finite. We define $f(\infty)$ to be the value of this limit.

Definition 3.28. Let T be an operator as in Definition 3.24. A function f is said to be locally regular on $\overline{\sigma}_S(T)$ if it is regular an open set $U \subset \mathbb{H}$ as in Definition 3.20 and at infinity.
We will denote by $\mathcal{R}_{\overline{\sigma}_S(T)}$ the set of locally regular functions on $\overline{\sigma}_S(T)$.

Consider $k \in \mathbb{H}$ and the homeomorphism

$$\Phi : \overline{\mathbb{H}} \to \overline{\mathbb{H}}$$

defined by

$$p = \Phi(s) = (s-k)^{-1}, \quad \Phi(\infty) = 0, \quad \Phi(k) = \infty.$$

Definition 3.29. Let $T : \mathcal{D}(T) \to V$ be a linear closed operator as in Definition 3.24 with $\rho_S(T) \cap \mathbb{R} \neq \emptyset$ and suppose that $f \in \mathcal{R}_{\overline{\sigma}_S(T)}$. Let us consider the function

$$\phi(p) := f(\Phi^{-1}(p))$$

and the operator

$$A := (T - k\mathcal{I})^{-1}, \quad \text{for some} \quad k \in \rho_S(T) \cap \mathbb{R}.$$

We define

$$f(T) = \phi(A). \tag{3.15}$$

Remark 3.30. Observe that, if $k \in \mathbb{R}$, we have that:

- the function ϕ is regular because it is the composition of the function f which is regular and $\Phi^{-1}(p) = p^{-1} + k$ which is regular with real coefficients;
- in the case $k \in \rho_S(T) \cap \mathbb{R}$ we have that $(T - k\mathcal{I})^{-1} = -S^{-1}(k,T)$.

We are now ready to state our main results (see [6]):

Theorem 3.31. *If $k \in \rho_S(T) \cap \mathbb{R} \neq \emptyset$ and Φ, ϕ are as above, then $\Phi(\overline{\sigma}_S(T)) = \sigma_S(A)$ and the relation $\phi(p) := f(\Phi^{-1}(p))$ determines a one-to-one correspondence between $f \in \mathcal{R}_{\overline{\sigma}_S(T)}$ and $\phi \in \mathcal{R}_{\overline{\sigma}_S(A)}$.*

Theorem 3.32. *Let $T : \mathcal{D}(T) \subset V \to V$ be a linear closed operator as in Definition 3.24 with $\rho_S(T) \cap \mathbb{R} \neq \emptyset$ and let $f \in \mathcal{R}_{\overline{\sigma}_S(T)}$. Then the operator $f(T)$ defined in (3.15) is independent of $k \in \rho_S(T) \cap \mathbb{R}$.*

Let W be an open set such that $\overline{\sigma}_S(T) \subset W$ and let f be a regular function on $W \cup \partial W$. Set $W_I = W \cap L_I$ for $I \in \mathbb{S}$ be such that its boundary ∂W_I is positively oriented and consists of a finite number of rectifiable Jordan curves. Then

$$f(T) = f(\infty)\mathcal{I} + \frac{1}{2\pi} \int_{\partial W_I} S^{-1}(s,T) ds_I f(s). \tag{3.16}$$

4. Slice monogenic functions and functional calculus for n-tuples of noncommuting operators

In this section we show how the theory of slice "holomorphicity" can be introduced also in the case of functions with values in a Clifford algebra (it will be called slice monogenicity). The theory of slice monogenic functions is the basis for a functional calculus dealing with n-tuples of operators (note that, in the quaternionic case, we proposed a functional calculus for a single quaternionic linear operator).

4.1. Slice monogenic functions

The real Clifford algebra $\mathbb{R}_n$ is generated by the n units $e_1, \dots, e_n$ satisfying $e_i e_j + e_j e_i = -2\delta_{ij}$. An element $a \in \mathbb{R}_n$ is of the form $a = \sum a_A e_A$, A is a subset of indices in $\{1, \dots, n\}$, $e_A = e_{i_1} \dots e_{i_r}$ and $e_\emptyset = 1$. A vector $(x_1, \dots, x_n)$ in $\mathbb{R}^n$ can be identified with an element in $\underline{x} \in \mathbb{R}_n$ by the map $(x_1, \dots, x_n) \mapsto \underline{x} = x_1 e_1 + \dots + x_n e_n$ and similarly a vector $x = (x_0, \dots, x_n) \in \mathbb{R}^{n+1}$ can be identified with $x_0 + \underline{x}$. The real part x_0 of x will be also denoted by $\mathrm{Re}[x]$.

Definition 4.1. By $\mathbb{S}$ we will denote the sphere of unit 1-vectors in $\mathbb{R}^n$, i.e.,

$$\mathbb{S} = \{\underline{x} = e_1 x_1 + \dots + e_n x_n \mid x_1^2 + \dots + x_n^2 = 1\}.$$

To each $x \in \mathbb{R}^{n+1}$ it is possible to associate the imaginary unit I_x defined as

$$I_x = \begin{cases} \dfrac{\underline{x}}{|\underline{x}|} & \text{if } \underline{x} \neq 0, \\ \text{any element of } \mathbb{S} \text{ otherwise.} \end{cases}$$

Given an element $I \in \mathbb{S}$, the complex plane $\mathbb{R} + I\mathbb{R}$ passing through 1 and $I \in \mathbb{S}$ is denoted by L_I, and an element belonging to it is denoted by $u + Iv$, for $u, v \in \mathbb{R}$.

Definition 4.2. Let $U \subseteq \mathbb{R}^{n+1}$ be a domain and let $f : U \to \mathbb{R}_n$ be a real differentiable function. Let $I \in \mathbb{S}$ and let f_I be the restriction of f to the complex plane L_I. We say that f is a left slice-monogenic function, in short s-monogenic, if for every $I \in \mathbb{S}$

$$\frac{1}{2}\left(\frac{\partial}{\partial u} + I \frac{\partial}{\partial v}\right) f_I(u + Iv) = 0.$$

A key fact is that any s-monogenic function can be developed into power series and also that it admits a Cauchy integral representation, as proved in [10]:

Proposition 4.3. *If $B = B(x_0, R) \subseteq \mathbb{R}^{n+1}$ is a ball centered in a real point x_0 with radius $R > 0$, then $f : B \to \mathbb{R}_n$ is s-monogenic if and only if it has a series expansion of the form*

$$f(x) = \sum_{m \geq 0} x^m \frac{1}{m!} \frac{\partial^m f}{\partial u^m}(x_0) \tag{4.1}$$

converging on B.

Theorem 4.4. *Let $B = B(0, R) \subseteq \mathbb{R}^{n+1}$ be a ball with center in 0 and radius $R > 0$ and let $f : B \to \mathbb{R}_n$ be an s-monogenic function. If $x \in B$ then*

$$f(x) = \frac{1}{2\pi} \int_{\partial \Delta_x(0,r)} (\zeta - x)^{-1} \, d\zeta_{I_x} f(\zeta)$$

where $\zeta \in L_{I_x} \bigcap B$, $d\zeta_{I_x} = -d\zeta I_x$ and $r > 0$ is such that

$$\overline{\Delta_x(0, r)} = \{u + I_x v \mid u^2 + v^2 \leq r^2\}$$

contains x and is contained in B.

As in the quaternionic case, the key ingredient to define a functional calculus is what we call noncommutative Cauchy kernel series.

Definition 4.5. Let $x = \mathrm{Re}[x] + \underline{x}$, $s = \mathrm{Re}[s] + \underline{s}$ be such that $sx \neq xs$. We will call noncommutative Cauchy kernel series the following expansion

$$S^{-1}(s, x) := \sum_{n \geq 0} x^n s^{-1-n} \tag{4.2}$$

defined for $|x| < |s|$.

We have the following result (cf. Section 3):

Theorem 4.6. *Let $x = \mathrm{Re}[x] + \underline{x}$, $s = \mathrm{Re}[s] + \underline{s}$ be such that $xs \neq sx$. Then*

$$\sum_{n \geq 0} x^n s^{-1-n} = -(x^2 - 2x\mathrm{Re}[s] + |s|^2)^{-1}(x - \overline{s})$$

for $|x| < |s|$. Moreover, $S^{-1}(s, x)$ is irreducible and $\lim_{x \to \overline{s}} S^{-1}(s, x)$ does not exist.

4.2. Functional calculus for n-tuples of bounded operators

In this section we will consider a Banach space V over $\mathbb{R}$ (the case of complex Banach spaces can be discussed in a similar fashion) with norm $\|\cdot\|$. It is possible to endow V with an operation of multiplication by elements of $\mathbb{R}_n$ which gives a two-sided module over $\mathbb{R}_n$. We recall that a two-sided module V over $\mathbb{R}_n$ is called a Banach module over $\mathbb{R}_n$, if there exists a constant $C \geq 1$ such that

$$\|va\| \leq C\|v\||a|, \quad \|av\| \leq C|a|\|v\|$$

for all $v \in V$ and $a \in \mathbb{R}_n$. By V_n we denote the two-sided Banach module over $\mathbb{R}_n$ corresponding to $V \otimes \mathbb{R}_n$. An element in V_n is of the type $\sum_A v_A \otimes e_A$ (where $A = i_1 \dots i_r$, $i_\ell \in \{1, 2, \dots, n\}$, $i_1 < \dots < i_r$ is a multi-index). The multiplications (right and left) of an element $v \in V_n$ with a scalar $a \in \mathbb{R}_n$ are defined as

$$va = \sum_A v_A \otimes (e_A a), \quad \text{and} \quad av = \sum_A v_A \otimes (a e_A).$$

For simplicity, we will write $\sum_A v_A e_A$ instead of $\sum_A v_A \otimes e_A$. We define

$$\|v\|^2_{V_n} = \sum_A \|v_A\|^2_V.$$

By $\mathcal{B}(V)$ we will denote the space of bounded $\mathbb{R}$-homomorphisms of the Banach space V into itself endowed with the natural norm denoted by $\|\cdot\|_{\mathcal{B}(V)}$. If $T_A \in \mathcal{B}(V)$, we can define the operator $T = \sum_A T_A e_A$ and its action on

$$v = \sum v_B e_B \in V_n$$

as

$$T(v) = \sum_{A,B} T_A(v_B) e_A e_B.$$

The set of all such bounded operators is denoted by $\mathcal{B}_n(V_n)$. Their norm is defined by

$$\|T\|^2_{\mathcal{B}_n(V_n)} = \sum_A \|T_A\|^2_{\mathcal{B}(V)}.$$

In the sequel, we will only consider operators of the form $T = T_0 + \sum_{j=1}^n e_j T_j$ where $T_\mu \in \mathcal{B}(V)$ for $\mu = 0, 1, \dots, n$. The set of such operators in $\mathcal{B}_n(V_n)$ will be denoted by $\mathcal{B}^{0,1}_n(V_n)$.

Definition 4.7. Let $T \in \mathcal{B}^{0,1}_n(V_n)$ and $s = \mathrm{Re}[s] + \underline{s}$. We define the S-resolvent operator series as

$$S^{-1}(s,T) := \sum_{n\geq 0} T^n s^{-1-n} \tag{4.3}$$

for $\|T\| < |s|$.

To have an operator defined also outside the ball $\|T\| < |s|$ it is necessary to compute the sum of the series (4.3). Despite the fact that the setting of Clifford algebras implies restrictions on the operations one can perform, as $\mathbb{R}_n$ is, in general, a non division algebra, we have the following:

Theorem 4.8. *Let $T \in \mathcal{B}^{0,1}_n(V_n)$ and $s = \mathrm{Re}[s] + \underline{s}$. Then*

$$\sum_{n\geq 0} T^n s^{-1-n} = -(T^2 - 2T\mathrm{Re}[s] + |s|^2\mathcal{I})^{-1}(T - \overline{s}\mathcal{I}), \tag{4.4}$$

for $\|T\| < |s|$.

When $Ts\mathcal{I} = sT$, the operator $S^{-1}(s,T)$ equals $(s\mathcal{I} - T)^{-1}$ where the series (4.3) *converges.*

This result leads to the following definition:

Definition 4.9. (The S-resolvent operator) Let $T \in \mathcal{B}_n^{0,1}(V_n)$ and $s = \mathrm{Re}[s] + \underline{s} \in \rho_S(T)$. We define the S-resolvent operator as

$$S^{-1}(s,T) := -(T^2 - 2\mathrm{Re}[s]T + |s|^2\mathcal{I})^{-1}(T - \bar{s}\mathcal{I}). \tag{4.5}$$

Theorem 4.10. *Let $T \in \mathcal{B}_n^{0,1}(V_n)$ and $s = \mathrm{Re}[s] + \underline{s} \in \rho_S(T)$. Let $S^{-1}(s,T)$ be the S-resolvent operator defined in* (4.5)*. Then $S^{-1}(s,T)$ satisfies the (S-resolvent) equation*

$$S^{-1}(s,T)s - TS^{-1}(s,T) = \mathcal{I}. \tag{4.6}$$

The S-resolvent operator $S^{-1}(s,T)$ gives rise to the definition of S-spectrum:

Definition 4.11. (The S-spectrum and the S-resolvent set) Let $T \in \mathcal{B}_n^{0,1}(V_n)$ and $s = \mathrm{Re}[s] + \underline{s}$. We define the S-spectrum $\sigma_S(T)$ of T as:

$$\sigma_S(T) = \{s \in \mathbb{R}^{n+1} \quad : \quad T^2 - 2\ \mathrm{Re}[s]T + |s|^2\mathcal{I} \quad \text{is not invertible}\}.$$

The S-resolvent set $\rho_S(T)$ is defined by

$$\rho_S(T) = \mathbb{R}^{n+1} \setminus \sigma_S(T).$$

Theorem 4.12. (*Structure of the S-spectrum*) *Let $T \in \mathcal{B}_n^{0,1}(V_n)$ and let $p = \mathrm{Re}[p] + \underline{p}$ belong to $\sigma_S(T)$ with $\underline{p} \neq 0$. Then all the elements of the sphere $s = \mathrm{Re}[s] + \underline{s}$ with $\mathrm{Re}[s] = \mathrm{Re}[p]$ and $|\underline{s}| = |\underline{p}|$ belong to $\sigma_S(T)$.*

As we did in the quaternionic case, we now describe the class of functions for which we can construct a functional calculus. This class is determined by the need to provide local series expansion for the s-monogenic functions.

Definition 4.13. Let $T = T_0 + \sum_{j=1}^n e_j T_j \in \mathcal{B}_n^{0,1}(V_n)$. Let $U \subset \mathbb{R}^{n+1}$ be an open set containing $\sigma_S(T)$ and such that

(i) $\partial(U \cap L_I)$ is union of a finite number of rectifiable Jordan curves for every $I \in \mathbb{S}$,

(ii) $\sigma_S(T)$ is contained in a finite union of open balls $B_i \subset U$ with center in real points and annular domains $A_j \subset U$ with center in real points whose boundaries do not intersect $\sigma_S(T)$.

A function f is said to be locally s-monogenic on $\sigma_S(T)$ if there exists an open set $U \subset \mathbb{R}^{n+1}$, as above, on which f is s-monogenic.
We will denote by $\mathcal{M}_{\sigma_S(T)}$ the set of locally s-monogenic functions on $\sigma_S(T)$.

Theorem 4.14. *Let $T \in \mathcal{B}_n^{0,1}(V_n)$ and $f \in \mathcal{M}_{\sigma_S(T)}$. Let $U \subset \mathbb{R}^{n+1}$ be an open set as in Definition* 4.13 *and let $U_I = U \cap L_I$ for $I \in \mathbb{S}$. Then the integral*

$$\frac{1}{2\pi} \int_{\partial U_I} S^{-1}(s,T)\ ds_I\ f(s) \tag{4.7}$$

does not depend on the choice of the imaginary unit I and on the open set U.

Definition 4.15. Let $T \in \mathcal{B}_n^{0,1}(V_n)$ and $f \in \mathcal{M}_{\sigma_S(T)}$. Let $U \subset \mathbb{R}^{n+1}$ be an open set as in Definition 4.13, and set $U_I = U \cap L_I$ for $I \in \mathbb{S}$. We define

$$f(T) = \frac{1}{2\pi} \int_{\partial U_I} S^{-1}(s,T)\, ds_I\, f(s). \tag{4.8}$$

In the case of monomials, and thus of polynomials, we have the following result:

Theorem 4.16. *Let $x = \mathrm{Re}[x] + \underline{x}$, $a = \mathrm{Re}[a] + \underline{a} \in \mathbb{R}^{n+1}$, $m \in \mathbb{N}$. Consider the monomial $x^m a$ and the operator $T \in \mathcal{B}_n^{0,1}(V_n)$. Let $U \subset \mathbb{R}^{n+1}$ be an open set as in Definition* 3.20, *and set $U_I = U \cap L_I$ for $I \in \mathbb{S}$. Then*

$$T^m a = \frac{1}{2\pi} \int_{\partial U_I} S^{-1}(s,T)\, ds_I\, s^m\, a. \tag{4.9}$$

Remark 4.17. For an extension of this functional calculus and further properties, see [7].

Remark 4.18. To compare our new functional calculus with the existing versions which the reader can find in the literature (see, e.g., [15] and its references) we will now consider the subset $\mathcal{B}_n^1(V_n) \subset \mathcal{B}_n^{0,1}(V_n)$ whose elements are operators of the form $T = \sum_{j=1}^n T_j e_j$ where T_j are linear operators acting on the Banach space V. When $n = 1$, this corresponds to considering a single operator T_1 and $T = T_1 e_1$. To compute the S-spectrum we have to consider the S-eigenvalue equation. Since in this case, the variable $s \in \mathbb{C}$ commutes with T, the S-eigenvalue equation reduces to the classical eigenvalue equation. Finally, since the theory of s-monogenic functions coincides with the theory of holomorphic functions in one complex variable, our calculus reduces to the Riesz-Dunford calculus. Let $f(z)$ be any function holomorphic on the spectrum of T_1. The Riesz-Dunford calculus allows to compute $f(T_1)$. In our case, to get exactly the function $f(T_1)$ we need to consider $\tilde{f}(z) = f(-ze_1) = f(x_1 - e_1 x_0)$ where we have denoted $z = x_0 + e_1 x_1$. This is not surprising, since we are considering not the given operator T_1 as in the classical case, but its tensor with the imaginary unit e_1. The two calculi are therefore equivalent up to this identification.

4.3. Functional calculus for n-tuples of unbounded operators

Definition 4.19. Let V be a Banach space and V_n be the two-sided Banach module over $\mathbb{R}_n$ corresponding to $V \otimes \mathbb{R}_n$. Let $T_\mu : \mathcal{D}(T_\mu) \subset V \to V$ be linear closed densely defined operators for $\mu = 0, 1, \ldots, n$. Let

$$\mathcal{D}(T) = \{v \in V_n \;:\; v = \sum_B v_B e_B,\;\; v_B \in \bigcap_{\mu=0}^{n} \mathcal{D}(T_\mu)\ \}. \tag{4.10}$$

be the domain of the operator

$$T = T_0 + \sum_{j=1}^{n} e_j T_j, \qquad T : \mathcal{D}(T) \subset V_n \to V_n.$$

Let us assume that

1) $\bigcap_{\mu=0}^{n} \mathcal{D}(T_\mu)$ is dense in V_n,
2) $T - \overline{s}\mathcal{I}$ is densely defined in V_n,
3) $\mathcal{D}(T^2) \subset \mathcal{D}(T)$ is dense in V_n,
4) $T^2 - 2T\mathrm{Re}[s] + |s|^2\mathcal{I}$ is one-to-one with range V_n.

The S-resolvent operator is defined by

$$S^{-1}(s,T) = -(T^2 - 2T\mathrm{Re}[s] + |s|^2\mathcal{I})^{-1}(T - \overline{s}\mathcal{I}). \tag{4.11}$$

Definition 4.20. Let $T : \mathcal{D}(T) \to V_n$ be a linear closed densely defined operator as in Definition 4.19. We define the S-resolvent set of T to be the set

$$\rho_S(T) = \{s \in \mathbb{R}^{n+1} \text{ such that } S^{-1}(s,T) \text{ exists and it is in } \mathcal{B}_n(V_n)\}. \tag{4.12}$$

We define the S- spectrum of T as the set

$$\sigma_S(T) = \mathbb{R}^{n+1} \setminus \rho_S(T). \tag{4.13}$$

Theorem 4.21. *(S-resolvent operator equation) Let $T : \mathcal{D}(T) \to V_n$ be a linear closed densely defined operator. Let $s \in \rho_S(T)$. Then $S^{-1}(s,T)$ satisfies the (S-resolvent) equation*

$$S^{-1}(s,T)s - TS^{-1}(s,T) = \mathcal{I}.$$

Let V be a Banach space and $T = T_0 + \sum_{j=1}^{m} e_j T_j$ where $T_\mu : \mathcal{D}(T_\mu) \to V$ are linear operators for $\mu = 0, 1, \dots, n$. If at least one of the T_j's is an unbounded operator then its resolvent is not defined at infinity. It is therefore natural to consider closed operators T for which the resolvent $S^{-1}(s,T)$ is not defined at infinity and to define the extended spectrum as

$$\overline{\sigma}_S(T) := \sigma_S(T) \cup \{\infty\}.$$

Let us consider $\overline{\mathbb{R}}^{n+1} = \mathbb{R}^{n+1} \cup \{\infty\}$ endowed with the natural topology: a set is open if and only if it is union of open discs $D(x,r)$ with center at points in $x \in \mathbb{R}^{n+1}$ and radius r, for some r, and/or union of sets the form $\{x \in \mathbb{R}^{n+1} \mid |x| > r\} \cup \{\infty\} = D'(\infty, r) \cup \{\infty\}$, for some r.

Definition 4.22. We say that f is s-monogenic function at ∞ if $f(x)$ is an s-monogenic function in a set $D'(\infty, r)$ and $\lim_{x\to\infty} f(x)$ exists and it is finite. We define $f(\infty)$ to be the value of this limit.

Definition 4.23. Let $T : \mathcal{D}(T) \to V_n$ be a linear closed operator as in Definition 4.19 . A function f is said to be locally s-monogenic on $\overline{\sigma}_S(T)$ if it is s-monogenic an open set $U \subset \mathbb{R}^{n+1}$ as in Definition 4.13 and at infinity.
We will denote by $\mathcal{M}_{\overline{\sigma}_S(T)}$ the set of locally s-monogenic functions on $\overline{\sigma}_S(T)$.

Consider $k \in \mathbb{R}^{n+1}$ and the homeomorphism

$$\Phi : \overline{\mathbb{R}}^{n+1} \to \overline{\mathbb{R}}^{n+1}$$

defined by

$$p = \Phi(s) = (s-k)^{-1}, \quad \Phi(\infty) = 0, \quad \Phi(k) = \infty.$$

Definition 4.24. Let $T : \mathcal{D}(T) \to V_n$ be a linear closed operator as in Definition 3.24 with $\rho_S(T) \cap \mathbb{R} \neq \emptyset$ and suppose that $f \in \mathcal{M}_{\overline{\sigma}_S(T)}$. Let us consider

$$\phi(p) := f(\Phi^{-1}(p))$$

and the operator

$$A := (T - k\mathcal{I})^{-1}, \quad \text{for some} \quad k \in \rho_S(T) \cap \mathbb{R}.$$

We define

$$f(T) = \phi(A). \tag{4.14}$$

Remark 4.25. Observe that, if $k \in \mathbb{R}$, we have that:

i) the function ϕ is s-monogenic because it is the composition of the function f which is s-monogenic and $\Phi^{-1}(p) = p^{-1} + k$ which is s-monogenic with real coefficients;
ii) in the case $k \in \rho_S(T) \cap \mathbb{R}$ we have that $(T - k\mathcal{I})^{-1} = -S^{-1}(k, T)$.

Theorem 4.26. *If $k \in \rho_S(T) \cap \mathbb{R} \neq \emptyset$ and Φ, ϕ are as above, then $\Phi(\overline{\sigma}_S(T)) = \sigma_S(A)$ and the relation $\phi(p) := f(\Phi^{-1}(p))$ determines a one-to-one correspondence between $f \in \mathcal{M}_{\overline{\sigma}_S(T)}$ and $\phi \in \mathcal{M}_{\overline{\sigma}_S(A)}$.*

Theorem 4.27. *Let $T : \mathcal{D}(T) \to V_n$ be a linear closed operator as in Definition* 3.24 *with $\rho_S(T) \cap \mathbb{R} \neq \emptyset$ and suppose that $f \in \mathcal{M}_{\overline{\sigma}_S(T)}$. Then operator $f(T)$ defined in* (4.14) *is independent of $k \in \rho_S(T) \cap \mathbb{R}$.*

Let W be an open set such that $\overline{\sigma}_S(T) \subset W$ and let f be an s-monogenic function on $W \cup \partial W$. Set $W_I = W \cap L_I$ for $I \in \mathbb{S}$ be such that its boundary ∂W_I is positively oriented and consists of a finite number of rectifiable Jordan curves. Then

$$f(T) = f(\infty)\mathcal{I} + \frac{1}{2\pi} \int_{\partial W_I} S^{-1}(s, T) ds_I f(s). \tag{4.15}$$

5. Monogenic functions and the monogenic functional calculus

A functional calculus which is based on the classical notion of monogenic functions was extensively studied by Jefferies, Kisil, MacIntosh and their coworkers. We mention here, with no claim of completeness, the works [16], [17], [18], [19], [20], the book [15] and the references therein. To start with, we will quickly recall the basic notions on monogenic functions.

5.1. Monogenic functions

The well-known notion of monogenic functions with values in a Clifford algebra (see [3]) is based on the so-called Dirac operator $\partial_{\underline{x}} = \sum_{j=1}^n e_j \partial_{x_j}$.

Definition 5.1. A real differentiable function $f : U \subseteq \mathbb{R}^n \to \mathbb{R}_n$ on an open set U is called (left) monogenic in U if it satisfies $\partial_{\underline{x}} f(\underline{x}) = 0$ on U.

Remark 5.2. A variation of the Dirac operator is the Weyl operator:

$$\partial_{x_0} + \partial_{\underline{x}},$$

whose nullsolutions $f : U \subseteq \mathbb{R}^{n+1} \to \mathbb{R}_n$ on an open set U are still called (left) monogenic. Moreover, in the literature, the Weyl operator is often called Dirac operator since it is possible to obtain one from the other by grouping the imaginary units and making some identifications in a suitable way.

Monogenic functions can be expanded in power series in terms of the symmetric polynomials $(e_j x_n + e_n x_j)$, $1 \leq j \leq n-1$. One has to consider these symmetric polynomials and the sum of all their possible permutations. For functions in the kernel of the Weyl operator the situation is similar and the symmetric polynomials are written in terms of the building blocks $(e_j x_0 - e_0 x_j)$, $1 \leq j \leq n$. It is natural to give the following:

Definition 5.3. Homogeneous monogenic polynomials of degree k are defined as

$$V^{\ell_1,\dots,\ell_k}(x) = \frac{1}{k!} \sum_{\ell_1,\dots,\ell_k} z_{\ell_1} \dots z_{\ell_k}, \tag{5.1}$$

where $z_j = x_j e_0 - x_0 e_j$ and the sum is taken over all different permutation of $\ell_1, \dots, \ell_k$.

Definition 5.4. Denote by Σ_n the volume $2\pi^{(n+1)/2}/\Gamma((n+1)/2)$ of the unit n-sphere in $\mathbb{R}^{n+1}$ and by $\bar{x} = x_0 - \underline{x}$ the conjugate of $x = x_0 + \underline{x}$. For each $x \in \mathbb{R}^{n+1}$, define the function $G(\cdot, x)$ as

$$G(\omega, x) = \frac{1}{\Sigma_n} \frac{\bar{\omega} - \bar{x}}{|\omega - x|^{n+1}}. \tag{5.2}$$

Note that $G(\omega, x)$, for $\omega \neq x$, is both left and right monogenic as a function of ω. It plays the role of the Cauchy kernel as shown in the following result (see [3]).

Theorem 5.5. *Let $\Omega \subset \mathbb{R}^{n+1}$ be a bounded open set with smooth boundary $\partial\Omega$ and exterior unit normal $n(\omega)$ defined for all $\omega \in \partial\Omega$. For any left monogenic function f defined in a neighborhood of U of $\overline{\Omega}$, we have the Cauchy formula*

$$\int_{\partial\Omega} G(\omega, x) n(\omega) f(\omega) d\mu(\omega) = \begin{cases} f(x), & \text{if } x \in \Omega, \\ 0, & \text{if } x \notin \Omega, \end{cases} \tag{5.3}$$

where μ is the surface measure of $\partial\Omega$.

5.2. Monogenic functional calculus

The Cauchy formula (5.3) is the starting point for the monogenic functional calculus. To this purpose, it is useful to consider a suitable series expansion of the kernel $G(\omega, x) = G_\omega(x)$:

$$G_\omega(x) = \sum_{k \geq 0} \left(\sum_{(\ell_1,\dots,\ell_k)} V^{\ell_1,\dots,\ell_k}(x) W_{\ell_1,\dots,\ell_k}(\omega) \right)$$

in the region $|x| > |\omega|$ (see [3]) where, for each $\omega \in \mathbb{R}^{n+1}$, $\omega \neq 0$,

$$W_{\ell_1,\ldots,\ell_k}(\omega) = (-1)^k \partial_{\omega_{\ell_1}} \ldots \partial_{\omega_{\ell_k}} G_\omega(0)$$

and $V^{\ell_1,\ldots,\ell_k}(x)$ are defined in (5.1).
Keeping in mind the definition of Banach modules, see Section 4, consider now an n-tuple $T = (T_1, \ldots, T_n)$ of bounded linear operator acting on a Banach space X and let

$$R > (1 + \sqrt{2}) \| \sum_{j=1}^{n} T_j e_j \|. \tag{5.4}$$

Suppose to formally replace z_j by T_j and 1 by the identity operator $\mathcal{I}$ in the Cauchy kernel series. It can be shown (see [18] Lemma 3.12, [15], Lemma 4.7) that

$$G_\omega(T) := \sum_{k \geq 0} \left(\sum_{(\ell_1,\ldots,\ell_k)} V^{\ell_1,\ldots,\ell_k}(T) W_{\ell_1,\ldots,\ell_k}(\omega) \right) \tag{5.5}$$

where

$$V^{\ell_1,\ldots,\ell_k}(T) = \frac{1}{k!} \sum_{\ell_1,\ldots,\ell_k} T_{\ell_1} \ldots T_{\ell_k},$$

converges uniformly for all $\omega \in \mathbb{R}^{n+1}$ such that $|\omega| \geq R$, where R is given in (5.4). We set the sum of the series (5.5) equal to $G_\omega(T)$ which turns out to be a bounded operator.

Remark 5.6. In the Clifford setting there are several possible notions of spectrum. In [18] the so-called resolvent set is the set of $\omega \in \mathbb{R}^n$ such that the series (5.5) converges. The spectral set $\sigma_C(T)$ of T is defined as the set complement of the resolvent set.

An important result is the following (see [18]):

Theorem 5.7. *Let $(T_1, \ldots, T_n)$ be an n-tuple of bounded self-adjoint operators. Let Ω be a domain with piecewise smooth boundary whose complement is connected, and suppose that $\sigma_C(T) \subseteq \Omega$. Then, for every $f \in \mathcal{M}(\Omega)$ the mapping*

$$f(x) \mapsto f(T) = \int_{\partial\Omega} G_\omega(T) n(\omega) f(\omega) d\mu(\omega)$$

defines a functional calculus.

Remark 5.8. If T is an n-tuple of bounded self adjoint operators and Ω is an open set with piecewise smooth boundary with connected complement containing the spectral set $\sigma_C(T)$ then according to [18] the map in Theorem 5.7 defines a functional calculus for functions monogenic on Ω. This fact is guaranteed by a Runge type approximation theorem.

In [15] the definition of spectrum is different:

Definition 5.9. The monogenic spectrum $\gamma(T)$ of the n-tuple T is the complement of the largest connected open set U in $\mathbb{R}^{n+1}$ in which the function $G_\omega(T)$ defined by the series above is the restriction of a monogenic function with domain U.

Let $\langle T, \underline{\xi}\rangle = \sum_{j=1}^n T_j\xi_j$ and suppose that $\sigma(\langle T, \underline{\xi}\rangle)$ is real for all $\underline{\xi} \in \mathbb{R}^n$ (here σ denotes the spectrum in the classical sense, i.e., the set of singularities of $(\lambda\mathcal{I} - \langle T, \underline{\xi}\rangle)^{-1}$). Then we have the following result (see [15]):

Theorem 5.10. *Let $T = (T_1, \dots, T_n)$ be an n-tuple of noncommuting bounded linear operator acting on a Banach space X and suppose that $\sigma(\langle T, \underline{\xi}\rangle) \subseteq \mathbb{R}$ for all $\underline{\xi} \in \mathbb{R}^n$. Then the $\mathcal{B}_n(X_n)$-valued function $\omega \to G_\omega(T)$ defined in* (5.5) *is the restriction to the region*

$$\Gamma = \{\omega \in \mathbb{R}^{n+1} \ : \ |\omega| > (1+\sqrt{2})\|\sum_{j=1}^n T_j e_j\|\}$$

of a function two-sided monogenic on $\mathbb{R}^{n+1}\backslash\mathbb{R}^n$.

This result guarantees that $G_\omega(T)$ is monogenic outside a ball. However, $G_\omega(T)$ can be monogenic in a larger set containing Γ. Denote with the same symbol $G_\omega(T)$ the maximal monogenic extension and let Ω be the union of all open sets containing the open set Γ on which is defined a two-sided monogenic function whose restriction on Γ equals the series $G_\omega(T)$. Then the extension is unique because the domain Ω is connected, contains Γ and the spectrum is a subset of $\mathbb{R}^n$ and hence it cannot disconnect a set in $\mathbb{R}^{n+1}$.

Definition 5.11. Let $T = (T_1, \dots, T_n)$ be an n-tuple of noncommuting bounded linear operators acting on a Banach space X and $\underline{\xi} \in \mathbb{R}^n$. Suppose that $\sigma(\langle T, \underline{\xi}\rangle)$ is real for all $\underline{\xi} \in \mathbb{R}^n$. Let $\Omega \subseteq \mathbb{R}^{n+1}$ be a bounded open neighborhood of $\gamma(T)$ with smooth boundary and exterior normal $n(\omega)$, for all $\omega \in \partial\Omega$. Let f be a monogenic function defined in an open neighborhood of $\overline{\Omega}$. By definition $f(T)$ is given by

$$f(T) = \int_{\partial\Omega} G_\omega(T) n(\omega) f(\omega) d\mu(\omega). \tag{5.6}$$

Denoting by $\mathcal{M}(\gamma(T), \mathbb{R}_n)$ the right module of monogenic functions defined in a neighborhood of $\gamma(T)$ in $\mathbb{R}^{n+1}$, we have that the map $f \mapsto f(T)$, $f \in \mathcal{M}(\gamma(T), \mathbb{R}_n)$ defines a functional calculus. It is a right-module homomorphism.

Remark 5.12. Let $p(\underline{x})$ be a complex-valued polynomial, $\underline{x}, \underline{\xi} \in \mathbb{R}^n$ and

$$p(\underline{x}) = p(x_1\xi_1 + \dots + x_n\xi_n).$$

Then

$$p(T_1, \dots, T_n) = p(T_1\xi_1 + \dots + T_n\xi_n).$$

Note that the function $G_\omega(T)$ admits a plane wave expansion as follows

Proposition 5.13. *Let $\omega \in \mathbb{R}^{n+1}$, $\omega = \omega_0 + \underline{\omega}$, $\omega_0 \neq 0$. Then*

$$G_\omega(T) = \frac{(n-1)!}{2} \left(\frac{i}{2\pi}\right)^n \operatorname{sgn}(\omega_0)^{n-1} \int_{S^{n-1}} (1+is)(\langle \omega\mathcal{I} - T, s\rangle - \omega_0 s)^{-n}\, ds. \quad (5.7)$$

When the condition $\sigma(\langle T, \underline{\xi}\rangle) \in \mathbb{R}$ is satisfied for all $\underline{\xi} \in \mathbb{R}^n$, then $\gamma(T) \subset \mathbb{R}^n$ is the complement in $\mathbb{R}^{n+1}$ of the points ω at which the function defined by the integral above is continuous. In the case of commuting bounded linear operators the spectrum can be determined directly as shown in the next result (see Theorem 3.3 and Corollary 3.4 in [20]).

Theorem 5.14. *Let $T = (T_1, \dots, T_n)$ be an n-tuple of commuting bounded linear operator acting on a Banach space X and suppose that $\sigma(T_j) \subseteq \mathbb{R}$ for all $j = 1, \dots, n$. Then $\gamma(T)$ is the complement in $\mathbb{R}^n$ of the set of all $\lambda \in \mathbb{R}^n$ for which the operator $\sum_{j=1}^n (\lambda_j \mathcal{I} - A_j)^2$ is invertible in $\mathcal{B}(X)$ (equivalently: $(\underline{\lambda}\mathcal{I} - T)$ is invertible in* End(X)*).*

5.3. Relation between the monogenic spectrum and the Weyl spectrum

There is a relation between the monogenic functional calculus and the so-called Weyl functional calculus that is based on the Fourier transform. More precisely, let $T = (T_1, \dots, T_n)$ be an n-tuple of bounded operators acting on a Banach space. Suppose the operators T_j are of Paley-Wiener type (r, s) (see [15] for the definition). The Weyl functional calculus is the map that associates to $f \in \mathcal{S}(\mathbb{R}^n)$ the operator

$$\mathcal{W}_T(f) = (2\pi)^{-n} \int_{\mathbb{R}^n} e^{i\langle T, \xi\rangle} \hat{f}(\xi) d\xi,$$

where $\hat{f}$ is the Fourier transform of f. This leads to the following definition:

Definition 5.15. Let $T = (T_1, \dots, T_n)$ be an n-tuple of bounded operators acting on a Banach space and let the operators T_j be of Paley-Wiener type (r, s). The support of the distribution $\mathcal{W}_T$ is called the joint spectrum of T and is denoted by $\tilde{\gamma}(T)$.

However, one can also introduce the definition of operators of Paley-Wiener type s (see [15]). For these operators the monogenic and the joint spectrum are related by the following result:

Proposition 5.16. *Let $s \geq 0$ and let T be an n-tuple of operators of Paley-Wiener type s. Then $\tilde{\gamma}(T) = \gamma(T)$.*

6. Cauchy-Fueter regularity and the difficulties of the related functional calculus

The most successful notion extending holomorphy to the quaternionic setting is the one of regularity in the sense of Cauchy-Fueter, so one may wonder whether it is possible to define a functional calculus associated to it. In this section we show the difficulties arising if one wishes to define a functional calculus in this setting.

We begin by recalling the definition of regularity and the fundamental properties of the functions regular in this sense.

6.1. Fueter regularity

Definition 6.1. Let $U \subseteq \mathbb{H}$ be an open set and let $f : U \to \mathbb{H}$ be a real differentiable function. We say that f is a left-regular function if

$$\frac{\partial f}{\partial \bar{q}} = \frac{\partial f}{\partial x_0} + i\frac{\partial f}{\partial x_1} + j\frac{\partial f}{\partial x_2} + k\frac{\partial f}{\partial x_3} = 0.$$

We will denote by $\mathcal{R}(U)$ the right quaternionic vector space of (left)-regular functions.

Definition 6.2. The function

$$G(q) = \frac{\bar{q}}{|q|^4} = \frac{q^{-1}}{|q|^2} = q^{-2}\bar{q}^{-1}$$

is called Cauchy-Fueter kernel. It is both left and right regular on $\mathbb{H}\setminus\{0\}$.

As it is well known, there is a Cauchy integral formula for regular functions (see [22]):

Theorem 6.3. *Let U be an open set in $\mathbb{H}$ and $f \in \mathcal{R}(U)$. Let $\Gamma = \partial U$ be a rectifiable 3-cell which is homologous, in the singular homology of $U\setminus\{q_0\}$, to a differentiable 3-cell whose image is ∂B for some ball $B \subset U$. Then*

$$f(q_0) = \frac{1}{2\pi^2}\int_\Gamma G(q-q_0)Dqf(q)$$

where $Dq = dx_1 \wedge dx_2 \wedge dx_3 - i dx_0 \wedge dx_2 \wedge dx_3 + j dx_0 \wedge dx_1 \wedge dx_3 - k dx_0 \wedge dx_1 \wedge dx_2$.

A polynomial function in the variable q is not regular and, in particular, the terms $a_n q^n$ or $a_0 q a_1 q \dots a_n q$ that is the natural generalization of $a_n z^n$, are not regular. The power series representing a regular function and the Laurent series representing a function with an isolated singularity can be expressed in terms of the special homogeneous regular polynomials similar to the ones introduced in the monogenic setting.

Definition 6.4. Let us consider a set $\nu = \{\lambda_1, \dots, \lambda_n\}$, where $1 \le \lambda_i \le 3$, $\forall i = 1, 2, \dots, n$. We can specify ν giving three integers n_1, n_2, n_3, such that $n_1 + n_2 + n_3 = n$ and n_λ is the number of λ's in the set ν. We will denote by σ_n the set of $\nu = [n_1, n_2, n_3]$. If $n = 0$, we set $\nu = \emptyset$. For every $\nu \in \sigma_n$, we define

$$P_\nu(q) = \frac{1}{n!}\sum_{1\le\lambda_1,\dots,\lambda_n\le 3}(x_0 e_{\lambda_1} - x_{\lambda_1})\dots(x_0 e_{\lambda_n} - x_{\lambda_n}),$$

where $e_\lambda = i, j, k$ when $\lambda = 1, 2, 3$ respectively and the sum is over all the different orderings of n elements λ_i, $(\lambda_i = 1, 2, 3)$.

We set the positions

$$\partial_\nu = \frac{\partial^n}{\partial x_1^{n_1} \partial x_2^{n_2} \partial x_3^{n_3}} \qquad \text{and} \qquad G_\nu(q) = \partial_\nu G(q), \tag{6.1}$$

where $G(q)$ is the Cauchy-Fueter kernel. We have the following proposition:

Proposition 6.5. *The expansions*

$$G(q,p) := G(q-p) = \sum_{n=0}^{+\infty} \sum_{\nu \in \sigma_n} P_\nu(p) G_\nu(q) = \sum_{n=0}^{+\infty} \sum_{\nu \in \sigma_n} G_\nu(q) P_\nu(p)$$

hold for $|p| < |q|$.

Theorem 6.6. *Let* $f : U \to \mathbb{H}$, $f \in \mathbb{R}_l(U)$. *Let* $q_0 \in U$ *and* $\delta < \text{dist } (q_0, \partial U)$. *Then there exists an open ball* $B = \{q \in \mathbb{H} : |q - q_0| < \delta\}$ *such that* $f(q)$ *can be represented by the uniformly convergent series*

$$f(q) = \sum_{n=0}^{+\infty} \sum_{\nu \in \sigma_n} P_\nu(q - q_0) a_\nu,$$

where

$$a_\nu = (-1)^n \partial_\nu f(q_0) = \frac{1}{2\pi^2} \int_{|q-q_0|=\delta} G_\nu(q - q_0) Dq f(q),$$

and $G_\nu(q)$ *is defined in* (6.1). *Moreover we have*

$$\int_S G_\mu(q)\, Dq\, P_\nu(q) = 2\pi^2\, \delta_{\mu\nu}$$

where S *is any sphere containing the origin, and* $\delta_{\mu\nu}$ *denotes the Kronecker delta.*

Definition 6.7. Let V be a left quaternionic Banach space. A function $f : \mathbb{H} \to V$ is said to be left regular in $q_0 \in \mathbb{H}$ if there exists an open ball $B(q_0, r)$ such that, for every point $q \in B(q_0, r)$ the function $f(q)$ can be represented by the following series

$$f(q) = \sum_{n=0}^{+\infty} \sum_{\nu \in \sigma_n} P_\nu(q - q_0) f_\nu(q_0), \qquad f_\nu(q_0) \in V,$$

uniformly converging in the norm of V for any q such that $|q - q_0| \leq r$. An analogous definition can be given for a vector right regular functions.

We now state a result whose proof is similar to the one given in the complex case.

Theorem 6.8. *Let* V *be a left quaternionic Banach space and let* $f : \mathbb{H} \to V$ *be a left regular function on* U. *Let* Γ *be a closed rectifiable* 3*-cell. Let* V *be a bounded closed set in* $\mathbb{H}$, *and let* $q_0 \in V$. *Suppose that* W, Γ *and* U *be such that* $W \cup \Gamma \subset U$ *and* $\Gamma = \partial W$. *Then*

$$f(q_0) = \frac{1}{2\pi^2} \int_\Gamma \mathcal{G}(q - q_0)\, Dq\, f(q),$$

where $\mathcal{G}(q - q_0)$ *is the Cauchy-Fueter kernel.*

From now on, we will consider only linear bounded operators and we will follow the ideas in [18]. We introduce a regular function which is related to the resolvent operator and is regular where defined. The idea is to generalize what happens in the complex setting: classically, one considers the Cauchy-Riemann kernel $g(z) = (z - \xi)^{-1}$ defined for $z \neq \xi$ and introduces $R(z,T) = (zI - T)^{-1}$ which is defined for z not in the spectrum of T. In our case, we consider the function $G(q,p)$ written in series expansion as (replacing p by T):

$$\mathcal{G}(q,T) = \sum_{n=0}^{+\infty} \sum_{\nu \in \sigma_n} P_\nu(T) G_\nu(q) = \sum_{n=0}^{+\infty} \sum_{\nu \in \sigma_n} G_\nu(q) P_\nu(T). \tag{6.2}$$

The expansions hold for $\|T\| < |q|$ (cf. Proposition 6.5) and define a bounded operator. It is natural to give the following definition:

Definition 6.9. The maximal open set $\rho(T)$ in $\mathbb{H}$ on which the series (6.2) converges in the operator norm topology to a bounded operator is called the resolvent set of T. The spectral set $\sigma(T)$ of T is defined as the complement set in $\mathbb{H}$ of the resolvent set.

Definition 6.10. A function $f : \mathbb{H} \to \mathbb{H}$ is said to be locally right-regular on the spectral set $\sigma(T)$ of an operator $T \in B(V)$ if there is an open set $U \subset \mathbb{H}$ containing $\sigma(T)$ whose boundary ∂U is a rectifiable 3-cell and such that f is regular in every connected component of U. We will denote by $\mathcal{R}_{r,\sigma(T)}$ the set of locally right regular functions on $\sigma(T)$.

Definition 6.11. Let $f \in \mathcal{R}_{r,\sigma(T)}$ and $T \in B(V)$ and set

$$f(T) := \frac{1}{2\pi^2} \int_{\partial U} f(q) Dq \mathcal{G}(q,T),$$

where U is an open set in $\mathbb{H}$ containing $\sigma(T)$.

Since we have that $F(f+g) = F(f) + F(g)$ and $F(pf) = pF(f)$ the following proposition holds.

Proposition 6.12. *The map $F : \mathcal{R}_{r,\sigma(T)} \to B(V)$ defined by $F(f) = f(T)$ is a left vector space homomorphism.*

Theorem 6.13. *Let*

$$f(q) = \sum_{n=0}^{N} \sum_{\nu \in \sigma_n} a_\nu P_\nu(q)$$

be a right regular polynomial and let U be a ball with center in the origin and radius $r > \|T\|$. Then

$$f(T) = \sum_{n=0}^{N} \sum_{\nu \in \sigma_n} a_\nu P_\nu(T).$$

Proof. Let U be an open set in $\mathbb{H}$ containing $\sigma(T)$. We have

$$f(T) = \frac{1}{2\pi^2} \int_{\partial U} \sum_{n=0}^{N} \sum_{\nu \in \sigma_n} a_\nu P_\nu(q) Dq \mathcal{G}(q,T)$$
$$= \frac{1}{2\pi^2} \sum_{n=0}^{N} \sum_{\nu \in \sigma_n} \int_{\partial U} a_\nu P_\nu(q) Dq \mathcal{G}(q,T)$$

We have, by Proposition 6.2:

$$\int_{\partial U} a_\nu P_\nu(q) Dq \mathcal{G}(q,T) = \int_{\partial U} a_\nu P_\nu(q) Dq \sum_{m\geq 0} \sum_{\mu \in \sigma_m} \mathcal{G}_\mu(q) P_\mu(T)$$
$$= \sum_{m\geq 0} \sum_{\mu \in \sigma_m} a_\nu \int_{\partial U} P_\nu(q) Dq \mathcal{G}_\mu(q) P_\mu(T)$$
$$= 2\pi^2 a_\nu P_\nu(T)$$

which gives $f(T) = \sum_{n=0}^{N} \sum_{\nu\in\sigma_n} a_\nu P_\nu(T)$. □

Proposition 6.14. *For any open set U with piecewise smooth boundary which does not contain $\sigma(T)$ and for any $f \in \mathcal{R}(U)$ we have*

$$\int_{\partial U} f(q) Dq \mathcal{G}(q,T) = 0.$$

Thanks to this proposition, we can replace a ball with center in the origin and suitable radius by any open set containing $\sigma(T)$ and, by the density of polynomials $P_\nu(Q)$ in the set of regular functions, we obtain:

Theorem 6.15. *If the right regular function*

$$f(q) = \sum_{n=0}^{+\infty} \sum_{\nu\in\sigma_n} \alpha_\nu P_\nu(q),$$

converges in a neighborhood U_0 of $\sigma(T)$, then

$$f(T) = \sum_{n=0}^{+\infty} \sum_{\nu\in\sigma_n} \alpha_\nu P_\nu(T),$$

converges in the operator norm topology.

Proof. As U_0 is an open set, it contains a circle

$$U_\delta = \{q : \, |q| \leq \rho(T) + \delta \ \}, \quad \delta > 0$$

in its interior. Hence the series (3.28) converges uniformly in the circle U_δ for some $\delta > 0$ thanks to Theorem 6.6. By the Cauchy integral formula we have

$$f(T) = \frac{1}{2\pi^2}\int_{\partial U_\delta} f(q)\, Dq\, \mathcal{G}(q,T) = \frac{1}{2\pi^2}\int_{\partial U_\delta} \sum_{n=0}^{+\infty}\sum_{\nu\in\sigma_n} \alpha_\nu P_\nu(q)\, Dq\, \mathcal{G}(q,T)$$

$$= \frac{1}{2\pi^2}\sum_{n=0}^{+\infty}\sum_{\mu\in\sigma_n}\int_{\partial U_\delta} \alpha_\nu P_\nu(q)\, Dq\, \mathcal{G}(q,T) = \sum_{n=0}^{+\infty}\sum_{\nu\in\sigma_n} \alpha_\nu P_\nu(T). \qquad \square$$

6.2. Some remarks and open problems

1. The properties which can be proved for the functional calculus defined in [18] and [19] can be demonstrated also in this case. One may also think to generalize the functional calculus as in [15]. However, this functional calculus possesses a strong limitation: even when considering the simplest case of regular function, i.e., a regular (symmetric) polynomial, we have that this function is formed by using the components of a given operator T not the operator T itself. For example, $P(q) = x_0 i - x_1$ is a regular polynomial and $P(T) = T_0 i - T_1$ for any bounded operator $T = T_0 + T_1 i + T_2 j + T_3 k$.
2. The above feature of the functional calculus does not seem to have physical interest when considering a linear quaternionic operator T.
3. As we have shown in Section 3, in the case of the S-resolvent operator the sum of the series $\sum_{n\geq 0} q^n s^{-1-n}$ equals $-(q^2 - 2q\mathrm{Re}[s] + |s|^2)^{-1}(q-\overline{s})$ for $|q| < |s|$ and it does not depend on the commutativity of the components of q so that when one replaces q by an operator T with noncommuting components the sum remains the same. In this case: what is the sum $G(q,T)$ of

$$G(q,p) = \sum_{n=0}^{+\infty}\sum_{\nu\in\sigma_n} P_\nu(p) G_\nu(q) \tag{6.3}$$

   when one replaces p by operator T with noncommuting components?
4. In the case in which the components of T commute, the sum $G(q,T)$ is

$$\mathcal{G}(q,T) = (q\mathcal{I} - T)^{-2}(\overline{q\mathcal{I} - T})^{-1}.$$

   The knowledge of the sum $G(q,T)$ in the general case would naturally lead to a notion of spectrum of the operator T in the case of Fueter regularity.
5. When we consider unbounded operators the series

$$\sum_{n=0}^{+\infty}\sum_{\nu\in\sigma_n} P_\nu(T) G_\nu(q)$$

   does not converge. So it is crucial to manage the sum of such a series so that one could extend the functional calculus to the case of unbounded operators.

References

[1] S. Adler, *Quaternionic Quantum Field Theory*, Oxford University Press (1995).

[2] R.F.V. Anderson, *The Weyl functional calculus*, J. Funct. Anal. **4** (1969), 240–267.

[3] F. Brackx, R. Delanghe, F. Sommen, *Clifford Analysis*, Pitman Res. Notes in Math., 76, 1982.

[4] F. Colombo, G. Gentili, I. Sabadini, D.C. Struppa, *A functional calculus in a non commutative setting*, Electron. Res. Announc. Math. Sci., **14** (2007), 60–68.

[5] F. Colombo, G. Gentili, I. Sabadini, D.C. Struppa, *Non commutative functional calculus: bounded operators*, preprint, 2007.

[6] F. Colombo, G. Gentili, I. Sabadini, D.C. Struppa, *Non commutative functional calculus: unbounded operators*, preprint, 2007.

[7] F. Colombo, I. Sabadini, *A structure formula for slice monogenic functions and some of its consequences*, this volume.

[8] F. Colombo, I. Sabadini, F. Sommen, D.C. Struppa, *Analysis of Dirac Systems and Computational Algebra*, Progress in Mathematical Physics, Vol. 39, Birkhäuser, Boston, 2004.

[9] F. Colombo, I. Sabadini, D.C. Struppa, *A new functional calculus for noncommuting operators*, J. Funct. Anal., **254** (2008), 2255–2274.

[10] F. Colombo, I. Sabadini, D.C. Struppa, *Slice monogenic functions*, to appear in Israel Journal of Mathematics.

[11] N. Dunford, J. Schwartz, *Linear operators, part I: general theory* , J. Wiley and Sons (1988).

[12] G. Gentili, C. Stoppato, *Zeros of regular functions and polynomials of a quaternionic variable*, to appear in Michigan Math. J.

[13] G. Gentili, D.C. Struppa, *A new approach to Cullen-regular functions of a quaternionic variable*, C.R. Acad. Sci. Paris, **342** (2006), 741–744.

[14] G. Gentili, D.C. Struppa, *A new theory of regular functions of a quaternionic variable*, Adv. Math., **216** (2007), 279–301.

[15] B. Jefferies, *Spectral properties of noncommuting operators*, Lecture Notes in Mathematics, 1843, Springer-Verlag, Berlin, 2004.

[16] B. Jefferies, A. McIntosh, *The Weyl calculus and Clifford analysis*, Bull. Austral. Math. Soc., **57** (1998), 329–341.

[17] B. Jefferies, A. McIntosh, J. Picton-Warlow, *The monogenic functional calculus*, Studia Math., **136** (1999), 99–119.

[18] V.V. Kisil, E. Ramirez de Arellano, *The Riesz-Clifford functional calculus for noncommuting operators and quantum field theory*, Math. Methods Appl. Sci., **19** (1996), 593–605.

[19] V.V. Kisil, E. Ramirez de Arellano, *A functional model for quantum mechanics: unbounded operators*, Math. Methods Appl. Sci., **20** (1997), 745–757.

[20] A. McIntosh, A. Pryde, A functional calculus for several commuting operators, Indiana U. Math. J., **36** (1987), 421–439.

[21] C. Stoppato, *Poles of regular quaternionic functions*, To appear in Complex Var. Elliptic Equ.

[22] A. Sudbery, *Quaternionic analysis*, Math. Proc. Camb. Phil. Soc. **85** (1979), 199–225.

[23] J.L. Taylor, *The analytic-functional calculus for several commuting operators*, Acta Math., **125** (1970), 1–38.

[24] J.L. Taylor, *A general framework for a multi-operator functional calculus*, Advances in Math., **9** (1972), 183–252.

[25] M.E. Taylor, *Functions of several self-adjoint operators*, Proc. Amer. Math. Soc., **19** (1968), 91–98.

Fabrizio Colombo and Irene Sabadini
Dipartimento di Matematica
Politecnico di Milano
Via Bonardi, 9
I-20133 Milano, Italy
e-mail: `fabrizio.colombo@polimi.it`
`irene.sabadini@polimi.it`

Graziano Gentili
Dipartimento di Matematica
Universitá di Firenze
Viale Morgagni, 67 A
Firenze, Italy
e-mail: `gentili@math.unifi.it`

Daniele C. Struppa
Department of Mathematics
and Computer Sciences
Chapman University
Orange, CA 92866, USA
e-mail: `struppa@chapman.edu`

Quaternionic and Clifford Analysis
Trends in Mathematics, 101–114

A Structure Formula for Slice Monogenic Functions and Some of its Consequences

Fabrizio Colombo and Irene Sabadini

Abstract. In this paper we show a structure formula for slice monogenic functions (see Lemma 2.2 and [1] for further details): we will show that this formula is a key tool to prove several results, among which we mention the Cauchy integral formula with slice monogenic kernel. This Cauchy formula allows us to extend the validity of the functional calculus for n-tuples of noncommuting operators introduced in [6]. In this wider setting, most of the properties which hold for the Riesz-Dunford functional calculus of a single operator, such as the Spectral Mapping Theorem and the Spectral Radius Theorem, still hold.

Mathematics Subject Classification (2000). Primary 30G35; Secondary 47A10.

Keywords. Slice monogenic functions, slice monogenic kernel, structure formula for slice monogenic functions, functional calculus for n-tuples of linear operators.

1. Introduction

In this paper we deepen the study of the slice monogenic functions introduced in [5] and of the related functional calculus for n-tuples of noncommuting operators. The literature on functional calculi is quite wide. We mention here, without claim completeness, the works [7], [8], [9], [10], [11] and the literature therein. The main purpose of this paper is to show a new version of the Cauchy integral formula obtained in [5] (see the following Theorem 3.8 and [1]). Then we show how to use it to generalize the functional calculus for n-tuples of noncommuting operators introduced in [6].

It is interesting to note that the Cauchy formula with slice monogenic kernel proved in [1] was suggested by the functional calculus formula (14) in [6] and that its proof is based on a structure formula for slice monogenic functions (see formula (2.1) in Lemma 2.2). Formula (2.1) is of independent interest, since it relates the values of a slice monogenic function in a given point $x_0 + I_x|\underline{x}|$ to the values it assumes in the two conjugate points $x = x_0 \pm I|\underline{x}|$ lying on the plane L_I.

Using Theorem 3.8, we can define a functional calculus for functions defined on more general domains then the ones studied in [6]. Moreover, most of the properties that hold for the Riesz-Dunford functional calculus can be proved also for our functional calculus (see Section 4 and [1]).

The authors have also studied the quaternionic functional calculus related to the Cauchy formula with slice regular quaternionic kernel, see [2]. This calculus is based on the analogue of the structure formula (2.1), on the analogue of the Cauchy formula with slice regular kernel (3.5) and on some of their consequences which have been deduced also in quaternionic case, see [3].

1.1. Notations and preliminaries

The setting in which we will work is the real Clifford algebra $\mathbb{R}_n$ over n imaginary units $e_1, \dots, e_n$ satisfying the relations

$$e_i e_j + e_j e_i = -2\delta_{ij}.$$

An element in the Clifford algebra will be denoted by $\sum_A e_A x_A$ where $A = i_1 \dots i_r$, $i_\ell \in \{1, 2, \dots, n\}$, $i_1 < \dots < i_r$ is a multi-index and $e_A = e_{i_1} e_{i_2} \dots e_{i_r}$. In the Clifford algebra $\mathbb{R}_n$, we can identify some specific elements with the vectors in the Euclidean space $\mathbb{R}^n$: an element $(x_1, x_2, \dots, x_n) \in \mathbb{R}^n$ can be identified with a so-called 1-vector in the Clifford algebra through the map

$$(x_1, x_2, \dots, x_n) \mapsto \underline{x} = x_1 e_1 + \dots + x_n e_n,$$

while an element $(x_0, x_1, \dots, x_n) \in \mathbb{R}^{n+1}$ will be identified with the element

$$x = x_0 + \underline{x} = x_0 + \sum_{j=1}^{n} x_j e_j$$

called, in short, vector.

Let us denote by $\mathbb{S}$ the sphere of unit 1-vectors in $\mathbb{R}^n$, i.e.,

$$\mathbb{S} = \{\underline{x} = e_1 x_1 + \dots + e_n x_n \mid x_1^2 + \dots + x_n^2 = 1\}.$$

The vector space $\mathbb{R} + I\mathbb{R}$ passing through 1 and $I \in \mathbb{S}$ will be denoted by L_I, while an element belonging to L_I will be denoted by $u + Iv$, for u, $v \in \mathbb{R}$. Observe that L_I, for every $I \in \mathbb{S}$, can be identified with the complex plane.

Given an element $x = x_0 + \underline{x} \in \mathbb{R}^{n+1}$ let us set

$$I_x = \begin{cases} \dfrac{\underline{x}}{|\underline{x}|} & \text{if } \underline{x} \neq 0, \\ \text{any element of } \mathbb{S} \text{ otherwise.} \end{cases}$$

Definition 1.1. For any element $x \in \mathbb{R}^{n+1}$, we define

$$[x] = \{y \in \mathbb{R}^{n+1} \ : \ y = \mathrm{Re}[x] + I|\underline{x}|, \quad I \in \mathbb{S}\}.$$

The set $[x]$ is a $(n-1)$-dimensional sphere in $\mathbb{R}^{n+1}$.

Remark 1.2. Note that the relation in $\mathbb{R}^{n+1}$ defined by $x \sim y$ if and only if $\mathrm{Re}[y] = \mathrm{Re}[x]$, $|y| = |x|$ is an equivalence relation. The $(n-1)$-sphere associated to $x \in \mathbb{R}^{n+1}$ can also be described as the equivalence class of x. When $x \in \mathbb{R}$, its equivalence class contains x only. In this case, the $(n-1)$-dimensional sphere has radius equal to zero.

We now come to the definition of slice monogenic function, see [5].

Definition 1.3. Let $U \subseteq \mathbb{R}^{n+1}$ be a domain and let $f : U \to \mathbb{R}_n$ be a function. Let $I \in \mathbb{S}$ and let f_I be the restriction of f to the complex plane L_I. We say that f is a (left) slice monogenic function, or s-monogenic function, if for every $I \in \mathbb{S}$, we have

$$\frac{1}{2}\left(\frac{\partial}{\partial u} + I\frac{\partial}{\partial v}\right) f_I(u+Iv) = 0.$$

Analogously, it is possible to define a notion of right s-monogenicity. Sometimes, we will write $\bar{\partial}_I f_I$ to denote the left s-monogenicity condition

$$\bar{\partial}_I f_I = \frac{1}{2}\left(\frac{\partial}{\partial u} + I\frac{\partial}{\partial v}\right) f_I$$

and $f_I \bar{\partial}_I$ instead of the right condition

$$f_I \bar{\partial}_I = \frac{1}{2}\left(\frac{\partial}{\partial u} f_I + \frac{\partial}{\partial v} f_I I\right).$$

For s-monogenic functions it is possible to give the definition of derivative and such a notion is well defined.

Definition 1.4. Let U be a domain in $\mathbb{R}^{n+1}$ and let $f : U \to \mathbb{R}_n$ be an s-monogenic function. Its s-derivative is defined by

$$\partial_s f(x) := \begin{cases} \partial_I(f)(x) & \text{if } x = u + Iv,\ v \neq 0 \\ \partial_u f(u) & \text{if } x = u \in \mathbb{R}, \end{cases} \tag{1.1}$$

where

$$\partial_I := \frac{1}{2}\left(\frac{\partial}{\partial u} - I\frac{\partial}{\partial v}\right).$$

In analogy with the case of holomorphic functions, for s-monogenic functions we have

$$\partial_s f(x) = \partial_u f(u+Iv)$$

and the derivatives of order $n \geq 1$ are given by $\partial_s^n f(x) = \partial_u^n f(u+Iv)$.

The plan of the paper is as follows: in Section 2 we introduce the structure formula for s-monogenic functions as well as some consequences on the zeroes of s-monogenic functions; in Section 3 we state the Cauchy formula with s-monogenic kernel and finally, in Section 4 we show its application to generalize the validity of the functional calculus for n-tuples of operators introduced in [6], also providing some properties of this functional calculus.

2. The structure of slice monogenic functions

In this section we announce and we sketch the proofs of some of the most important results proved by the authors in [1] for the theory of slice monogenic functions. We begin by recalling the Identity Principle for s-monogenic functions (see [5]) which will be crucial to prove our structure formula:

Theorem 2.1 (Identity principle). *Let U be a domain in $\mathbb{R}^{n+1}$ such that $U \cap \mathbb{R} \neq \emptyset$ and $U \cap L_I$ is a domain for all $I \in \mathbb{S}$. Let $f : U \to \mathbb{R}_n$ be an s-monogenic function, and Z the set of its zeroes. If there is an imaginary unit I such that $L_I \cap Z$ has an accumulation point, then $f \equiv 0$ on U.*

The structure formula for s-monogenic functions is a powerful tool which relates the values of a function in a given point $x = x_0 + I_x|\underline{x}|$ to the values it assumes in two conjugate points $x = x_0 \pm I|\underline{x}|$ lying on a plane L_I. The fact that there should be a relation between the values of a function f in a point x and in $x = x_0 \pm I|\underline{x}|$ is suggested by the functional calculus formula in [6]. Indeed, if the operator T is the multiplication by the variable x, we get the value $f(x)$ independently on the plane L_I chosen to integrate. The only condition to be satisfied is that the integration contour has to strictly contain the section of the S-spectrum of the operator T on the plane L_I, for all $I \in \mathbb{S}$ (see Definition 4.1). We point out that the functional calculus formula in [6] can be applied to functions admitting power series expansion. The general validity of the structure formula, will allow us to extend both the Cauchy formula and the functional calculus to a wider class of s-monogenic functions, depending only on the domains on which they are defined.

Lemma 2.2 (The structure formula for s-monogenic functions). *Let $U \subseteq \mathbb{R}^{n+1}$ be a domain such that $U \cap \mathbb{R} \neq \emptyset$, $U \cap L_I$ is a domain for all $I \in \mathbb{S}$ and U contains the $(n-1)$-sphere $[x]$ defined by x whenever $x \in U$. Let $f : U \to \mathbb{R}_n$ be an s-monogenic function. Then for all $x \in U$ and $I \in \mathbb{S}$ the following formula holds:*

$$f(x) = \frac{1}{2}\Big[1 - I_x I\Big] f(x_0 + I|\underline{x}|) + \frac{1}{2}\Big[1 + I_x I\Big] f(x_0 - I|\underline{x}|). \tag{2.1}$$

Proof. If x is a real vector the formula in Lemma 2.2 holds trivially. Otherwise, set the positions, $u = x_0$, $v = |\underline{x}|$ and define the functions

$$\eta_I : U \cap L_I \to \mathbb{R}_n, \qquad \eta_I(u,v) := f(u + Iv) + f(u - Iv),$$

and

$$\theta_I : U \cap L_I \to \mathbb{R}_n, \qquad \theta_I(u,v) := I[f(u - Iv) - f(u + Iv)].$$

We observe that we have the following identity:

$$\frac{1}{2}\Big[1 - I_x I\Big] f(u + Iv) + \frac{1}{2}\Big[1 + I_x I\Big] f(u - Iv) = \frac{1}{2}\Big[\eta_I(u,v) + I_x \theta_I(u,v)\Big]. \tag{2.2}$$

Moreover, for every $I_x \in \mathbb{S}$ the function $\eta_I(u,v) + I_x\theta_I(u,v)$ satisfies the Cauchy-Riemann equation $\bar{\partial}_{I_x}(\eta_I(u,v) + I_x\theta_I(u,v)) = 0$, thus the function $\frac{1}{2}\big[\eta_I(u,v) +$

$I_x\theta_I(u,v)\big]$ is s-monogenic. Finally, observe that if $I = I_x$, we have

$$\frac{1}{2}\big[\eta_{I_x}(u,v) + I_x\theta_{I_x}(u,v)\big] = f(x)$$

so formula (2.1) holds on L_{I_x}. For the Identity Principle the formula holds in the whole domain. □

Remark 2.3. The class of functions of the form $\eta_I(u,v) + I_x\theta_I(u,v)$ satisfying a suitable system of differential equations is interesting per se. For example, consider functions of the form $\eta_I(u,v) + I_x\theta_I(u,v)$ satisfying the Vekua system, instead of the Cauchy-Riemann system. If the functions are also monogenic (in the classical sense), then one obtains the so-called axially monogenic functions studied by Sommen and others, see [12].

The following results are direct consequences of the structure formula (2.2):

Corollary 2.4. *Let $U \subseteq \mathbb{R}^{n+1}$ be a domain such that $U \cap \mathbb{R} \neq \emptyset$, $U \cap L_I$ is a domain for all $I \in \mathbb{S}$ and U contains the $(n-1)$-sphere $[x]$ defined by x whenever $x \in U$. Let $f : U \to \mathbb{R}_n$ be an s-monogenic function. If $f(u_0 + Iv_0) = f(u_0 - Iv_0) = a$ then $f(x) = a$ for all $x \in [u_0 + Iv_0]$. In particular, if $a = 0$, i.e., $f(u_0 + Iv_0) = 0$ and $f(u_0 - Iv_0) = 0$, the $(n-1)$-sphere $[u_0 + Iv_0]$ belongs to the zero set of f.*

Corollary 2.5. *Let $U \subseteq \mathbb{R}^{n+1}$ be a domain such that $U \cap \mathbb{R} \neq \emptyset$, $U \cap L_I$ is a domain for all $I \in \mathbb{S}$ and U contains the $(n-1)$-sphere $[x]$ defined by x whenever $x \in U$. Let $f : U \to \mathbb{R}_n$ be an s-monogenic function. If $f(u_0 + Iv_0) \neq 0$ and $f(u_0 - Iv_0) = 0$, then for all $x \in [u_0 + Iv_0]$ we have*

$$f(x) = \frac{1}{2}(1 - I_x I)f(u_0 + Iv_0).$$

In particular, we can describe the nature of the zeroes of polynomials and power series in the variable $x \in \mathbb{R}^{n+1}$ with coefficients in the Clifford algebra $\mathbb{R}_n$.

Corollary 2.6. *Let $\sum_{m\geq 0} x^m a_m$ be a power series converging in the ball $B(0,R) \subset \mathbb{R}^{n+1}$. If $u_0 \pm Iv_0$ are both solutions to*

$$\sum_{m\geq 0} x^m a_m = 0,$$

then all the elements in the $(n-1)$-sphere $[u_0 + Iv_0]$ are solutions to the equation.

The result holds in particular for polynomial equations:

Corollary 2.7. *Let $p(x) = \sum_{m=0}^{N} x^m a_m$ be a polynomial. If $u_0 \pm Iv_0$ are both solutions to*

$$\sum_{m=0}^{N} x^m a_m = 0,$$

then all the elements in the $(n-1)$-sphere $[u_0 + Iv_0]$ are solutions to the equation.

Note that, in general, given a polynomial $p(x)$ in the variable $x \in \mathbb{R}^{n+1}$, the equation $p(x) = 0$ does not admit solutions (see [5]): sometimes the solutions exist but they are not vectors in the Clifford algebra $\mathbb{R}_n$, sometimes the solutions do not exist at all. It is sufficient to think, for example, to the polynomial $(1 - e_{123})x = 1$ in the Clifford algebra $\mathbb{R}_3$. However, there are examples in which a polynomial equation admits an infinite number of solutions, for example the zero set of $x^2 - 2x + 2 = 0$ in $\mathbb{R}_n$, for any fixed n, is the whole $(n-1)$-sphere $[1 + I]$.

3. The Cauchy formula with slice monogenic kernel

In this section we state and give a sketch of the proof of the Cauchy formula for s-monogenic functions which is more general then the one proved in [5]. It is interesting to note that this new formula has been stimulated by the functional calculus for n-tuples of noncommuting operators.

Let us begin by recalling the notion of noncommutative Cauchy kernel series for $x, s \in \mathbb{R}^{n+1}$ and of noncommutative Cauchy kernel.

Definition 3.1. Let $x = \mathrm{Re}[x] + \underline{x}$, $s = \mathrm{Re}[s] + \underline{s}$ be such that $sx \neq xs$. We will call noncommutative Cauchy kernel series the following expansion

$$S^{-1}(s, x) := \sum_{n \geq 0} x^n s^{-1-n} \tag{3.1}$$

defined for $|x| < |s|$.

We will use the sum of the series (3.1) to prove the new Cauchy formula in the sequel.

Theorem 3.2. *(See [5]) Let $x = \mathrm{Re}[x] + \underline{x}$, $s = \mathrm{Re}[s] + \underline{s}$ be such that $xs \neq sx$. Then*

$$\sum_{n \geq 0} x^n s^{-1-n} = -(x^2 - 2x\mathrm{Re}[s] + |s|^2)^{-1}(x - \overline{s})$$

for $|x| < |s|$.

We will call the expression

$$S^{-1}(s, x) = -(x^2 - 2x\mathrm{Re}[s] + |s|^2)^{-1}(x - \overline{s}), \tag{3.2}$$

defined for $x^2 - 2x\mathrm{Re}[s] + |s|^2 \neq 0$, *noncommutative Cauchy kernel.* With an abuse of notation, we denote the noncommutative Cauchy kernel series and the noncommutative Cauchy kernel with the same symbol $S^{-1}(s, x)$. In fact they coincide where they are both defined by virtue of their monogenicity (see Proposition 3.3) and of the Identity Principle, see [5]. Therefore note that the noncommutative Cauchy kernel is defined on a set which is larger then the set $\{(x, s) \in \mathbb{R}^{n+1} \times \mathbb{R}^{n+1} \mid |x| < |s|\}$ where the noncommutative Cauchy kernel series converges.

The following result concerns the fact that the new kernel has slice monogenicity properties.

Proposition 3.3. *The function $S^{-1}(s,x)$ is left s-monogenic in the variable x and right s-monogenic in the variable s in its domain of definition.*

Proof. The proof follows by direct computations. □

Proposition 3.4. *Let $x = \mathrm{Re}[x] + \underline{x}$, $s = \mathrm{Re}[s] + \underline{s}$ be such that $x \neq \bar{s}$. Then the following identity holds:*

$$(x - \bar{s})^{-1} s (x - \bar{s}) - x = -(s - \bar{x}) x (s - \bar{x})^{-1} + s, \tag{3.3}$$

or, equivalently,

$$-(x - \bar{s})^{-1}(x^2 - 2x\mathrm{Re}[s] + |s|^2) = (s^2 - \mathrm{Re}[x]s + |x|^2)(s - \bar{x})^{-1}. \tag{3.4}$$

Proof. One may prove the identities by direct computations. □

To obtain the new version of the Cauchy formula, we recall some results which are proved in [1]:

Proposition 3.5. *Let $x \in \mathbb{R}^{n+1} \backslash \mathbb{R}$. If $I \neq I_x$, the function $S^{-1}(s,x) = S_{\bar{x}}^{-1}(s)$ has the two singularities $\mathrm{Re}[x] \pm I|\underline{x}|$ on the plane L_I. On the plane L_{I_x} the function $S_{\bar{x}}^{-1}(s) = (x-s)^{-1}$ has only one singularity at the point x. If $x \in \mathbb{R}$, the function $S_{\bar{x}}^{-1}(s)$ has only one singularity at the point x.*

The previous proposition states that $S_{I_x}^{-1}(s,x)$, i.e., the restriction of $S^{-1}(s,x)$ to the plane L_{I_x}, has a removable singularity at the point $s = \bar{x}$. However, equality (3.4) and the proof of Theorem 2.11 in [6] show that the function $S^{-1}(s,x)$ still has a singularity at the point $s = \bar{x}$.

We now state a version of the Stokes' theorem for s-monogenic functions which will allow us to prove the Cauchy formula with the kernel defined in (3.2).

Lemma 3.6. *Let f, g be continuously differentiable functions on an open set $U_I = U \cap L_I$ of the plane L_I. Then for every 2-chain $C \subset U_I$ we have*

$$\int_{\partial C} g(s) ds_I f(s) = 2 \int_C ((g(s)\bar{\partial}_I) f(s) + g(s)(\bar{\partial}_I f(s))) d\sigma$$

where $s = u + Iv$ is the variable on L_I, $ds_I = -Ids$, $d\sigma = du \wedge dv$.

An immediate consequence of the above lemma is the following:

Corollary 3.7. *Let f and g be left s-monogenic and right s-monogenic, respectively, on an open set U. For any $I \in \mathbb{S}$ and any 2-chain C in $U \cap L_I$ we have:*

$$\int_{\partial C} g(s) ds_I f(s) = 0.$$

We are now ready to state the Cauchy formula with s-monogenic kernel. Here we just give a sketch of its proof. All the details are in Theorem 2.16 in [1].

Theorem 3.8. *Let $U \subset \mathbb{R}^{n+1}$ be a domain, such that $U \cap L_I$ is a domain for all $I \in \mathbb{S}$, $U \cap \mathbb{R} \neq \emptyset$ and $[x] \subset U$ whenever $x \in U$. Suppose that $\partial(U \cap L_I)$ is finite union of rectifiable Jordan curves for every $I \in \mathbb{S}$. Let f be a (left) s-monogenic function on U and set $ds_I = ds/I$. Then*

$$f(x) = \frac{1}{2\pi} \int_{\partial(U\cap L_I)} S^{-1}(s,x) ds_I f(s) \tag{3.5}$$

where $S^{-1}(s,x)$ is defined in (3.2) and the integral does not depend on U and on the imaginary unit $I \in \mathbb{S}$.

Proof. First of all, the integral at the right-hand side of (3.5) does not depend on the open set U: it follows from the fact that $S^{-1}(s,x)$ is right s-monogenic in s, and Corollary 3.7.

Let us show that the integral (3.5) does not depend on the choice of the imaginary unit $I \in \mathbb{S}$.

The zeroes of the function $x^2 - 2s_0x + |s|^2$ consist either of a real point x or a 2-sphere $[x]$. On L_{I_x} we find only the point x as a singularity and the result follows from the Cauchy formula on the plane L_{I_x}. When the singularity is a real number, the integral reduces again to a Cauchy integral of the complex analysis. If the zero is not real, on any complex plane L_I we find the two zeroes

$$s_{1,2} = x_0 \pm I|\underline{x}|.$$

In this case, we calculate the residues about the points s_1 e s_2 on the plane L_I for $I \neq I_x$. Let us start with s_1 by setting the positions

$$s = x_0 + I|\underline{x}| + \varepsilon e^{I\theta}, \qquad s_0 = x_0 + \varepsilon\cos\theta, \qquad \overline{s} = x_0 - I|\underline{x}| + \varepsilon e^{-I\theta},$$
$$ds_I = -[\varepsilon I e^{I\theta}] I d\theta = \varepsilon e^{I\theta} d\theta, \quad |s|^2 = x_0^2 + 2x_0\varepsilon\cos\theta + \varepsilon^2 + |\underline{x}|^2 + 2\varepsilon\sin\theta|\underline{x}|,$$

we have

$$\rho_1^\varepsilon = \int_0^{2\pi} -(-2x\varepsilon\cos\theta + 2x_0\varepsilon\cos\theta + \varepsilon^2 + 2\varepsilon\sin\theta|\underline{x}|)^{-1}$$
$$\times (x - [x_0 - I|\underline{x}| + \varepsilon e^{-I\theta}])\varepsilon e^{I\theta} d\theta f(x_0 + I|\underline{x}| + \varepsilon e^{I\theta})$$

for $\varepsilon \to 0$ we get

$$\rho_1^0 = \int_0^{2\pi} (2x\cos\theta - 2x_0\cos\theta - 2\sin\theta|\underline{x}|)^{-1}(\underline{x} + I|\underline{x}|)e^{I\theta} d\theta f(x_0 + I|\underline{x}|)$$
$$= -\frac{1}{2|\underline{x}|^2} \int_0^{2\pi} [(\underline{x})^2\cos\theta + \sin\theta|\underline{x}|\underline{x}$$
$$+ \underline{x}I|\underline{x}|\cos\theta + \sin\theta|\underline{x}|^2 I][\cos\theta + I\sin\theta] d\theta f(x_0 + I|\underline{x}|).$$

With some calculations we obtain

$$\rho_1^0 = \frac{\pi}{|\underline{x}|}\Big[|\underline{x}| - \underline{x}I\Big] f(x_0 + I|\underline{x}|).$$

Recalling that $\underline{x}/|\underline{x}| = I_x$ we get the first residue

$$\rho_1^0 = \frac{1}{2}\Big[1 - I_x I\Big] f(x_0 + I|\underline{x}|).$$

With analogous calculations we prove that the residue about s_2 is

$$\rho_2^0 = \frac{1}{2}\Big[1 + I_x I\Big] f(x_0 - I|\underline{x}|).$$

So by the residues theorem we get:

$$\frac{1}{2\pi}\int_{\partial(U\cap L_I)} S^{-1}(s,x) ds_I f(s) = \rho_1^0 + \rho_2^0.$$

The statement now follows from Lemma 2.2. □

We conclude this section with the formula for the derivatives of an s-monogenic function using the s-monogenic Cauchy kernel. To this aim, we define a product between s-monogenic polynomials which preserves the s-monogenicity:

Definition 3.9. Let $f(x) = \sum_{i=0}^{n} x^i a_i$ and $g(x) = \sum_{i=0}^{m} x^i b_i$, where a_i, $b_i \in \mathbb{R}_n$. We define the s-monogenic product of f and g as

$$f * g(x) := \sum_{j=0}^{n+m} x^j c_j$$

with $c_j = \sum_{i+k=j} a_i b_k$. We will denote by f^{n*} the product $f * \cdots * f$, n-times.

This product is computed by taking the coefficients of the polynomials on the right, like in the case in which the variables and the coefficients commute. When the coefficients of a polynomial f are real numbers, the s-monogenic product coincides with the usual product, i.e., $f * g = fg$.

The following result shows that s-monogenic functions are infinitely differentiable.

Theorem 3.10. *Let $U \subset \mathbb{R}^{n+1}$ be a domain, such that $U \cap L_I$ is a domain for all $I \in \mathbb{S}$, $U \cap \mathbb{R} \neq \emptyset$ and $[x] \subset U$ whenever $x \in U$. Suppose $\partial(U \cap L_I)$ is a finite union of rectifiable Jordan curves for every $I \in \mathbb{S}$. Let f be an s-monogenic function on U and set $ds_I = ds/I$. Let $x = x_0 + \underline{x}$, and $s = s_0 + \underline{s}$. Then*

$$\partial_{x_0}^n f(x) = \frac{n!}{2\pi}\int_{\partial(U\cap L_I)} (x^2 - 2s_0 x + |s|^2)^{-n-1}(x - \overline{s})^{(n+1)*} ds_I f(s)$$

$$= \frac{n!}{2\pi}\int_{\partial(U\cap L_I)} [S^{-1}(s,x)(x-\overline{s})^{-1}]^{n+1}(x-\overline{s})^{(n+1)*} ds_I f(s) \qquad (3.6)$$

where

$$(x - \overline{s})^{n*} = \sum_{k=0}^{n} \frac{n!}{(n-k)!k!} x^{n-k}\overline{s}^k, \qquad (3.7)$$

and $S^{-1}(s,x)$ is defined in (3.2). Moreover, the integral does not depend on U and on the imaginary unit $I \in \mathbb{S}$.

Proof. First of all, we recall that the s-derivative defined in (1.1) coincides, for s-monogenic functions, with the partial derivative with respect to the scalar coordinate x_0. To compute $\partial^n_{x_0} f(x)$, we can compute the derivative of the integrand, since f and its derivatives with respect to x_0 are continuous functions on $\partial(U\cap L_I)$. Thus we get

$$\partial^n_{x_0} f(x) = \frac{1}{2\pi}\int_{\partial(U\cap L_I)} \partial^n_{x_0}[S^{-1}(s,x)]ds_I f(s).$$

To prove the statement, it is sufficient to compute by recurrence $\partial^n_{x_0}[S^{-1}(s,x)]$. For the derivative of $\partial_{x_0} S^{-1}(s,x)$ we have

$$\partial_{x_0} S^{-1}(s,x) = (x^2 - 2s_0 x + |s|^2)^{-2}(x-\bar{s})^{2*}.$$

We now assume that

$$\partial^n_{x_0} S^{-1}(s,x) = (-1)^{n+1} n!(x^2 - 2s_0 x + |s|^2)^{-(n+1)}(x-\bar{s})^{(n+1)*},$$

holds and we compute $\partial^{n+1}_{x_0} S^{-1}(s,x)$. We have:

$$\begin{aligned}\partial^{n+1}_{x_0} S^{-1}(s,x) &= \partial_{x_0}[(-1)^{n+1} n!(x^2 - 2s_0 x + |s|^2)^{-(n+1)}(x-\bar{s})^{(n+1)*}]\\ &= (-1)^{n+2}(n+1)!(x^2 - 2s_0 x + |s|^2)^{-(n+2)}(2x-2s_0)(x-\bar{s})^{(n+1)*}\\ &\quad + (-1)^{n+1}(n+1)!(x^2 - 2s_0 x + |s|^2)^{-(n+1)}(x-\bar{s})^{n*}\\ &= (-1)^{n+2}(n+1)!(x^2 - 2s_0 x + |s|^2)^{-(n+2)}\\ &\quad \times [(2x-2s_0)(x-\bar{s}) - (x^2 - 2s_0 x + |s|^2)] * (x-\bar{s})^{n*}\end{aligned}$$

here we have used the fact that the s-monogenic product coincides with the usual one when the coefficients a real numbers, so

$$\partial^{n+1}_{x_0} S^{-1}(s,x) = (-1)^{n+2}(n+1)!(x^2 - 2s_0 x + |s|^2)^{-(n+2)}[x^2 - 2x\bar{s} + \bar{s}^2] * (x-\bar{s})^{n*}.$$

We get the last equality in (3.6) by recalling that

$$S^{-1}(s,x)(x-\bar{s})^{-1} = (x^2 - 2s_0 x + |s|^2)^{-1}. \qquad \square$$

In the next section we introduce the functional calculus for n-tuples of noncommuting operators based on the Cauchy formula (3.5).

4. The functional calculus for n-tuples of noncommuting operators

In the sequel, we will consider a Banach space V over $\mathbb{R}$ (the case of complex Banach spaces can be discussed in a similar fashion) with norm $\|\cdot\|$ and let $V_n = V \otimes \mathbb{R}_n$. We will denote by $\mathcal{B}(V)$ the space of bounded $\mathbb{R}$-homomorphisms of the Banach space V to itself endowed with the natural norm denoted by $\|\cdot\|_{\mathcal{B}(V)}$. Given $T_A \in \mathcal{B}(V)$, we can introduce the operator $T = \sum_A T_A e_A$ and its action on $v = \sum v_B e_B \in V_n$ as $T(v) = \sum_{A,B} T_A(v_B) e_A e_B$. The operator T is a right-module homomorphism which is a bounded linear map on V_n. The set of all such bounded operators is denoted by $\mathcal{B}_n(V_n)$. We define $\|T\|^2_{\mathcal{B}_n(V_n)} = \sum_A \|T_A\|^2_{\mathcal{B}(V)}$. In the sequel, we will consider operators of the form

$$T = T_0 + \sum_{j=1}^{n} e_j T_j$$

where $T_\mu \in \mathcal{B}(V)$ for $\mu = 0, 1, \dots, n$. The subset of such operators in $\mathcal{B}_n(V_n)$ will be denoted by $\mathcal{B}^{0,1}_n(V_n)$.

The S-spectrum of T, defined below, generalizes the definition of spectral set in the case of a single operator.

Definition 4.1 (The S-spectrum and the S-resolvent set). Let $T \in \mathcal{B}^{0,1}_n(V_n)$ and $s \in \mathbb{R}^{n+1}$. We define the S-spectrum $\sigma_S(T)$ of T as:

$$\sigma_S(T) = \{s \in \mathbb{R}^{n+1} \;\;:\;\; T^2 - 2\,\mathrm{Re}[s]T + |s|^2\mathcal{I} \quad \text{is not invertible}\}.$$

The S-resolvent set $\rho_S(T)$ is defined by

$$\rho_S(T) = \mathbb{R}^{n+1} \setminus \sigma_S(T).$$

We recall some properties of the spectrum which are originally proven in [6] and can be found also in [4]. First of all, we recall that given an operator $T \in \mathcal{B}^{0,1}_n(V_n)$ and $p = \mathrm{Re}[p] + \underline{p} \in \sigma_S(T)$, then all the elements of the $(n-1)$-sphere $[p]$ belongs to $\sigma_S(T)$. This fact implies that if $x \in \sigma_S(T)$ then either x is a real point or the whole $(n-1)$-sphere $[x]$ belongs to $\sigma_S(T)$. Moreover, we have

Theorem 4.2 (Compactness of S-spectrum). *Let $T \in \mathcal{B}^{0,1}_n(V_n)$. Then the S-spectrum $\sigma_S(T)$ is a compact nonempty set. Moreover $\sigma_S(T)$ is contained in $\{s \in \mathbb{R}^{n+1} : |s| \leq \|T\| \}$.*

The following is a fundamental definition:

Definition 4.3 (The S-resolvent operator). Let $T \in \mathcal{B}^{0,1}_n(V_n)$ and $s \in \rho_S(T)$. We define the S-resolvent operator as

$$S^{-1}(s,T) := -(T^2 - 2\mathrm{Re}[s]T + |s|^2\mathcal{I})^{-1}(T - \overline{s}\mathcal{I}). \tag{4.1}$$

In the case of a single operator the S-resolvent operator (4.1) becomes the classical one used in the Riesz-Dunford functional calculus. We now describe the class of functions for which we can define our functional calculus.

Definition 4.4. Let $T = T_0 + \sum_{j=1}^{n} e_j T_j \in \mathcal{B}^{0,1}_n(V_n)$. Let $U \subset \mathbb{R}^{n+1}$ be an open set containing the $(n-1)$-sphere $[x]$ for every $x \in U$ and such that

(i) $\partial(U \cap L_I)$ is union of a finite number of rectifiable Jordan curves for every $I \in \mathbb{S}$,

(ii) U contains the S-spectrum $\sigma_S(T)$, $U \cap \mathbb{R} \neq \emptyset$.

A function f is said to be locally s-monogenic on $\sigma_S(T)$ if there exists an open set $U \subset \mathbb{R}^{n+1}$ as above, on which f is s-monogenic.

We will denote by $\mathcal{M}_{\sigma_S(T)}$ the set of locally s-monogenic functions on $\sigma_S(T)$.

The following result is a consequence of the Cauchy formula (3.5) and of a well-known corollary of the Hahn-Banach theorem.

Theorem 4.5. *Let $T \in \mathcal{B}_n^{0,1}(V_n)$ and $f \in \mathcal{M}_{\sigma_S(T)}$. Let $U \subset \mathbb{R}^{n+1}$ be any open set as in Definition 4.4 and let $U_I = U \cap L_I$ for $I \in \mathbb{S}$. Then the integral*

$$\frac{1}{2\pi} \int_{\partial U_I} S^{-1}(s,T)\, ds_I\, f(s) \tag{4.2}$$

does not depend on the open set U and on the choice of the imaginary unit $I \in \mathbb{S}$.

Note that we can formally replace the variable x in $S^{-1}(s,x)$ by an operator T by Theorem 3.2 in [6]. Theorem 4.5 states that the operator (4.2) defined by the integral does not depend on the open set U and on the choice of the imaginary unit $I \in \mathbb{S}$, thus the following definition is well posed.

Definition 4.6. Let $T \in \mathcal{B}_n^{0,1}(V_n)$ and $f \in \mathcal{M}_{\sigma_S(T)}$. Let $U \subset \mathbb{R}^{n+1}$ be any open set as in Definition 4.4, and set $U_I = U \cap L_I$ for $I \in \mathbb{S}$. We define

$$f(T) = \frac{1}{2\pi} \int_{\partial U_I} S^{-1}(s,T)\, ds_I\, f(s). \tag{4.3}$$

4.1. Some properties of the functional calculus

The product and the composition of two s-monogenic functions is not, in general, an s-monogenic function. Here we give sufficient conditions to guarantee the s-monogenicity of the product and of the composition of s-monogenic functions.

Definition 4.7. Let U be a domain in $\mathbb{R}^{n+1}$. Let $I = I_1 \in \mathbb{S}$ let $I_2, \dots, I_n$ be a completion to an orthonormal basis of $\mathbb{R}_n$ and let

$$f_I(z) = \sum_{|A|=0}^{n-1} F_A(z) I_A, \quad I_A = I_{i_1} \dots I_{i_s}, \quad z = u + Iv$$

(where F_A are holomorphic functions, $A = i_1 \dots i_s$ is a subset of $\{2, \dots, n\}$, with $i_1 < \dots < i_s$, or, when $|A| = 0$, $I_\emptyset = 1$) be the corresponding splitting. The subclass of functions $f \in \mathcal{M}(U)$ such that

$$f_I(z) = \sum_{|A|=0, |A| even}^{n-1} F_A(z) I_A, \quad I_A = I_{i_1} \dots I_{i_s},$$

for all $I \in \mathbb{S}$ will be denoted by $\widetilde{\mathcal{M}}(U)$.

For the product of s-monogenic functions the subset $\widetilde{\mathcal{M}}(U)$ is sufficient to guarantee the validity of the following proposition:

Proposition 4.8. *Let U be a domain in $\mathbb{R}^{n+1}$. Let $f \in \widetilde{\mathcal{M}}(U)$, $g \in \mathcal{M}(U)$, then $fg \in \mathcal{M}(U)$.*

To consider the composition of s-monogenic functions we define the following subset of $\mathcal{M}(U)$.

Definition 4.9. Let U be a domain in $\mathbb{R}^{n+1}$. For $\mathbb{R}_n$-valued functions, we define

$$\mathcal{N}(U) = \{f \in \mathcal{M}(U) \mid f(L_I) \subseteq L_I, \forall I \in \mathbb{S}\}.$$

The relation among the sets $\mathcal{N}(U)$, $\widetilde{\mathcal{M}}(U)$ and $\mathcal{M}(U)$ is given by the following lemma.

Lemma 4.10. *Let U be a domain in $\mathbb{R}^{n+1}$. We have $\mathcal{N}(U) \subseteq \widetilde{\mathcal{M}}(U) \subseteq \mathcal{M}(U)$.*

Lemma 4.11. *Let U, U' be two domains in $\mathbb{R}^{n+1}$, $U' \cap \mathbb{R} \neq \emptyset$ and let $f \in \mathcal{N}(U')$, $g \in \mathcal{N}(U)$ with $g(U) \subseteq U'$. Then $f(g(x))$ is s-monogenic in U.*

Now becomes natural the definition to follow.

Definition 4.12. In Definition 4.4 consider instead of s-monogenic functions, the subset of functions belonging to $\widetilde{\mathcal{M}}$ (resp. $\mathcal{N}$). This subclass of $\mathcal{M}_{\sigma_S(T)}$ will be denoted by $\widetilde{\mathcal{M}}_{\sigma_S(T)}$ (resp. $\mathcal{N}_{\sigma_S(T)}$).

Here we have the algebraic properties of our functional calculus:

Theorem 4.13. *Let $T \in \mathcal{B}_n^{0,1}(V_n)$.*

(a) *If f and $g \in \mathcal{M}_{\sigma_S(T)}$ then $(f+g)(T) = f(T) + g(T)$, $(f\lambda)(T) = f(T)\lambda$, for all $\lambda \in \mathbb{R}_n$.*
(b) *If $\phi \in \widetilde{\mathcal{M}}_{\sigma_S(T)}$ and $g \in \mathcal{M}_{\sigma_S(T)}$. Then $(\phi g)(T) = \phi(T)g(T)$.*
(c) *If $f(s) = \sum_{n\geq 0} s^n p_n$, $p_n \in \mathbb{R}_n$, belongs to $\mathcal{M}_{\sigma_S(T)}$, then $f(T) = \sum_{n\geq 0} T^n p_n$.*
(d) *Let $f_m \in \mathcal{M}_{\sigma_S(T)}$, $m \in \mathbb{N}$ and let $U \supset \sigma_S(T)$ be an open set as in Definition 4.4. Then if f_m converges uniformly to f on $U_I = U \cap L_I$, for some $I \in \mathbb{S}$, then $f_m(T)$ converges to $f(T)$ in $\mathcal{B}(V)$.*

The Spectral Mapping Theorem holds for our functional calculus in the following form (see [1]):

Theorem 4.14. *Let $T \in \mathcal{B}_n^{0,1}(V_n)$, $f \in \widetilde{\mathcal{M}}_{\sigma_S(T)}$. Then*

$$\sigma_S(f(T)) = f(\sigma_S(T)) = \{f(s) : s \in \sigma_S(T)\}.$$

As a consequence of the Spectral Mapping Theorem we can prove the theorem of composition of functions:

Theorem 4.15. *Let $T \in \mathcal{B}_n^{0,1}(V_n)$, $f \in \mathcal{N}_{\sigma_S(T)}$, $\phi \in \mathcal{N}_{\sigma_S(f(T))}$ and let $F(s) = \phi(f(s))$. Then $F \in \mathcal{M}_{\sigma_S(T)}$ and $F(T) = \phi(f(T))$.*

We finally conclude with the S-spectral radius theorem. We define the spectral radius related to the S-spectrum of T, that is the real nonnegative number

$$r_S(T) := \sup\{\, |s| \ : \ s \in \sigma_S(T)\,\},$$

so we have:

Theorem 4.16 (S-spectral radius theorem). *Let $T \in \mathcal{B}_n^{0,1}(V_n)$ and let $r_S(T)$ be the S-spectral radius of T. Then*

$$r_S(T) = \lim_{m\to\infty} \|T^m\|^{1/m}.$$

References

[1] F. Colombo, I. Sabadini, *The Cauchy formula with s-monogenic kernel and a functional calculus for noncommuting operators*, preprint, 2008.

[2] F. Colombo, I. Sabadini, *On some properties of the quaternionic functional calculus*, preprint, 2008.

[3] F. Colombo, G. Gentili, I. Sabadini, *A Cauchy kernel for slice regular functions*, preprint, 2008.

[4] F. Colombo, G. Gentili, I. Sabadini, D.C. Struppa, *An overview on functional calculus in different settings*, this volume.

[5] F. Colombo, I. Sabadini, D.C. Struppa, *Slice monogenic functions*, to appear in Israel Journal of Mathematics.

[6] F. Colombo, I. Sabadini, D.C. Struppa, *A new functional calculus for noncommuting operators*, J. Funct. Anal., **254** (2008), 2255–2274.

[7] B. Jefferies, *Spectral properties of noncommuting operators*, Lecture Notes in Mathematics, 1843, Springer-Verlag, Berlin, 2004.

[8] B. Jefferies, A. McIntosh, *The Weyl calculus and Clifford analysis*, Bull. Austral. Math. Soc., **57** (1998), 329–341.

[9] B. Jefferies, A. McIntosh, J. Picton-Warlow, *The monogenic functional calculus*, Studia Math., **136** (1999), 99–119.

[10] V.V. Kisil, E. Ramirez de Arellano, *The Riesz-Clifford functional calculus for noncommuting operators and quantum field theory*, Math. Methods Appl. Sci., **19** (1996), 593–605.

[11] A. McIntosh, A. Pryde, *A functional calculus for several commuting operators*, Indiana Univ. Math. J., **36** (1987), 421–439.

[12] F. Sommen, *Special functions in Clifford analysis and axial symmetry*, J. Math. Anal. Appl., **130** (1988), 110–133.

Fabrizio Colombo and Irene Sabadini
Dipartimento di Matematica
Politecnico di Milano
Via Bonardi, 9
I-20133 Milano, Italy
e-mail: fabrizio.colombo@polimi.it
irene.sabadini@polimi.it

Quaternionic and Clifford Analysis
Trends in Mathematics, 115–124

On the CK-extension for a Special Overdetermined System in Complex Clifford Analysis

Bram De Knock, Dixan Peña Peña and Frank Sommen

Abstract. In this paper we investigate a new overdetermined system in $\mathbb{R}^{m+1}$, called RicSom system, arising from adding one extra real dimension to the Hermitian Dirac system in $\mathbb{R}^m$, $m = 2n$, that uses the complex structure of $\mathbb{C}^n$. For this new system we consider a CK-extension type problem.

Mathematics Subject Classification (2000). Primary 30G35.

Keywords. Clifford algebras, Dirac operator, CK-extension problem.

1. Introduction and preliminaries

We first briefly present the basic definitions and some results of Hermitian Clifford analysis which are necessary for our purpose. For an in-depth study of this higher-dimensional function theory we refer to, e.g., [6, 7, 4, 1, 2].

Hermitian Clifford analysis focuses on the simultaneous null solutions of the orthogonal Dirac operators $\partial_{\underline{X}}$ and its twisted counterpart $\partial_{\underline{X}|}$, introduced below. Both Dirac operators being linked to each other by means of a so-called complex structure, the dimension of the Euclidean space, from which our complex Clifford algebra stems, is forced to be even (see, e.g., [1]). So, let $\mathbb{R}^{0,2n}$ be endowed with a non-degenerate quadratic form of signature $(0, 2n)$, let $(e_1, \ldots, e_{2n})$ be an orthonormal basis for $\mathbb{R}^{0,2n}$ and let $\mathbb{C}_{2n}$ be the complex Clifford algebra constructed over $\mathbb{R}^{0,2n}$. The non-commutative multiplication in $\mathbb{C}_{2n}$ is governed by

$$e_j e_k + e_k e_j = -2\,\delta_{jk} \quad , \quad j, k = 1, \ldots, 2n\ .$$

A basis for $\mathbb{C}_{2n}$ is obtained by considering for a set

$$A = \{j_1, \ldots, j_h\} \subset \{1, \ldots, 2n\} = M$$

the element $e_A = e_{j_1} \ldots e_{j_h}$, with $1 \leq j_1 < j_2 < \cdots < j_h \leq 2n$. For the empty set $\emptyset$ one puts $e_\emptyset = 1$, the identity element. Any complex Clifford number a in $\mathbb{C}_{2n}$

may thus be written as $a = \sum_A e_A a_A$, where $a_A \in \mathbb{C}$, or still as $a = \sum_{k=0}^{2n} [a]_k$, where $[a]_k = \sum_{|A|=k} e_A a_A$ is the so-called k-vector part of a ($k = 0, 1, \ldots, 2n$). Denoting then by $\mathbb{C}_{2n}^k$ the subspace of all k-vectors in $\mathbb{C}_{2n}$, i.e., the image of the projection operator $[\cdot]_k$, one has the multivector decomposition $\mathbb{C}_{2n} = \bigoplus_{k=0}^{2n} \mathbb{C}_{2n}^k$ leading to the identification of $\mathbb{C}$ with $\mathbb{C}_{2n}^0$ and of $\mathbb{R}^{0,2n}$ with the subspace of real Clifford vectors

$$\mathbb{R}_{0,2n}^1 = \{\underline{X} = \sum_{j=1}^{n} (e_j\, x_j + e_{n+j}\, y_j)\ , x_j, y_j \in \mathbb{R}\} \subset \mathbb{C}_{2n}^1 \,.$$

At the same time we introduce for each real vector $\underline{X}$ its twisted counterpart

$$\underline{X}| = \sum_{j=1}^{n} (e_j\, y_j - e_{n+j}\, x_j) \,.$$

Note that the square of a vector $\underline{X}$ (or $\underline{X}|$) is scalar valued and equals the norm squared up to a minus sign: $\underline{X}^2 = -\langle \underline{X}, \underline{X} \rangle = -|\underline{X}|^2 = -|\underline{X}||^2 = \underline{X}|^2$. Also observe that the vectors $\underline{X}$ and $\underline{X}|$ are orthogonal with respect to the standard Euclidean scalar product, which implies that the Clifford vectors $\underline{X}$ and $\underline{X}|$ anti-commute.

The Fischer dual of the vector $\underline{X}$ is the real vector valued first-order differential operator

$$\partial_{\underline{X}} = \sum_{j=1}^{n} (e_j\, \partial_{x_j} + e_{n+j}\, \partial_{y_j})\ ,$$

called Dirac operator. It is precisely this Dirac operator which underlies the notion of monogenicity of a function, a notion which is the higher-dimensional counterpart of holomorphy in the complex plane (see [3, 5]). In what follows we denote by Ω an open subset of $\mathbb{R}^{2n}$. A continuously differentiable function $f : \Omega \to \mathbb{C}_{2n}$ is called (left) monogenic in Ω if $\partial_{\underline{X}} f = 0$ in Ω. Analogously, also a notion of monogenicity can be associated to the Fisher dual of the vector $\underline{X}|$, given by

$$\partial_{\underline{X}|} = \sum_{j=1}^{n} (e_j\, \partial_{y_j} - e_{n+j}\, \partial_{x_j}) \,.$$

As the Dirac operator $\partial_{\underline{X}}$ (respectively $\partial_{\underline{X}|}$) factorizes the Laplacian, i.e.,

$$-\partial_{\underline{X}}^2 = \Delta = -\partial_{\underline{X}|}^2\ ,$$

monogenicity with respect to $\partial_{\underline{X}}$ (respectively $\partial_{\underline{X}|}$) can be regarded as a refinement of harmonicity.

Further, a continuously differentiable function $f : \Omega \to \mathbb{C}_{2n}$ is called a (left) Hermitian monogenic (or h-monogenic) function in Ω if and only if it satisfies in Ω the system

$$\partial_{\underline{X}} f = 0 = \partial_{\underline{X}|} f \,. \tag{1.1}$$

The theory of the above system being equivalent with the system

$$\partial_{\underline{X}}\,[fI] = 0 = \partial_{\underline{X}|}\,[fI]\,,$$

with $I = I_1 I_2 \ldots I_n$ the primitive idempotent for which

$$I_j = \frac{1}{2}(1 - i e_j e_{n+j})\,, \quad j = 1, \ldots, n\,,$$

a study of the latter system is considered in [4]. Since moreover it is shown there that $\mathbb{C}_{2n} I \cong \mathbb{C}_n I$, the h-monogenic system to be examined reads

$$\partial_{\underline{X}}\,[FI] = 0 = \partial_{\underline{X}|}\,[FI] \tag{1.2}$$

for functions F with values in $\mathbb{C}_n = \mathrm{Alg}_{\mathbb{C}}\{e_1, \ldots, e_n\}$. Introducing the main involution $\tilde{\cdot}$ which leaves the multivector structure invariant, i.e.,

$$(ab)\tilde{} \;=\; \tilde{a}\tilde{b}\,, \qquad (e_A a_A)\tilde{} \;=\; \widetilde{e_A} a_A \;\; (A \subset M) \qquad \text{and} \qquad \widetilde{e_j} \;=\; -e_j$$

and writing in a natural notation

$$\partial_{\underline{x}} = \sum_{j=1}^{n} e_j\,\partial_{x_j} \qquad \text{and} \qquad \partial_{\underline{y}} = \sum_{j=1}^{n} e_j\,\partial_{y_j}\,,$$

following alternative formulation for the Hermitian monogenic system (1.2) is given in [4].

Proposition 1.1. *Let Ω be an open subset of $\mathbb{R}^{2n}$ and let $F : \Omega \to \mathbb{C}_n$ be a continuously differentiable function in Ω. Then, the function $FI : \Omega \to \mathbb{C}_n I$ is h-monogenic in Ω if and only if F satisfies in Ω the system*

$$\partial_{\underline{x}}\,F - i\,\widetilde{F}\,\partial_{\underline{y}} = 0 = -\widetilde{F}\,\partial_{\underline{x}} + i\,\partial_{\underline{y}}\,F\,. \tag{1.3}$$

Now take a (1-)vector $\underline{v} \in \mathbb{C}_n^1$ and a k-vector $a^{(k)} \in \mathbb{C}_n^k$, $k = 0, \ldots, n$. The products $\underline{v}a^{(k)}$ and $a^{(k)}\underline{v}$ can then be decomposed into a $(k-1)$-vector and a $(k+1)$-vector part as follows:

$$\underline{v}a^{(k)} \;=\; \underline{v} \cdot a^{(k)} + \underline{v} \wedge a^{(k)}\,, \tag{1.4}$$

$$a^{(k)}\underline{v} \;=\; (-1)^{k-1}\,\underline{v} \cdot a^{(k)} + (-1)^k\,\underline{v} \wedge a^{(k)}\,, \tag{1.5}$$

with

$$\underline{v} \cdot a^{(k)} \;\equiv\; \left[\underline{v}a^{(k)}\right]_{k-1} \;=\; \frac{1}{2}\left(\underline{v}a^{(k)} - (-1)^k\,a^{(k)}\underline{v}\right) \;=\; (-1)^{k-1}\,a^{(k)} \cdot \underline{v}\,,$$

$$\underline{v} \wedge a^{(k)} \;\equiv\; \left[\underline{v}a^{(k)}\right]_{k+1} \;=\; \frac{1}{2}\left(\underline{v}a^{(k)} + (-1)^k\,a^{(k)}\underline{v}\right) \;=\; (-1)^k\,a^{(k)} \wedge \underline{v}\,.$$

When considering the multivector decomposition of F into its k-vector parts, i.e., $F = \sum_{k=0}^{n} [F]_k$, $[F]_k : \Omega \to \mathbb{C}_n^k$, we arrive at another characterization of h-monogenic functions with values in $\mathbb{C}_n I$.

Proposition 1.2. *Let Ω be an open subset of $\mathbb{R}^{2n}$ and let $F : \Omega \to \mathbb{C}_n$ be a continuously differentiable function in Ω. Then, the function $FI : \Omega \to \mathbb{C}_n I$ is h-monogenic in Ω if and only if F satisfies in Ω the system*

$$\left(\partial_{\underline{x}} + i\,\partial_{\underline{y}}\right) \cdot [F]_k \ = \ 0 \ = \ \left(\partial_{\underline{x}} - i\,\partial_{\underline{y}}\right) \wedge [F]_k \ , \quad k = 0, \ldots, n \, . \tag{1.6}$$

Proof. The Hermitian system (1.3) can be reformulated in terms of the multivector decomposition of F as follows:

$$\begin{cases} \sum_{k=0}^{n} \left(\partial_{\underline{x}}\,[F]_k + (-1)^{k-1}\, i\,[F]_k\,\partial_{\underline{y}}\right) = 0 \\ \sum_{k=0}^{n} \left((-1)^{k-1}\,[F]_k\,\partial_{\underline{x}} + i\,\partial_{\underline{y}}\,[F]_k\right) = 0 \end{cases} .$$

The above system is equivalent with

$$\begin{cases} \sum_{k=0}^{n} \left[\left(\partial_{\underline{x}}\,[F]_k + (-1)^{k-1}\,[F]_k\,\partial_{\underline{x}}\right) + i\left((-1)^{k-1}\,[F]_k\,\partial_{\underline{y}} + \partial_{\underline{y}}\,[F]_k\right)\right] = 0 \\ \sum_{k=0}^{n} \left[\left(\partial_{\underline{x}}\,[F]_k - (-1)^{k-1}\,[F]_k\,\partial_{\underline{x}}\right) + i\left((-1)^{k-1}\,[F]_k\,\partial_{\underline{y}} - \partial_{\underline{y}}\,[F]_k\right)\right] = 0 \end{cases}$$

and can on account of (1.4) and (1.5) be rewritten as

$$\begin{cases} \sum_{k=0}^{n} \left(\partial_{\underline{x}} + i\,\partial_{\underline{y}}\right) \cdot [F]_k = 0 \\ \sum_{k=0}^{n} \left(\partial_{\underline{x}} - i\,\partial_{\underline{y}}\right) \wedge [F]_k = 0 \end{cases}$$

which is equivalent with (1.6). □

2. The RicSom system

In this section we present an extension of one dimension of the h-monogenic system (1.1), which will be called RicSom system, named after Richard Sommen. Our motivation for introducing and studying the latter system stems from the problem of constructing an h-monogenic Cauchy-like integral, the non-tangential boundary limits of which give rise to a Hilbert-like operator in $\mathbb{R}^{2n}$. Unfortunately, such a Hermitian Cauchy integral would have to be defined on $\mathbb{R}^{2n+2}$ (causing a jump of two dimensions at once), since the Hermitian framework requires all involved vector spaces to be even dimensional. Moreover, it is by no means clear how to construct a mutual fundamental solution of both the Dirac operator and its twisted counterpart, which would then act as a Cauchy kernel. In future work we hope to tackle those problems making use of the RicSom system.

In what follows we denote by Γ an open subset of

$$\mathbb{R}^{2n+1} = \{(x_0, x_1, \ldots, x_n, y_1, \ldots, y_n) : x_0, x_j, y_j \in \mathbb{R}, j = 1, \ldots, n\}$$

and we introduce the Clifford algebra $\mathbb{C}_{2n+2} = \mathrm{Alg}_{\mathbb{C}}\{e_0, e_0|, e_1, \ldots, e_{2n}\}$, where two extra basis vectors e_0 and $e_0|$ are introduced, following the usual multiplication rules

$$\begin{aligned} e_0^2 &= -1 \,, & e_0 e_j + e_j e_0 &= 0 \,, & j &= 1, \dots, 2n \,, \\ e_0|^2 &= -1 \,, & e_0| e_j + e_j e_0| &= 0 \,, & j &= 0, \dots, 2n \,. \end{aligned}$$

Definition 2.1. For functions $g : \Gamma \to \mathbb{C}_{2n+2}$, the *RicSom system* is given by

$$\left\{ \begin{array}{l} e_0 \ \partial_{x_0} g + \partial_{\underline{X}} \ g = 0 \\ e_0| \ \partial_{x_0} g + \partial_{\underline{X}|} \ g = 0 \end{array} \right. . \tag{2.1}$$

We remark that it can be shown that the RicSom system (2.1) is invariant under the same invariance group of the h-monogenic system (1.1), viz

$$\widetilde{\mathrm{U}}(n) \;=\; \{ s \in \mathrm{Spin}(2n) \mid \exists\, \theta \geq 0 : sI = \exp(i\theta) I \}$$

(see [1]).

Now, analogously as has been done in, e.g., [1] for the Clifford algebra $\mathbb{C}_{2n}$, here, a decomposition of $\mathbb{C}_{2n+2}$ will be obtained in terms of complex spinor spaces. Therefore we introduce the idempotents

$$\begin{aligned} I_0 &= \frac{1}{2}(1 - i\, e_0 e_0|) \,, & I_j &= \frac{1}{2}(1 - i\, e_j e_{n+j}) \,, & j &= 1, \dots, n \,, \\ K_0 &= \frac{1}{2}(1 + i\, e_0 e_0|) \,, & K_j &= \frac{1}{2}(1 + i\, e_j e_{n+j}) \,, & j &= 1, \dots, n \,, \end{aligned}$$

being mutually commuting and self-adjoint for which it moreover holds that

$$I_j + K_j = 1 \,, \quad j = 0, \dots, n$$

and thus

$$\prod_{j=0}^{n} (I_j + K_j) = 1 \,,$$

the left-hand side consisting of 2^{n+1} terms, each one being a self-adjoint idempotent annihilating all other terms. In this way, we may write

$$\mathbb{C}_{2n+2} \equiv \mathbb{C}_{2n+2} \prod_{j=0}^{n} (I_j + K_j) \,,$$

yielding a decomposition of $\mathbb{C}_{2n+2}$ as a direct sum of 2^{n+1} components, all of them being mutually isomorphic minimal left ideals, and thus constituting realizations of complex spinor space. Considering the primitive idempotent $J = I_0 I_1 \dots I_n$, we will now use $\mathbb{C}_{2n+2} J$ as a standard model for complex spinor space, since properties of solutions of the RicSom system (2.1) can be studied by examining properties of solutions with values in one chosen minimal left ideal, identified with $\mathbb{C}_{2n+2} J$. Since the primitive idempotent J acts as a translator, i.e.,

$$e_0| \, J = -i\, e_0 \, J \qquad \text{and} \qquad e_{n+j} \, J = -i\, e_j \, J \,, \quad j = 1, \dots, n \,, \tag{2.2}$$

we moreover have that $\mathbb{C}_{2n+2} J \cong \mathbb{C}_{n+1} J$ with $\mathbb{C}_{n+1} = \mathrm{Alg}_{\mathbb{C}}\{e_0, e_1, \dots, e_n\}$.

Hence, to study the RicSom system in a general way, it is sufficient to consider the system

$$\begin{cases} \left(e_0\ \partial_{x_0} + \partial_{\underline{X}}\ \right)[GJ] = 0 \\ \left(e_0|\ \partial_{x_0} + \partial_{\underline{X}|}\right)[GJ] = 0 \end{cases} \tag{2.3}$$

for functions $G : \Gamma \to \mathbb{C}_{n+1}$, since GJ then takes values in the complex spinor space $\mathbb{C}_{n+1}J \cong \mathbb{C}_{2n+2}J$. On account of (2.2), we then have that

$$\left(e_0\,\partial_{x_0} + \partial_{\underline{X}}\ \right)[GJ] \quad = \quad \left((e_0\partial_{x_0})\,G + \partial_{\underline{x}}\,G - i\,\widetilde{G}\,\partial_{\underline{y}}\right)\,J$$

and

$$\left(e_0|\,\partial_{x_0} + \partial_{\underline{X}|}\right)[GJ] \quad = \quad \left(-i\,\widetilde{G}\,(e_0\partial_{x_0}) + \partial_{\underline{y}}\,G + i\,\widetilde{G}\,\partial_{\underline{x}}\right)\,J\ ,$$

which leads to following alternative formulation of the RicSom system (2.3) where only the algebra $\mathbb{C}_{n+1}$ plays a role.

Proposition 2.2. *Let Γ be an open subset of $\mathbb{R}^{2n+1}$ and let $G : \Gamma \to \mathbb{C}_{n+1}$ be a continuously differentiable function in Γ. Then, G satisfies the RicSom system* (2.3) *if and only if*

$$\begin{cases} (e_0\partial_{x_0})\,G + \partial_{\underline{x}}\,G - i\,\widetilde{G}\,\partial_{\underline{y}} = 0 \\ G\,(e_0\partial_{x_0}) - G\,\partial_{\underline{x}} + i\,\partial_{\underline{y}}\,\widetilde{G} = 0 \end{cases}\ . \tag{2.4}$$

Notice that system (2.4) can be considered as an extension of one dimension of the h-monogenic system (1.3).

Let us specialize further to k-vector valued functions $G^{(k)} : \Gamma \to \mathbb{C}^k_{n+1}$ with $k \in \{0, \ldots, n+1\}$. Splitting each of the two equations of the system (2.4) into a $(k-1)$-vector and a $(k+1)$-vector part leads to

$$\begin{aligned} &(e_0\partial_{x_0})\,G^{(k)} + \partial_{\underline{x}}\,G^{(k)} - i\,\widetilde{G^{(k)}}\,\partial_{\underline{y}} = 0 \\ \Longleftrightarrow\ &\begin{cases} \left(e_0\partial_{x_0} + \partial_{\underline{x}} + i\,\partial_{\underline{y}}\right)\cdot G^{(k)} &= \quad 0 \\ \left(e_0\partial_{x_0} + \partial_{\underline{x}} - i\,\partial_{\underline{y}}\right)\wedge G^{(k)} &= \quad 0 \end{cases} \end{aligned} \tag{2.5}$$

and

$$\begin{aligned} &G^{(k)}\,(e_0\partial_{x_0}) - G^{(k)}\,\partial_{\underline{x}} + i\,\partial_{\underline{y}}\,\widetilde{G^{(k)}} = 0 \\ \Longleftrightarrow\ &\begin{cases} \left(-e_0\partial_{x_0} + \partial_{\underline{x}} + i\,\partial_{\underline{y}}\right)\cdot G^{(k)} &= \quad 0 \\ \left(-e_0\partial_{x_0} + \partial_{\underline{x}} - i\,\partial_{\underline{y}}\right)\wedge G^{(k)} &= \quad 0 \end{cases}\ . \end{aligned} \tag{2.6}$$

Thus we arrive at following characterization of the RicSom system (2.3) for k-vector valued functions in $\mathbb{C}_{n+1}$, $k = 0, \ldots, n+1$.

Proposition 2.3. *Let Γ be an open subset of $\mathbb{R}^{2n+1}$ and let $G^{(k)} : \Gamma \to \mathbb{C}^k_{n+1}$ be a continuously differentiable function in Γ. Then, $G^{(k)}$ satisfies the RicSom system* (2.3) *if and only if $G^{(k)}$ is independent of x_0 and h-monogenic in Γ with respect to $\partial_{\underline{X}}$ and $\partial_{\underline{X}|}$.*

Proof. The function $G^{(k)}$ satisfies the RicSom system (2.3) if and only if it simultaneously satisfies the systems (2.5) and (2.6). Taking now deliberate combinations of the equations of those systems, $G^{(k)}$ thus has to fulfil

$$\left\{ \begin{array}{l} (e_0\partial_{x_0}) \cdot G^{(k)} = 0 = (e_0\partial_{x_0}) \wedge G^{(k)} \\ \left(\partial_{\underline{x}} + i\,\partial_{\underline{y}}\right) \cdot G^{(k)} = 0 = \left(\partial_{\underline{x}} - i\,\partial_{\underline{y}}\right) \wedge G^{(k)} \end{array} \right. .$$

The first line is then equivalent with $G^{(k)}$ being independent of the variable x_0. On account of Proposition 1.2, the second line is equivalent with $G^{(k)}$ being h-monogenic with respect to $\partial_{\underline{X}}$ and $\partial_{\underline{X}|}$. □

Finally, decomposing a function $G : \Gamma \to \mathbb{C}_{n+1}$ into its k-vector parts then leads to following system which can be considered as an extension of the Hermitian monogenic system (1.6) with one extra dimension.

Proposition 2.4. *Let Γ be an open subset of $\mathbb{R}^{2n+1}$ and let $G : \Gamma \to \mathbb{C}_{n+1}$ be a continuously differentiable function in Γ. Then, G satisfies the RicSom system* (2.3) *if and only if for each $k = 0, \ldots, n+1$*

$$\left\{ \begin{array}{l} (e_0\partial_{x_0}) \cdot G^{(k+1)} + \left(\partial_{\underline{x}} - i\,\partial_{\underline{y}}\right) \wedge G^{(k-1)} = 0 \\ (e_0\partial_{x_0}) \wedge G^{(k-1)} + \left(\partial_{\underline{x}} + i\,\partial_{\underline{y}}\right) \cdot G^{(k+1)} = 0 \end{array} \right. , \tag{2.7}$$

where $G^{(k)}$ denotes the projection of G onto the space $\mathbb{C}^k_{n+1}$ of k-vectors in $\mathbb{C}_{n+1}$.

Proof. The proof runs along similar lines as the proof of Proposition 1.2. □

Notice that for $k = 0$ the system (2.7) reduces to

$$(e_0\partial_{x_0}) \cdot G^{(1)} = 0 = \left(\partial_{\underline{x}} + i\,\partial_{\underline{y}}\right) \cdot G^{(1)} ,$$

while for $k = n+1$ the same system reads

$$(e_0\partial_{x_0}) \wedge G^{(n)} = 0 = \left(\partial_{\underline{x}} - i\,\partial_{\underline{y}}\right) \wedge G^{(n)} .$$

We have now studied the theory of the original RicSom system (2.1) for functions with values in the specific spinor space $\mathbb{C}_{2n+2}J$, which we had chosen as standard model. Let us consider another primitive idempotent, e.g., $K = K_0 I_1 \ldots I_n$, instead of J. Then of course the theory of the RicSom system

$$\left\{ \begin{array}{l} \left(e_0\ \partial_{x_0} + \partial_{\underline{X}}\ \right) [GK] = 0 \\ \left(e_0|\,\partial_{x_0} + \partial_{\underline{X}|}\right) [GK] = 0 \end{array} \right. \tag{2.8}$$

for functions $G : \Gamma \to \mathbb{C}_{n+1}$ is mathematically equivalent with the RicSom system (2.3). However, since the newly chosen primitive idempotent K yields the slightly different conversion relations

$$e_0| K = i\, e_0\, K \qquad \text{and} \qquad e_{n+j}\, K = -i\, e_j\, K\ , \quad j = 1, \ldots, n\ ,$$

other characterizations of the RicSom system are obtained then the ones mentioned in Proposition 2.2–2.4. We summarize our results in the following propositions.

Proposition 2.5. *Let Γ be an open subset of $\mathbb{R}^{2n+1}$ and let $G : \Gamma \to \mathbb{C}_{n+1}$ be a continuously differentiable function in Γ. Then, G satisfies the RicSom system (2.8) if and only if*

$$\begin{cases} (e_0\partial_{x_0})\, G + \partial_{\underline{x}}\, G - i\, \widetilde{G}\, \partial_{\underline{y}} = 0 \\ G\, (e_0\partial_{x_0}) + G\, \partial_{\underline{x}} - i\, \partial_{\underline{y}}\, \widetilde{G} = 0 \end{cases} .$$

Proposition 2.6. *Let Γ be an open subset of $\mathbb{R}^{2n+1}$ and let $G^{(k)} : \Gamma \to \mathbb{C}^k_{n+1}$ be a continuously differentiable function in Γ. Then, $G^{(k)}$ satisfies the RicSom system (2.8) if and only if it fulfils one of the following conditions:*

(i) $\left(e_0\, \partial_{x_0} + \partial_{\underline{X}}\right) [G^{(k)} K] = 0,$

(ii) $\left(e_0|\, \partial_{x_0} + \partial_{\underline{X}|}\right) [G^{(k)} K] = 0.$

Proposition 2.7. *Let Γ be an open subset of $\mathbb{R}^{2n+1}$ and let $G : \Gamma \to \mathbb{C}_{n+1}$ be a continuously differentiable function in Γ. Then, G satisfies the RicSom system (2.8) if and only if it fulfils one of the following conditions:*

(i) $\left(e_0\, \partial_{x_0} + \partial_{\underline{X}}\right) [G^{(k)} K] = 0,\ k = 0, \ldots, n+1,$

(ii) $\left(e_0|\, \partial_{x_0} + \partial_{\underline{X}|}\right) [G^{(k)} K] = 0,\ k = 0, \ldots, n+1.$

where $G^{(k)}$ denotes the projection of G onto the space $\mathbb{C}^k_{n+1}$ of k-vectors in $\mathbb{C}_{n+1}$.

3. CK-extension for the RicSom system

Let f be a $\mathbb{C}_{2n+2}$ valued analytic function in $\mathbb{R}^{2n}$. In this section we deal with the following problem. Does there exist a $\mathbb{C}_{2n+2}$-valued function g which satisfies the RicSom system in $\mathbb{R}^{2n+1}$ such that $g|_{\mathbb{R}^{2n}} = f$? This problem will be called the Cauchy–Kowalewski extension (or CK-extension) problem for the RicSom system. The following result will be very useful.

Lemma 3.1. *Assume that g satisfies the equation $e_0\, \partial_{x_0} g + \partial_{\underline{X}} g = 0$ in $\mathbb{R}^{2n+1}$ and let $f = g|_{\mathbb{R}^{2n}}$. Then g satisfies the RicSom system in $\mathbb{R}^{2n+1}$ if and only if $e_0 \partial_{\underline{X}} f = e_0| \partial_{\underline{X}|} f$.*

Proof. Let us first suppose that g satisfies the RicSom system in $\mathbb{R}^{2n+1}$. Then we get that

$$\begin{cases} \partial_{x_0} g - e_0 \partial_{\underline{X}} g = 0 \\ \partial_{x_0} g - e_0| \partial_{\underline{X}|} g = 0 \end{cases} .$$

From the above it follows that $e_0\partial_{\underline{X}}g = e_0|\partial_{\underline{X}|}g$ in $\mathbb{R}^{2n+1}$ and hence also that $e_0\partial_{\underline{X}}f = e_0|\partial_{\underline{X}|}f$.

Conversely, assume that $e_0\partial_{\underline{X}}f = e_0|\partial_{\underline{X}|}f$ and define $h = e_0|\partial_{x_0}g + \partial_{\underline{X}|}g$. It is easy to check that h vanishes in $\mathbb{R}^{2n}$. Indeed,

$$\begin{aligned} h|_{\mathbb{R}^{2n}} &= e_0|(\partial_{x_0}g - e_0|\partial_{\underline{X}|}g)|_{\mathbb{R}^{2n}} \\ &= e_0|((\partial_{x_0}g)|_{\mathbb{R}^{2n}} - e_0|\partial_{\underline{X}|}f) \\ &= e_0|((\partial_{x_0}g)|_{\mathbb{R}^{2n}} - e_0\partial_{\underline{X}}f) \\ &= e_0e_0|(e_0\partial_{x_0}g + \partial_{\underline{X}}g)|_{\mathbb{R}^{2n}} \\ &= 0\,. \end{aligned}$$

As $e_0\,\partial_{x_0}h + \partial_{\underline{X}}h = 0$ in $\mathbb{R}^{2n+1}$ it follows immediately (see [3]) that $h \equiv 0$ in $\mathbb{R}^{2n+1}$. □

In what follows, CKf stands for the function defined by

$$\mathrm{CK}f(x_0,\underline{X}) = \sum_{k=0}^{\infty} \frac{x_0^k}{k!}\,(e_0\partial_{\underline{X}})^k f(\underline{X})\,.$$

Clearly, $\mathrm{CK}f(0,\underline{X}) = f(\underline{X})$ and it may be proved that CKf satisfies the equation $e_0\partial_{x_0}\mathrm{CK}f + \partial_{\underline{X}}\mathrm{CK}f = 0$ in $\mathbb{R}^{2n+1}$ (see [3]).

Similarly, the function

$$\mathrm{CK}|f(x_0,\underline{X}) = \sum_{k=0}^{\infty} \frac{x_0^k}{k!}\,(e_0|\partial_{\underline{X}|})^k f(\underline{X})$$

satisfies the equation $e_0|\partial_{x_0}\mathrm{CK}|f + \partial_{\underline{X}|}\mathrm{CK}|f = 0$ in $\mathbb{R}^{2n+1}$ and $\mathrm{CK}|f(0,\underline{X}) = f(\underline{X})$.

We can now formulate the main result of the section.

Theorem 3.2. *The CK-extension problem for the RicSom system is solvable if and only if it holds that $e_0\partial_{\underline{X}}f = e_0|\partial_{\underline{X}|}f$. Moreover, its unique solution is given by $g = \mathrm{CK}f = \mathrm{CK}|f$.*

Proof. Suppose that there exists a function g satisfying the RicSom system in $\mathbb{R}^{2n+1}$ such that $g|_{\mathbb{R}^{2n}} = f$. Let h be given by $h = g - \mathrm{CK}f$. Notice that then one has $e_0\,\partial_{x_0}h + \partial_{\underline{X}}h = 0$ in $\mathbb{R}^{2n+1}$ and $h \equiv 0$ in $\mathbb{R}^{2n}$. It follows that $g = \mathrm{CK}f$ in $\mathbb{R}^{2n+1}$. Lemma 3.1 now yields $e_0\partial_{\underline{X}}f = e_0|\partial_{\underline{X}|}f$.

Conversely, if $e_0\partial_{\underline{X}}f = e_0|\partial_{\underline{X}|}f$, then Lemma 3.1 shows that the function CKf is a solution of the CK-extension problem for the RicSom system.

To end the proof we remark that $\mathrm{CK}f = \mathrm{CK}|f$ iff $e_0\partial_{\underline{X}}f = e_0|\partial_{\underline{X}|}f$. □

References

[1] F. Brackx, J. Bureš, H. De Schepper, D. Eelbode, F. Sommen and V. Souček, *Fundaments of Hermitean Clifford Analysis. Part I: Complex structure*, Complex Analysis and Operator Theory **1(3)** (2007), 341–365.

[2] F. Brackx, J. Bureš, H. De Schepper, D. Eelbode, F. Sommen and V. Souček, *Fundaments of Hermitean Clifford Analysis. Part II: Splitting of h-monogenic equations*, Complex Variables and Elliptic Equations **52(10–11)** (2007), 1063-1079.

[3] F. Brackx, R. Delanghe and F. Sommen, *Clifford Analysis*, Research Notes in Mathematics **76**, Pitman Advanced Publishing Program, Boston-London-Melbourne, 1982.

[4] F. Brackx, H. De Schepper and F. Sommen, *The Hermitean Clifford analysis toolbox*, to appear in Advances in Applied Clifford Algebras.

[5] R. Delanghe, F. Sommen and V. Souček, *Clifford Algebra and Spinor-Valued Functions. A Function Theory for the Dirac Operator*, Mathematics and its Applications **53**, Kluwer Academic Publishers, Dordrecht, 1992.

[6] R. Rocha-Chávez, M. Shapiro and F. Sommen, *Integral theorems for functions and differential forms in* $\mathbb{C}^m$, Research Notes in Mathematics **428**, Chapman & Hall/ CRC, New York, 2002.

[7] I. Sabadini and F. Sommen, *Hermitian Clifford analysis and resolutions*, Mathematical Methods in the Applied Sciences **25(16–18)** (2002), 1395–1414.

Bram De Knock, Dixan Peña Peña and Frank Sommen
Department of Mathematical Analysis
Faculty of Engineering
Ghent University
Galglaan 2
9000 Gent
Belgium
e-mail: bdk@cage.UGent.be
dixan@cage.UGent.be
fs@cage.UGent.be

Quaternionic and Clifford Analysis
Trends in Mathematics, 125–135

Polynomial Invariants for the Rarita-Schwinger Operator

David Eelbode and Dalibor Šmíd

Abstract. We show that polynomial invariant operators on functions with values in the Spin(n)-representation with highest weight $\left(\frac{3}{2}, \frac{1}{2}, \ldots, \frac{1}{2}\right)$ are spanned by powers of the symbols of the Laplace and Rarita-Schwinger operators. This result generalizes the well-known description of polynomial invariants on the scalar and spinor-valued functions. We describe the operators in the language of Clifford analysis.

Mathematics Subject Classification (2000). 15A66.

Keywords. Rarita-Schwinger operator, Fischer decomposition, Invariant polynomials.

1. Why using Clifford analysis?

Clifford analysis, nowadays regarded as a broadly accepted branch of classical analysis, is usually described as a generalization of classical complex analysis in the plane to a higher-dimensional setting. The Dirac operator, playing the role of the generalized Cauchy-Riemann operator, lies at the heart of the theory and the study of its null solutions has always been the main topic of research in Clifford analysis. We refer the reader to the standard references [1, 9, 12]. During the last decade however, it became clear that Clifford analysis techniques can be used to study more general invariant differential operators. The reason for this is that arbitrary representations $\mathbb{V}_\lambda$ for $\mathfrak{so}(m)$, defined by their highest weight λ containing, e.g., half-integers only, can be defined as function spaces containing spinor-valued polynomials satisfying certain conditions expressed in terms of Dirac operators, see [8]. This means that sections of associated vector bundles, taking values in $\mathbb{V}_\lambda$, can be seen as functions $F(\underline{x}; \underline{u}_1, \ldots, \underline{u}_k)$ in several vector variables. This leads

David Eelbode was supported as a postdoctoral fellow by the F.W.O. Vlaanderen (Belgium)
Dalibor Šmíd was supported by GAČR 201/06/P267 and MSM 0021620839.

to an elegant framework to study higher spin operators from a function theoretical point of view, and yields a valuable alternative for the standard multi-index notation used in physics and abstract representation theory.

In the present paper, techniques from both classical Clifford analysis and representation theory will be used to investigate polynomial invariants in the space $\mathrm{End}(\mathbb{S}_1)$, where $\mathbb{S}_1$ denotes the irreducible $\mathfrak{so}(m)$-module whose highest weight is given by $(\frac{3}{2}, \frac{1}{2}, \ldots, \frac{1}{2})$. This module can be defined as the vector space consisting of 1-homogeneous spinor-valued null solutions for the Dirac operator $\underline{\partial}$ on $\mathbb{R}^m$. Our motivation for investigating these invariants is inspired by the recent applications of the theory of Howe dual pairs [13, 14] in the setting of Clifford analysis, see, e.g., [2], which turned out to be crucial to find a multiplicity-free decomposition for spaces of polynomial solutions for certain invariant operators.

2. Clifford algebras and Clifford analysis

Let $(e_1, \ldots, e_m)$ be an orthonormal basis for the complex vector space $\mathbb{C}^m$ endowed with the standard bilinear form $\mathcal{B}(e_i, e_j) = \delta_{ij}$, and let $\mathbb{C}_m$ be the Clifford algebra generated by this basis, together with the multiplication rules $e_ie_j + e_je_i = -2\delta_{ij}$. For the general theory of Clifford algebras we refer to, e.g., [9, 12, 16]. In the present paper, the so-called spinor spaces carrying the basic spinor representations for the orthogonal Lie algebra will be of crucial importance. In order to define these representation(s) for $\mathfrak{so}(m)$ with half-integer highest weight, we need the so-called Witt basis for $\mathbb{C}^m$. Consider the case of an even dimension $m = 2n$ first, and consider the Witt decomposition into maximal isotropic subsets, given by

$$\mathbb{C}^{2n} = \mathrm{span}_{\mathbb{C}}(\mathfrak{f}_j, \mathfrak{f}_j^\dagger)_j = \mathrm{span}_{\mathbb{C}}\left(\frac{e_{2j-1} - ie_{2j}}{2}, -\frac{e_{2j-1} + ie_{2j}}{2}\right)_j ,$$

where $1 \leq j \leq n$. The Witt basis vectors satisfy the multiplication rules

$$\mathfrak{f}_j\mathfrak{f}_k + \mathfrak{f}_k\mathfrak{f}_j = \mathfrak{f}_j^\dagger\mathfrak{f}_k^\dagger + \mathfrak{f}_k^\dagger\mathfrak{f}_j^\dagger = 0 \quad , \quad \mathfrak{f}_j\mathfrak{f}_k^\dagger + \mathfrak{f}_k^\dagger\mathfrak{f}_j = \delta_{jk} \ .$$

Introducing n idempotents by means of $I_k = \mathfrak{f}_k\mathfrak{f}_k^\dagger$, one can define a primitive idempotent $I = I_1 \cdots I_n$. The space $\mathbb{C}_{2n}I$ is a minimal left ideal which serves as an irreducible representation space for the simple algebra $\mathbb{C}_{2n}$ under left multiplication, i.e., $\mathbb{C}_{2n} \cong \mathrm{End}(\mathbb{C}_{2n}I)$. Restricting the multiplicative action to the even subalgebra $\mathbb{C}_{2n}^+$, the complex vector space $\mathbb{C}_{2n}I$ splits into two irreducible subspaces $\mathbb{C}_{2n}^+I$ and $\mathbb{C}_{2n}^-I$. Introducing the Grassman algebra $\Lambda^\dagger$ spanned by the daggered Witt basis vectors, we then have that $\mathbb{C}_{2n}^\pm I \cong \Lambda_\pm^\dagger I$, with $\Lambda_\pm^\dagger := \mathbb{C}_{2n}^\pm \cap \Lambda^\dagger$. In the following definitions, we will realize the spinor space(s) as subspaces of the Clifford algebra. In case of an odd dimension $m = 2n + 1$, we therefore add $I_0 = \frac{1}{2}(1 + ie_{2n+1})$ to the set of idempotents I_j, leading to $I' = I_1 \cdots I_n I_0$.

Definition 1. *In case the dimension $m = 2n$ is even, the spinor spaces $\mathbb{S}^\pm$ are given by $\mathbb{S}^\pm = \mathbb{C}_m^\pm I = \Lambda_\pm^\dagger I$. In case the dimension $m = 2n + 1$ is odd, the spinor space $\mathbb{S}$ is given by $\mathbb{S} = \mathbb{C}_m I' = \Lambda^\dagger I'$.*

The following is crucial for what follows:

Proposition 1. *In case $m = 2n$, the spaces $\mathbb{S}^{\pm}$ provide models for the irreducible $\mathfrak{so}(m)$-modules with highest weight $(\frac{1}{2}, \ldots, \frac{1}{2}, \pm\frac{1}{2})$. In case $m = 2n+1$, the space $\mathbb{S}$ defines a model for the irreducible $\mathfrak{so}(m)$-module with highest weight $(\frac{1}{2}, \ldots, \frac{1}{2})$.*

Remark: For the sake of convenience, we will restrict ourselves to the case of odd dimension in what follows.

In classical Clifford analysis, the Dirac operator is defined as $\partial_{\underline{x}} = \sum_i e_i \partial_{x_i}$ and acts on functions $f \in \mathcal{C}^\infty(\mathbb{R}^m, \mathbb{S})$ on $\mathbb{R}^m$ with values in the spinor space $\mathbb{S}$. This operator factorizes the Laplacian Δ_m in m dimensions, i.e., $\partial_{\underline{x}}^2 = -\Delta_m$, and null solutions for this operator are called monogenic functions. Of particular interest for what follows are spaces of homogeneous monogenic polynomials. Let us denote by $\mathcal{M}_k(\mathbb{S})$ the space of spinor-valued k-homogeneous monogenic polynomials on $\mathbb{R}^m$. Defining the action of $\mathfrak{so}(m)$ as the one derived from the classical L-representation of $\mathrm{Spin}(m)$ given by $L(s)[f(\underline{x})] = sf(\bar{s}\underline{x}s)$, one obtains a module for $\mathfrak{so}(m)$:

Proposition 2. *The space $\mathbb{S}_k := \mathcal{M}_k(\mathbb{S})$ yields a model for the $\mathfrak{so}(m)$-module with highest weight $(k + \frac{1}{2}, \frac{1}{2}, \ldots, \frac{1}{2})$, for all $k \in \mathbb{N}$.*

A crucial property is the so-called Fischer decomposition on $\mathbb{R}^m$, which describes the decomposition of spaces of homogeneous spinor-valued polynomials in terms of these irreducible modules containing homogeneous monogenic polynomials:

Theorem 1. *The space $\mathcal{P}_k(\mathbb{S})$ of k-homogeneous spinor-valued polynomials decomposes as*

$$\mathcal{P}_k(\mathbb{S}) = \bigoplus_{j=0}^{k} \underline{x}^j \mathcal{M}_{k-j}(\mathbb{S}) \, .$$

3. Dimensions of the space of invariants

Let $\mathfrak{g}$ be a complex Lie algebra of the type $\mathfrak{so}(m)$ and let us denote by G the corresponding Lie group $\mathrm{Spin}(m)$. The space of polynomial operators acting on functions with values in a $\mathfrak{g}$-representation $\mathbb{V}$ is $\mathcal{P}(\mathrm{End}\,\mathbb{V})$. This is itself a $\mathfrak{g}$-representation. The space of invariants $\mathcal{P}(\mathrm{End}\,\mathbb{V})^G$ consists from the point of view of representation theory precisely of the trivial summands occurring in its decomposition into irreducibles ([11], Lecture 14). We will treat the odd-dimensional case $\mathfrak{g} = B_n$, $m = 2n+1$, in detail and make some remarks on the even-dimensional case afterwards.

Let us fix some notation. $\mathbb{V} \equiv \mathbb{S}_1$ is the representation with highest weight $(\frac{3}{2}, (\frac{1}{2})_{n-1})$, as expressed in the standard basis e_i. By $(\alpha)_j$ we mean a shortcut for $\alpha, \ldots, \alpha$, j-times. We denote by $\mathbb{V}_k$ the representation of $\mathfrak{g}$ with highest weight $(k, (0)_{n-1}) \equiv ke_1$ and use a shortcut (k) for the highest weight of $\mathbb{V}_k$.

The following proposition tells us something about the representation in which the polynomial operators take values:

Proposition 3.

$$\operatorname{End}\mathbb{S}_1 = \mathbb{V}_0 \oplus \mathbb{V}_1 \oplus \mathbb{V}_2 \oplus \mathbb{V}_3 \oplus \mathbb{W},$$

where the representation $\mathbb{W}$ *contains no irreducible summand of type* $\mathbb{V}_k$, $k \geq 0$.

Proof. Since $\mathbb{S}_1$ is a selfdual representation, we are actually decomposing $\operatorname{End}\mathbb{S}_1 = \mathbb{S}_1 \otimes \mathbb{S}_1$. First we use Freudenthal formula for calculating multiplicities of weights of $\mathbb{S}_1$ and then Brauer-Klimyk formula for identifying all irreducible summands of type $\mathbb{V}_k$ in the tensor product. Since low-dimensional cases can be easily checked by direct calculation, we will assume $n > 3$.

Multiplicities of weights: Let μ be a weight and $m_\lambda(\mu)$ its multiplicity in a representation $\mathbb{V}_\lambda$. Denote by Δ^+ the set of positive roots and by δ the sum of fundamental weights of $\mathfrak{g}$. Then multiplicities satisfy a recurrent expression called Freudenthal formula [15].

$$m_\lambda(\mu) = \frac{2}{|\lambda+\delta|^2 - |\mu+\delta|^2} \sum_{\alpha\in\Delta^+} \sum_{k=1}^{\infty} m_\lambda(\mu + k\alpha)(\mu + k\alpha, \alpha),$$

where the scalar product may be any suitable multiple of the Killing form – let us choose it such that its matrix with respect to the standard basis $\{e_i\}$ is the identity matrix. We know that $m_\lambda(\lambda) = 1$ and $\Delta^+ = \{e_i, 1 \leq i \leq n\} \cup \{e_i - e_j, 1 \leq i < j \leq n\} \cup \{e_i + e_j, 1 \leq i < j \leq n\}$.

The Weyl group W of B_n acts on the set of weights by permutations and sign reversals of coordinates. The dominant Weyl chamber Λ_W consists of weights $\mu = (\mu_1, \ldots, \mu_n)$ satisfying $\mu_1 \geq \mu_2 \geq \cdots \geq \mu_n \geq 0$. For each weight μ of $\mathbb{S}_1$ there is a unique element $w \in W$ such that $w.\mu \in \Lambda_W$. Any weight of $\mathbb{V}_\lambda$ is expressed as $\lambda - \sum k_\alpha \alpha$ where k_α are non-negative integers and the sum is over simple roots α. The only dominant weights of this type are λ and $\lambda - e_1 \equiv \left(\left(\frac{1}{2}\right)_n\right)$. Since the multiplicity function is constant on orbits, we need only to calculate $m_\lambda(\lambda - e_1)$. Freudenthal formula yields

$$\begin{aligned} m_\lambda(\lambda - e_1) &= \frac{2}{2n+1}\left(\sum_{i=1}^{n}((\lambda - e_1) + e_i, e_i) + \sum_{1\leq i<j\leq n}((\lambda - e_1) + e_i - e_j, e_i - e_j)\right) \\ &= \frac{2}{2n+1}\left(\frac{3}{2}n + 2\frac{n(n-1)}{2}\right) = n\,. \end{aligned}$$

In the first equality we used that $(\lambda - e_1) + e_i$ and $(\lambda - e_1) + e_i - e_j$ are on the orbit of λ and have thus multiplicity 1 and that $(\lambda - e_1) + e_i + e_j$, $(\lambda - e_1) + ke_i$, $k > 1$ and $(\lambda - e_1) + k(e_i - e_j)$, $k > 1$ are neither on the orbit of λ nor $\lambda - e_1$ and so none of them is a weight of $\mathbb{V}_\lambda \equiv \mathbb{S}_1$.

Decomposition of the tensor product: For a weight μ not fixed by any element of W denote by $[\mu]$ the unique element on the orbit of μ such that $[\mu] \in \Lambda_W$ and by $w(\mu)$ the unique element of W such that $w(\mu).\mu = [\mu]$. Then Brauer-Klimyk formula [15] says that for every pair μ, ρ of dominant weights

$$\mathbb{V}_\mu \otimes \mathbb{V}_\rho = \bigoplus_{\mu' \in \tilde{\Pi}_\mu} (-1)^{|w(\rho+\mu'+\delta)|} m_\mu(\mu') \mathbb{V}_{[\rho+\mu'+\delta]-\delta}, \tag{1}$$

where $\tilde{\Pi}_\mu$ is the set of all weights of $\mathbb{V}_\mu$ not fixed by any element of W and $|w|$ is the length of $w \in W$.

In our case $\mu = \rho = \lambda$, $\delta = \left(\frac{2n-1}{2}, \frac{2n-3}{2}, \dots, \frac{1}{2}\right)$ and we are looking for the occurrences of representations $(k), k \geq 0$. This means that

$$[\lambda + \lambda' + \delta] = (k) + \delta, \tag{2}$$

where $\lambda' = w'.\lambda$ or $w'.(\lambda - e_1)$ for some $w' \in W$ and we are of course interested in the $w \in W$ that is implicit in the square brackets. We shall denote $\nu := \lambda + \lambda' + \delta$.

The weight ν must differ from $(k)+\delta$ only by a permutation and sign changes, i.e., the set of absolute values of coordinates must be the same for both. Two cases can occur:

1) *Some coordinates of ν are negative.* It is clear that there can be only one such coordinate, $\nu_n = -\frac{1}{2}$. This happens precisely for $\lambda'_n = -\frac{3}{2}$. All the other coordinates of λ' are $\pm\frac{1}{2}$ and the only way to obtain the set equality $\{\nu_2, \dots, \nu_{n-1}\} = \{\frac{2n-3}{2}, \dots, \frac{3}{2}\}$ is to have $\lambda'_j = -\frac{1}{2}$ for $2 \leq j \leq n-1$. We get

$$\left[\left(\frac{3}{2}, \left(\frac{1}{2}\right)_{n-1}\right) + \left(\pm\frac{1}{2}, \left(-\frac{1}{2}\right)_{n-2}, -\frac{3}{2}\right) + \delta\right] = (1) \text{ or } (2) + \delta,$$

The multiplicity is 1 and the sign is -1, since the w of the square brackets is just the sign reversal of the last coordinate, which has length 1.

2) *All coordinates of ν are positive.* If ν_n is $\frac{3}{2}$, then $\lambda'_n = \frac{1}{2}$. Now coordinates of ν must be a permutation of coordinates of $(k) + \delta$, hence some coordinate of ν must be $\frac{1}{2}$. The only way how to do it is $\lambda'_{n-1} = -\frac{3}{2}$, $\nu_{n-1} = \frac{1}{2}$. As before, the remaining coordinates allow no freedom and we have

$$\left[\left(\frac{3}{2}, \left(\frac{1}{2}\right)_{n-1}\right) + \left(\pm\frac{1}{2}, \left(-\frac{1}{2}\right)_{n-3}, -\frac{3}{2}, \frac{1}{2}\right) + \delta\right] = (1) \text{ or } (2) + \delta,$$

where the multiplicity is 1 and the sign is again -1, since ν is moved to Λ_W by a transposition of the last two coordinates.

The value $\frac{1}{2}$ must appear among coordinates of ν and it can be only at ν_n or ν_{n-1}, since $|\lambda'_i| \leq \frac{3}{2}$. In the latter case then $\nu_n = \frac{3}{2}$. This is because $\nu_n > \frac{3}{2}$ requires $\lambda'_n > \frac{1}{2}$, so $\lambda'_n = \frac{3}{2}$, $\nu_n = \frac{5}{2}$, all other λ'_i are $\pm\frac{1}{2}$ and so none other ν_i can be $\frac{3}{2}$. We dealt with $\nu_{n-1} = \frac{1}{2}$, $\nu_n = \frac{3}{2}$ in the previous paragraph, therefore let us set $\nu_n = \frac{1}{2}$ which means $\lambda'_n = -\frac{1}{2}$. Thus any permutation of coordinates of ν pushing it into the dominant chamber fixes the last coordinate. Repeating this argument now for the possible positions of the value $\frac{3}{2}$, we see that either the permutation fixes also the $(n-1)$th coordinate, or it is the transposition of

$(n-1)$th and $(n-2)$th coordinate. If we go on, we get $n-2$ contributions with a transposition:

$$\left[\left(\frac{3}{2},\left(\frac{1}{2}\right)_{n-1}\right)+\left(\pm\frac{1}{2},\left(-\frac{1}{2}\right)_p,-\frac{3}{2},\frac{1}{2},\left(-\frac{1}{2}\right)_q\right)+\delta\right] = (1) \text{ or } (2)+\delta,$$

where $p+q=n-3$, $p\geq 0$, $q\geq 1$ (we already considered the case with $q=0$) and then the contribution with ν already dominant:

$$\left[\left(\frac{3}{2},\left(\frac{1}{2}\right)_{n-1}\right)+\left(\pm\frac{1}{2},\left(-\frac{1}{2}\right)_{n-1}\right)+\delta\right] = (1) \text{ or } (2)+\delta$$

$$\left[\left(\frac{3}{2},\left(\frac{1}{2}\right)_{n-1}\right)+\left(\pm\frac{3}{2},\left(-\frac{1}{2}\right)_{n-1}\right)+\delta\right] = (0) \text{ or } (3)+\delta\,.$$

The former gives $\mathbb{V}_1$ and $\mathbb{V}_2$ with multiplicity n, since now $[\lambda'] = \lambda - e_1$, and the latter $\mathbb{V}_0$ and $\mathbb{V}_3$ with multiplicity 1; signs are $+1$ in both cases.

To summarize, we get $\mathbb{V}_0$ and $\mathbb{V}_3$ with multiplicity 1 and $\mathbb{V}_1$ and $\mathbb{V}_2$ with multiplicity $n-(n-2)-1=1$ as well. □

Remark: In the even-dimensional case, $\mathfrak{g}=D_n$, the only difference is that we have two representations:

$$\mathbb{S}_1^+ := \left(\frac{3}{2},\left(\frac{1}{2}\right)_{n-2},\frac{1}{2}\right)$$

$$\mathbb{S}_1^- := \left(\frac{3}{2},\left(\frac{1}{2}\right)_{n-2},-\frac{1}{2}\right).$$

We shall denote by $\mathbb{S}_1$ the reducible representation $\mathbb{S}_1^+ \oplus \mathbb{S}_1^-$.

The analogue of Proposition 3 is

Proposition 4.

$$\mathrm{Hom}(\mathbb{S}_1^+,\mathbb{S}_1^-) = \mathbb{V}_0 \oplus \mathbb{V}_2 \oplus \mathbb{W}^{+-}$$

$$\mathrm{Hom}(\mathbb{S}_1^-,\mathbb{S}_1^+) = \mathbb{V}_0 \oplus \mathbb{V}_2 \oplus \mathbb{W}^{-+}$$

$$\mathrm{End}(\mathbb{S}_1^+) = \mathbb{V}_1 \oplus \mathbb{V}_3 \oplus \mathbb{W}^{++}$$

$$\mathrm{End}(\mathbb{S}_1^-) = \mathbb{V}_1 \oplus \mathbb{V}_3 \oplus \mathbb{W}^{--}$$

where the representations $\mathbb{W}^{+-}$, $\mathbb{W}^{-+}$, $\mathbb{W}^{++}$, $\mathbb{W}^{--}$ *contain no irreducible summand of type* $\mathbb{V}_k$. *From this we easily see that*

$$\mathrm{End}\,\mathbb{S}_1 = (\mathbb{V}_0 \oplus \mathbb{V}_1 \oplus \mathbb{V}_2 \oplus \mathbb{V}_3)^{\oplus 2} \oplus \mathbb{W},$$

where $\mathbb{W}$ *contains no summand of type* $\mathbb{V}_k$.

Let us return to the odd-dimensional case. The space of polynomials $\mathcal{P}(\operatorname{End}\mathbb{S}_1)$ is graded by homogeneity

$$\bigoplus_{j=0}^{\infty}\mathcal{P}^j(\operatorname{End}\mathbb{S}_1)=\bigoplus_{j=0}^{\infty}\left(\odot^j\mathbb{V}_1\otimes\operatorname{End}\mathbb{S}_1\right)=\bigoplus_{j=0}^{\infty}\left(\bigoplus_{i=0}^{\lfloor\frac{j}{2}\rfloor}\mathbb{V}_{j-2i}\otimes\operatorname{End}\mathbb{S}_1\right). \tag{3}$$

Here by $\odot^j\mathbb{V}_1$ we mean the jth symmetric tensor power of $\mathbb{V}_1$. The last equality comes from the fact that $\bigodot^p\mathbb{V}_1=\mathbb{V}_p\oplus\bigodot^{p-2}\mathbb{V}_1$, i.e., a symmetric tensor can be split into its tracefree (Cartan) part and its contraction in any two indices. This corresponds to the classical Fischer decomposition of polynomials into harmonic polynomials: a homogeneous polynomial $P(x)$ of degree p is written as a sum of a harmonic polynomial of degree p and $x^2R(x)$, where $R(x)$ is a polynomial of degree $p-2$.

Lemma 1. *Let $\mathfrak{g}$ be of type B_n, $\mathbb{V}_\mu$ an irreducible representation of $\mathfrak{g}$, $i\geq 0$. Then $\mathbb{V}_k\otimes\mathbb{V}_\mu$ contains a trivial irreducible summand $\mathbb{V}_0$ if and only if $\mu=(k)$. In such a case, its multiplicity is one.*

Proof. By Brauer-Klimyk formula (1) a weight μ' of $\mathbb{V}_\mu$ can contribute to a nonzero coefficient of a trivial summand only if

$$[(k)+\mu'+\delta]=\delta \tag{4}$$

The coordinates of all weights of a given finite-dimensional representation are either all integral or all half-integral. If μ is half-integral, then $(k)+\mu'+\delta$ and δ cannot be at the same time integral or half-integral. Action of the Weyl group implicit in the operation $[\cdot]$ preserves integrality, so the resulting dominant weight cannot equal δ.

Hence only for μ integral there can be a $\mathbb{V}_0$ in $\mathbb{V}_k\otimes\mathbb{V}_\mu$. Let us denote by $H(\mu')$ the non-negative number $\sum_{i=1}^n|\mu_i'|$. If μ' is a weight of $\mathbb{V}_\mu$, then $H(\mu')\leq H(\mu)$. If μ' satisfies (4), then

$$H(\delta)=H((k)+\mu'+\delta)\geq H((k))+H(\delta)-H(\mu'),$$

i.e., $H(\mu)\geq H(\mu')\geq k$. If $H(\mu)=k$ and $\mu_1'>-k$, then the first coordinate of $(k)+\mu'+\delta$ is greater than any coordinate of δ and so (4) cannot hold. Thus $H(\mu)=k$ means that any contributing μ' is $(-k,(0)_{n-1})$ and such a weight exists only in a representation with highest weight $\mu=(k)$, provided $H(\mu)=k$. Clearly this gives one trivial summand in $\mathbb{V}_k\otimes\mathbb{V}_k$.

Hence any other μ satisfying the statement must be integral with $m\equiv H(\mu)>k$. Such a representation can be realised inside the tensor power $\mathbb{V}_1^m$ (since $\mathbb{V}_1$ is the defining or vector representation) as a subspace of tracefree tensors satisfying certain symmetry [10]. Let us denote by π_m the projection from $\mathbb{V}_1^m$ to $\mathbb{V}_\mu$ and by π_k the projection $\mathbb{V}_1^k$ to $\mathbb{V}_k$. If there is a projection $\pi:\mathbb{V}_k\otimes\mathbb{V}_\mu\to(0)$, then $\pi\circ(\pi_k\otimes\pi_m)$ is a projection from $\mathbb{V}_1^k\otimes\mathbb{V}_1^m$ to a trivial summand. By classical invariant theory [11] the space of orthogonal invariants in $\mathbb{V}_1^p$ is zero for p odd and spanned by complete contractions for p even. A complete contraction for $p=k+m$,

$m > k$ must involve a contraction in two indices of $\mathbb{V}_1^m$, which is projected to zero by π_m. Thus for $H(\mu) > k$ there are no invariants in $\mathbb{V}_k \otimes \mathbb{V}_\mu$. □

Theorem 2. *The space of invariant k-homogeneous polynomials with values in* $\operatorname{End}\mathbb{S}_1$ *is one-dimensional for* $k < 2$ *and two-dimensional for* $k \geq 2$.

Proof. The invariants correspond to trivial summands in the right-hand side of expression (3). Lemma 1 implies that tensor product of $\mathbb{V}_{j-2i}$ with the representation $\mathbb{W} \subset \operatorname{End}\mathbb{S}_1$ of Proposition 3 contains no trivial summands and that there is precisely one trivial summand in $\mathbb{V}_{j-2i} \otimes \mathbb{V}_l$, $l \in \{0,1,2,3\}$ if and only if $j - 2i = l$. This holds for $l = 0$ in homogeneity 0, for $l = 1$ in homogeneity 1, for $l \in \{0,2\}$ in nonzero even homogeneities and for $l \in \{1,3\}$ for odd homogeneities higher than one. □

Remark: In the even-dimensional case, Lemma 1 and its proof remain the same. The dimensions of the spaces of invariants are twice as big in each homogeneity.

4. Bases of the spaces of invariants

Within the framework of Clifford analysis, a higher-spin Dirac operator (or sometimes higher-spin Rarita-Schwinger operator) can be defined as an operator acting on functions $F(\underline{x};\underline{u})$ depending on two vector variables. Indeed, if this function belongs for fixed $\underline{x}$ to $\mathcal{M}_k(\mathbb{S})$, it can be seen as a function on $\mathbb{R}^m$ taking values in the module $\mathbb{S}_k$, $k > 0$. The natural invariant operator acting between such functions is defined as the higher-spin Dirac operator, given by

$$\mathcal{R}'_k := \left(\frac{\underline{u}\,\partial_{\underline{u}}}{m+2k-2} + 1\right)\partial_{\underline{x}} : \mathcal{C}^\infty(\mathbb{R}^m, \mathcal{M}_k) \mapsto \mathcal{C}^\infty(\mathbb{R}^m, \mathcal{M}_k)\,.$$

From the point of view of representation theory, this operator is a Stein-Weiss gradient corresponding to the unique summand $\mathbb{S}_k$ in $\mathbb{V}_1 \otimes \mathbb{S}_k$. We shall denote by $\mathcal{R}_k$ the symbol of $\mathcal{R}'_k$, which is a 1-homogeneous polynomial operator with values in $\operatorname{End}(\mathbb{S}_k)$. Stein-Weiss gradients are invariant differential operators and so $\mathcal{R}_k$ is an invariant polynomial in $\mathcal{P}(\operatorname{End}\mathbb{S}_k)$.

The function theory for higher-spin Dirac operators is in full development, we refer to, e.g., [6, 7, 18]. We consider here the case of $k = 1$, the Rarita-Schwinger operator. According to Theorem 2, there ought to be two linearly independent invariant polynomials in each order of homogeneity $k \geq 2$ and one invariant for homogeneities 0 and 1. The latter ones are clearly a constant and $\mathcal{R}_1$. In order to find the former ones, the most obvious way to proceed is to start calculating powers of the operator $\mathcal{R}_1$. We have the following:

$$(\mathcal{R}_1)^2 = -|\underline{x}|^2 + \frac{4}{m^2}\left(m\langle \underline{u}, \underline{x}\rangle + \underline{u}\,\underline{x}\right)\langle \partial_{\underline{u}}, \underline{x}\rangle \ =: -|\underline{x}|^2 + \gamma_m T_1,$$

where we denoted $\gamma_m := \frac{4}{m^2}$ and defined T_1, a 2-homogeneous $(\operatorname{End}\mathbb{S}_1)$-valued polynomial. Since $|\underline{x}|^2$ is clearly an invariant polynomial, then so is T_1. It is then easily verified that

$$T_1^2 = (m-1)|\underline{x}|^2 T_1.$$

Hence

$$\mathrm{span}_{\mathbb{C}}\left(\{|\underline{x}|^{2k}(\mathcal{R}_1)^j \,|k \geq 0, 0 \leq j \leq 3\}\right)$$

is a subalgebra of $\mathcal{P}(\mathrm{End}\,\mathbb{S}_1)$, consisting only of invariants. Moreover, it has the same number of generators in each homogeneity as the algebra of invariants should have basis elements according to the previous section. We only need to show that the generators are linearly independent in order to conclude that the two algebras are isomorphic as vector spaces.

Referring to, e.g., [4, 6], we recall the definition for the twistor operator

$$\begin{array}{rcl} \mathcal{T}_k' : \mathcal{C}^\infty(\mathbb{R}^m, \underline{u}\mathcal{M}_{k-1}) & \mapsto & \mathcal{C}^\infty(\mathbb{R}^m, \mathcal{M}_k) \\ \underline{u}f(\underline{x};\underline{u}) & \mapsto & \left(\dfrac{\underline{u}\,\underline{\partial}_u}{m+2k-2}+1\right)\underline{\partial}_x \underline{u} f \end{array}$$

and the dual twistor operator

$$\begin{array}{rcl} (\mathcal{T}_k^*)' : \mathcal{C}^\infty(\mathbb{R}^m, \mathcal{M}_k) & \mapsto & \mathcal{C}^\infty(\mathbb{R}^m, \underline{u}\mathcal{M}_{k-1}) \\ f(\underline{x};\underline{u}) & \mapsto & -\dfrac{1}{m+2k-2}\underline{u}\,\underline{\partial}_u \underline{\partial}_x f\ . \end{array}$$

In terms of these operators, an alternative description for the polynomial invariant T_1 can be given:

$$\mathcal{T}_1\mathcal{T}_1^* = \gamma_m T_1\ .$$

Here $\mathcal{T}_1$ maps $\mathbb{S}_1$-valued functions into $\mathbb{S}_0$-valued functions and its adjoint the other way round. Since $\dim \mathbb{S}_0 < \dim \mathbb{S}_1$, the operator T_1 cannot be an isomorphism. On the other hand, $|\underline{x}|^2$ is an isomorphism for any nonzero x, hence $|\underline{x}|^2$ and T_1 are linearly independent. The operators $\mathcal{R}_1$, $|\underline{x}|^{2k}$ are isomorphisms for any nonzero x. This shows that we have two linearly independent generators in each homogeneity greater than one, namely $|\underline{x}|^{2k+2}$, $|\underline{x}|^{2k}T_1$ for even homogeneity and $|\underline{x}|^{2k+2}\mathcal{R}_1, |\underline{x}|^{2k}T_1\mathcal{R}_1$ for odd homogeneity. We can summarize our result:

Theorem 3. *Let $\mathfrak{g}$ be a simple complex Lie algebra of type B_n and $\mathbb{S}_1$ its representation with highest weight $\left(\frac{3}{2}, \frac{1}{2}, \ldots, \frac{1}{2}\right)$. The algebra of $\mathfrak{g}$-invariant polynomials with values in End($\mathbb{S}_1$) has as a vector space the following structure:*

$$\mathrm{End}(\mathbb{S}_1) \quad = \quad \mathrm{span}_{\mathbb{C}}\left(\{|\underline{x}|^{2k}(\mathcal{R}_1)^j \,|k \geq 0, 0 \leq j \leq 3\}\right).$$

Remark: In the even-dimensional case we have operators

$$\begin{aligned} (|\underline{x}|^2)^+ &: \mathbb{S}_1^+ \to \mathbb{S}_1^+ \\ (|\underline{x}|^2)^- &: \mathbb{S}_1^- \to \mathbb{S}_1^- \\ \mathcal{R}_1^+ &: \mathbb{S}_1^+ \to \mathbb{S}_1^- \\ \mathcal{R}_1^- &: \mathbb{S}_1^- \to \mathbb{S}_1^+\ . \end{aligned}$$

We may see them also as elements of $\mathrm{End}\,\mathbb{S}_1$, extending them by a zero operator to the whole $\mathbb{S}_1 = \mathbb{S}_1^+ \oplus \mathbb{S}_1^-$. By this extension we just choose a basis in the space of invariants: any linear combination of for instance $(|\underline{x}|^2)^+$ and $(|\underline{x}|^2)^-$ is still an invariant operator. Note that the products $\mathcal{R}^+\mathcal{R}^+$, $\mathcal{R}^-\mathcal{R}^-$, $(|\underline{x}|^2)^-\mathcal{R}^+$, $(|\underline{x}|^2)^+\mathcal{R}^-$, $(|\underline{x}|^2)^+(|\underline{x}|^2)^-$ and $(|\underline{x}|^2)^-(|\underline{x}|^2)^+$ are zero.

The operators have formally the same expression and all other statements in this section hold for them too. Thus we get

Theorem 4. *Let $\mathfrak{g}$ be a simple complex Lie algebra of type D_n and $\mathbb{S}_1 = \mathbb{S}_1^+ \oplus \mathbb{S}_1^-$ its representation where highest weights of $\mathbb{S}_1^\pm$ are $\left(\frac{3}{2}, \frac{1}{2}, \dots, \pm\frac{1}{2}\right)$. The algebra of $\mathfrak{g}$-invariant polynomials with values in End($\mathbb{S}_1$) has as a vector space the following structure:*

$$\mathrm{End}(\mathbb{S}_1) = \mathrm{span}_{\mathbb{C}}\left(\{(|\underline{x}|^{2k})^\pm(\mathcal{R}_1^\pm)^j | k \geq 0, 0 \leq j \leq 3\}\right).$$

Here by $(\mathcal{R}_1^\pm)^j$, $j > 0$ we mean the product $\Pi_{i=1}^j \mathcal{R}^{s(i)}$, where $s(i)$ is either $+$ or $-$ and the signs alternate.

Acknowledgments

The authors would like to acknowledge many motivating and enlightening discussions on the subject with Vladimír Souček. They also would like to thank the referee for the many useful remarks and suggestions.

References

[1] F. Brackx, R. Delanghe, F. Sommen, *Clifford Analysis*, Research Notes in Mathematics **76**, Pitman, London, 1982.

[2] F. Brackx, H. De Schepper, D. Eelbode, V. Souček, *Howe dual pairs in Hermitean Clifford analysis*, submitted.

[3] T. Branson, O. Hijazi, *Bochner-Weitzenböck formulas associated with the Rarita-Schwinger operator*, Internat. J. Math., **13** (2002), No. 2, pp. 137–182

[4] J. Bureš, *The Rarita-Schwinger operator and spherical monogenic forms*, Complex Variables Theory Appl. **43** No. 1 (2000), pp. 77–108.

[5] J. Bureš, V. Souček, *Eigenvalues for conformally invariant operators.*

[6] J. Bureš, F. Sommen, V. Souček, P. Van Lancker, *Rarita-Schwinger type operators in Clifford analysis*, Journal of Funct. Anal. **185** (2001), pp. 425–456.

[7] J. Bureš, F. Sommen, V. Souček, P. Van Lancker, *Symmetric analogues of Rarita-Schwinger equations*, Annals of Global Analysis Geometry **21** No. 3 (2001), pp. 215–240.

[8] Constales, D., Sommen, F., Van Lancker, P., *Models for irreducible Spin(m)-modules.*

[9] R. Delanghe, F. Sommen, V. Souček, *Clifford analysis and spinor valued functions*, Kluwer Academic Publishers, Dordrecht, 1992.

[10] W. Fulton, J. Harris, *Representation Theory: A First Course*, Graduate Texts in Mathematics **129**, Springer-Verlag, New York (1999).

[11] W. Goodman, *Representations and invariants of the classical groups*, Cambridge Univ. Press, Cambridge (1998)

[12] J. Gilbert, M.A.M. Murray, *Clifford algebras and Dirac operators in harmonic analysis*, Cambridge University Press, Cambridge, 1991.

[13] R. Howe, *Transcending classical invariant theory*, J. Am. Math. Soc. **2** No. 3 (1989), pp. 535–552.

[14] R. Howe, *Remarks on classical invariant theory*, Trans. Am. Math. Soc. **313** (1989), pp. 539–570.

[15] J.E. Humphreys, *Introduction to Lie Algebras and Representation Theory*, Graduate Texts in Mathematics **7**, Springer-Verlag, New York (1972).

[16] I. Porteous, *Clifford Algebras and the Classical groups*, Cambridge University Press (1995).

[17] V. Severa, *Invariant differential operators between spinor-valued forms*, PhD-thesis, Charles University, Praha (1998).

[18] P. Van Lancker, *Rarita-Schwinger fields in the half space*, Complex Variables and Elliptic Equations **51** No. 5-6 (2006), pp. 563–579.

David Eelbode
Clifford Research Group
Dept. of Mathematical Analysis
Ghent University
Galglaan 2, 9000 Ghent
Belgium
e-mail: deef@cage.ugent.be

Dalibor Šmíd
Mathematical Institute
Charles University
Sokolovska 83, 186 75 Prague 8
Czech Republic
e-mail: smid@karlin.mff.cuni.cz

Quaternionic and Clifford Analysis
Trends in Mathematics, 137–149

Hypermonogenic Functions and Their Dual Functions

Sirkka-Liisa Eriksson

Abstract. In this paper we present a new integral formulas for hypermonogenic functions where the kernels are also hypermonogenic functions. We also introduce dual k-hypermonogenic functions. If $k = 0$, then k-hypermonogenic functions are monogenic functions and their dual functions are also monogenic. If k is nonzero the only function that is k-hypermonogenic function and dual hypermogenic is zero function.

The theory of dual functions is very similar to the theory of hypermonogenic functions. We present their integral formula and use it to present the integral formula for $(1 - n)$-hypermonogenic functions.

Mathematics Subject Classification (2000). Primary 30G35; Secondary 30A05.

Keywords. Monogenic, hypermonogenic, Dirac operator, hyperbolic metric.

1. Introduction

There exist two generalizations of classical complex analysis to higher dimensions using geometric algebras. The first one is the theory of monogenic functions or regular functions introduced by R. Delanghe around 1970 based on Euclidean metric (see for example [1]) and the second one is the theory of hypermonogenic functions based on hyperbolic metric initiated by H. Leutwiler around 1990 ([13], [14]) and continued jointly with the author by [7], [9], [10] and [4]. The advantage of hypermonogenic functions is that positive and negative powers of hypercomplex variables are included to the theory which is not in the monogenic case. Hence elementary functions can be defined similarly as in classical complex analysis.

Hypermonogenic functions are solutions of the system

$$M_{n-1}f = Df + (n-1)\frac{Q'f}{x_n} = 0,$$

where $'$ is the main involution and Qf is given by the decomposition $f(x) = Pf(x) + Qf(x)e_n$ with $Pf(x), Qf(x) \in C\ell_{n-1}$. A Cauchy-type formula for hypermonogenic functions was proved in [4]. A key concept in the proof was k-hypermonogenic functions introduced in [5]. K-hypermonogenic functions are related to harmonic functions with respect to the Riemannian metric

$$ds^2 = \frac{\sum_{i=0}^n dx_i^2}{x_n^{\frac{2k}{n-1}}}.$$

They satisfy the equation $M_k f = Df + k\frac{Q'f}{x_n} = 0$.

The Dirac operator decomposes the usual Laplace operator to the first-order operators. An important property of the $M_k f$ operator is that it may be used to decompose the Laplace Beltrami operator $\triangle f - \frac{k}{x_n}\frac{\partial f}{\partial x_n}$ (see [7]). An example of a k-hypermonogenic is the function $|x|^{k-n+1} x^{-1}$. They are also related to polyharmonic functions. Indeed if a function H satisfies $\triangle^k H = 0$ then locally there exists $2i$-hypermonogenic functions $g_{ij} : \Omega \to Cl_n$ $(j = 1, 2)$ such that $\sum_{i=1}^k \left(g_{i1} + \frac{\partial g_{i2}}{\partial x_n} e_n\right) = H$ ([5]).

In this paper we review main properties of k-hypermonogenic functions. We also present a new integral formula for hypermonogenic functions (Theorem 3.10) where the kernels are also hypermonogenic functions. In Section 4 we introduce a new concept of a dual k-hypermonogenic function. If $k = 0$, then k-hypermonogenic functions are monogenic functions and their dual functions are also monogenic. If k is nonzero the only function that is k-hypermonogenic and dual hypermogenic is zero function. The theory of dual functions is very similar to the theory of hypermonogenic functions. We present their integral formula and use it to present the integral formula for $(1-n)$-hypermonogenic functions.

2. Preliminaries

We consider the Clifford algebra $C\ell_{0,n}$ generated by $e_1, \ldots, e_n$ satisfying the relation

$$e_i e_j + e_j e_i = -2\delta_{ij},$$

where δ_{ij} is the usual Kronecker delta.

The elements

$$x = x_0 + x_1 e_1 + \cdots + x_n e_n$$

for $x_0, \ldots, x_n \in \mathbb{R}$ are called *paravectors*. The set $\mathbb{R}^{n+1}$ is identified with the set of paravectors.

The *main involution* is the mapping $a \to a'$ defined for the generating elements by $e_i' = -e_i$ for $i = 1, \ldots, n$ and extended to the total algebra by linearity and the product rule $(ab)' = a'b'$. Similarly the reversion is the mapping $a \to a^*$ defined for the generating elements by $e_i' = -e_i$ for $i = 1, \ldots, n$ and extended to the total algebra by linearity and the product rule $(ab)^* = b^* a^*$. The conjugation is the mapping $a \to \overline{a}$ defined by $\overline{a} = (a')^* = (a^*)'$.

Any element $a \in C\ell_{0,n}$ may be uniquely decomposed as

$$a = b + ce_n$$

for $b, c \in C\ell_{0,n-1}$(the Clifford algebra generated by $e_1, \ldots, e_{n-1}$). The mappings $P : C\ell_{0,n} \to C\ell_{0,n-1}$ and $Q : C\ell_{0,n} \to C\ell_{0,n-1}$ are defined in [7] by

$$Pa = b, \qquad Qa = c.$$

In order to compute the P- and Q-parts we use the involution $a \to \widehat{a}$ defined for the generating elements by

$$\hat{e}_i = (-1)^{\delta_{in}} e_i$$

and extended to the total algebra by linearity and the product rule $\widehat{ab} = \widehat{a}\widehat{b}$. Then we obtain the formulas

$$Pa = \frac{1}{2}(a + \widehat{a}) \tag{2.1}$$

and

$$Qa = -\frac{1}{2}(a - \widehat{a})\, e_n. \tag{2.2}$$

The following calculation rules ([7]) hold

$$P(ab) = (Pa)\, Pb + (Qa)\, Q(b'), \tag{2.3}$$

$$\begin{aligned} Q(ab) &= (Pa)\, Qb + (Qa)\, P'(b) \\ &= aQb + (Qa)\, b'. \end{aligned} \tag{2.4}$$

Note that if $a \in Cl_{0,n}$ then

$$a' e_n = e_n \widehat{a}.$$

Moreover if $a \in Cl_{0,n-1}$ then

$$ae_n = e_n a'. \tag{2.5}$$

An element $I = e_1 e_2 \ldots e_n$ is called a *pseudo scalar*. Note that

$$IA_r = (-1)^{r(n-1)} A_r I$$

if $A = e_{j_1} e_{j_2} \ldots e_{j_r}$ for $1 \leq j_1 < j_2 < \cdots < j_r \leq n$. Moreover if n is odd then I belongs to the center of the $Cl_{0,n}$.

3. Hypermonogenic functions

We briefly recall the definition of hypermonogenic functions. Let Ω be an open subset of $\mathbb{R}^{n+1}$. We consider functions $f : \Omega \to C\ell_{0,n}$ whose components are continuously differentiable. The left Dirac operator in $C\ell_{0,n}$ is defined by

$$D_l f = \sum_{i=0}^{n} e_i \frac{\partial f}{\partial x_i}$$

and the right Dirac operator by

$$D_r f = \sum_{i=0}^{n} \frac{\partial f}{\partial x_i} e_i.$$

The operators $\overline{D_l}$ and $\overline{D_r}$ are defined by

$$\overline{D_l} f = \sum_{i=0}^{n} \overline{e_i} \frac{\partial f}{\partial x_i}, \quad \overline{D_r} f = \sum_{i=0}^{n} \frac{\partial f}{\partial x_i} \overline{e_i}.$$

Let Ω be an open subset of $\mathbf{R}^{n+1} \backslash \{x_n = 0\}$. The modified Dirac operators M_k^l, $\overline{M}_k^l$, M_k^r and $\overline{M}_k^r$ are introduced in [7] and [3] by

$$M_k^l f(x) = D_l f(x) + k \frac{Q' f}{x_n}$$
$$M_k^r f(x) = D_r f(x) + k \frac{Q f}{x_n}$$

and

$$\overline{M}_k^l f(x) = \overline{D}_l f(x) - k \frac{Q' f}{x_n},$$
$$\overline{M_k}^r f(x) = \overline{D}_r f(x) - k \frac{Q f}{x_n},$$

where $f \in \mathcal{C}^1(\Omega, C\ell_{0,n})$ and

$$(Qf)' = Q' f$$
$$(Pf)' = P' f.$$

The operator M_{n-1}^l is also denoted by M.

Definition 3.1. Let $\Omega \subset \mathbb{R}^{n+1}$ be open. A function $f : \Omega \to C\ell_{0,n}$ is **left k-hypermonogenic** if $f \in \mathcal{C}^1(\Omega)$ and

$$M_k^l f(x) = 0$$

for any $x \in \Omega \backslash \{y \in \mathbf{R}^{n+1} \mid y_n \neq 0\}$. The **right k-hypermonogenic functions** are defined similarly. The $(n-1)$-left hypermonogenic functions are called **hypermonogenic functions**. The set of left k-hypermonogenic functions in Ω is denoted by $\mathcal{M}_k(\Omega)$.

Paravector-valued hypermonogenic functions are *H-solutions* introduced by H. Leutwiler in [13] and [14]. Total Clifford algebra $C\ell_{0,n}$-valued hypermonogenic functions were introduced by the author and H. Leutwiler in [7]. Their theory is further developed in [2], [3], [5], [6], [8], [9], [10], [11], [12] and [16]. We state some main properties of hypermonogenic functions.

Lemma 3.2 (Generalized Cauchy-Riemann equations [7]**).** *Let Ω be an open subset of $\mathbb{R}^{n+1}$ and $f : \Omega \to C\ell_{0,n}$ be a mapping with continuous partial derivatives. The equation $M_k f = 0$ is equivalent with the following system of equations*

$$D_{n-1}(Pf) - \frac{\partial (Q' f)}{\partial x_n} + k \frac{Q' f}{x_n} = 0,$$
$$D_{n-1}(Qf) + \frac{\partial P'(f)}{\partial x_n} = 0.$$

There is an interplay between k-hypermonogenic and $-k$-hypermonogenic functions.

Lemma 3.3 ([3]). *If $k \in \mathbb{R}$ then*

$$M_k^l f = -x_n^k M_{-k}^l \left(x_n^{-k} f e_n \right) e_n.$$

Moreover the function f is k-hypermonogenic if and only if $x_n^{-k} f e_n$ is $-k$-hypermonogenic.

Lemma 3.4 ([7]). *Let $f : \Omega \to C\ell_{0,n}$ be twice continuously differentiable. Then*

$$P\left(M_k \overline{M}_k f\right) = \triangle Pf - \frac{k}{x_n} \frac{\partial Pf}{\partial x_n}$$

$$Q\left(M_k \overline{M}_k f\right) = \triangle Qf - \frac{k}{x_n} \frac{\partial Qf}{\partial x_n} + k \frac{Qf}{x_n^2}.$$

These are the Laplace-Beltrami equations with respect to the Riemannian metric

$$ds^2 = \frac{\sum_{i=0}^n dx_i^2}{x_n^{\frac{2k}{n-1}}}.$$

If $k = n-1$, the metric is the hyperbolic metric of the Poincaré model of the upper half-space. In case $n = k-1$ harmonic functions with respect to the above metric have been studied by Leutwiler in [15].

Example. Let $k \in \mathbb{N}$ and $1 \leq k \leq n-1$. Assume that $1 \leq i_1 < i_2 < \cdots < i_k \leq n-1$. Set $w = w_0 + w_{i_1} e_{i_1} + \cdots + w_{i_k} e_{i_k} + w_n e_n$.Then

1. w^m is k-hypermonogenic, if $m \in \mathbf{Z}$.
2. $e^w = \sum_{j=0}^\infty \frac{1}{j!} w^j$ is k-hypermonogenic.
3. $\sin w = \sum_{j=0}^\infty \frac{1}{(2j+1)!} (-1)^j w^{2j+1}$ is k-hypermonogenic.
4. $\cos w = \sum_{j=0}^\infty \frac{1}{(2j)!} (-1)^j w^{2j}$ is k-hypermonogenic.
5. If $f(z) = \sum_{j=0}^\infty a_j z^j$ is holomorphic and $a_j \in \mathbb{R}$ then $f(w) = \sum_{j=0}^\infty a_j w^j$ is k-hypermonogenic.
6. (Fueter construction) If $f = u + iv$ is holomorphic in an open set $\Omega \subset \mathbb{C}$ then

$$\widetilde{f}(w) = u\left(w_0, \sqrt{w_{i_1}^2 + \cdots + w_{i_k}^2 + w_n^2}\right) + \frac{w_{i_1} e_{i_1} + \cdots + w_{i_k} e_{i_k} + w_n e_n}{\sqrt{w_{i_1}^2 + \cdots + w_{i_k}^2 + w_n^2}} v\left(w_0, \sqrt{w_{i_1}^2 + \cdots + w_{i_k}^2 + w_n^2}\right)$$

is k-hypermonogenic.

Note that in the special case $k = n-1$ all the preceding functions are hypermonogenic ([13], [14]).

Using the definition it is easy to see the following result ([7], [10]).

Proposition 3.5. *Let f be a k-hypermonogenic function. Then*

$$\frac{\partial f}{\partial x_i}, \quad i = 1, \ldots, n-1$$

is k-hypermonogenic. Moreover the function $\frac{\partial f}{\partial x_n}$ *is k-hypermonogenic if and only if* $\frac{\partial f}{\partial x_n} = 0$.

A k-hypermonogenic function remains k-hypermonogenic if it is multiplied from the right with a constant belonging to $C\ell_{0,n-1}$.

Proposition 3.6. *The k-hypermonogenic functions in an open subset* Ω *of* $\mathbb{R}^{n+1}$ *form a right* $C\ell_{0,n-1}$*-module.*

In this paper we consider integral formulas for hypermonogenic functions for simplicity in the upper half-space

$$\mathbb{R}^{n+1}_{+} = \left\{(x_0, \ldots, x_n) \in \mathbb{R}^{n+1} \mid x_n > 0\right\}.$$

Some results hold also in whole $\mathbb{R}^{n+1}$ but they are more technical. The similar results holds also in lower half-space, since the following result holds.

Theorem 3.7. *Let* Ω *be an open subset of* $\mathbb{R}^{n+1}_{+}$ *and* $f : \Omega \to C\ell_n$ *be continuously differentiable. If* f *is k-hypermonogenic then* $\widehat{f}\left(\widehat{x}\right)$ *is k-hypermonogenic in* $\widehat{\Omega} = \left\{x \in \mathbb{R}^{n+1} \mid \widehat{x} \in \Omega\right\}$.

The integral formulas hypermonogenic functions were first proved for the P-part in [9] and then for the Q-part in [3]. We review them using slightly different notations. They can also be proved for $n+1$-chains satisfying $\overline{K} \subset \Omega$. If $k \neq n-1$ the integral formulas for k-hypermonogenic functions are presented in [4], but their kernels are more complicated.

Theorem 3.8. *Let* Ω *be an open subset of* $\mathbb{R}^{n+1}_{+}$ *and* $K \subset \Omega$ *a smoothly bounded compact set with outer unit normal field* ν. *If* f *is hypermonogenic in* Ω *and* $y \in K$ *then*

$$Pf\left(y\right) = \frac{2^n}{\omega_{n+1}} \int_{\partial K} P\left(y_n^n p\left(x,y\right) \nu\left(x\right) f\left(x\right)\right) \frac{d\sigma}{x_n^{n-1}} \tag{3.1}$$

$$Qf\left(y\right) = \frac{2^n}{\omega_{n+1}} \int_{\partial K} Q\left(y_n^{n-1} q\left(x,y\right) \nu\left(x\right) f\left(x\right)\right) d\sigma \tag{3.2}$$

where ω_{n+1} *is the surface measure of the unit ball in* $\mathbb{R}^{n+1}$ *and the kernel given by*

$$\begin{aligned} p\left(x,y\right) &= x_n^{n-1} \frac{\left(x-y\right)^{-1}}{\left|x-y\right|^{n-1}} e_n \frac{\left(x-\widehat{y}\right)^{-1}}{\left|x-\widehat{y}\right|^{n-1}} \\ &= \frac{x_n^{n-1}}{2y_n} \left(\frac{\left(x-y\right)^{-1}}{\left|x-y\right|^{n-1}} - \frac{\left(x-\widehat{y}\right)^{-1}}{\left|x-\widehat{y}\right|^{n-1}} \right) \end{aligned}$$

is paravector valued hypermonogenic and the other kernel

$$\begin{aligned} q\left(x,y\right) &= \frac{\left(x-\widehat{y}\right)^{-1}}{\left|x-\widehat{y}\right|^{n-1}} \left(x - Py\right) \frac{\left(x-y\right)^{-1}}{\left|x-y\right|^{n-1}} \\ &= \frac{\left(x-y\right)^{-1} + \left(x-\widehat{y}\right)^{-1}}{2\left|x-y\right|^{n-1} \left|x-\widehat{y}\right|^{n-1}}. \end{aligned}$$

is paravector valued $(1-n)$*-hypermonogenic.*

Note that the coordinate functions of $p(x,y)$ and $q(x,y)$ are the following

$$p(x,y) = \frac{2x_n^n \overline{P(x-y)}}{|y-x|^{n+1}|\widehat{y}-x|^{n+1}} - \frac{x_n^{n-1}\left(|P(y-x)|^2 - x_n^2 + y_n^2\right)}{|y-x|^{n+1}|\widehat{y}-x|^{n+1}} e_n,$$

$$q(x,y) = \frac{\left(|P(x-y)|^2 + x_n^2 + y_n^2\right)\overline{P(x-y)}}{|y-x|^{n+1}|\widehat{y}-x|^{n+1}} - \frac{x_n\left(y_n^2 - x_n^2 - |P(y-x)|^2\right)}{|y-x|^{n+1}|\widehat{y}-x|^{n+1}} e_n.$$

If using (2.1) and (2.2) we combine the formulas (3.1) and (3.2) we obtain the integral formula.

Theorem 3.9 ([3]). *Let Ω be an open subset of $\mathbb{R}_+^{n+1}$ and $\overline{K} \subset \Omega$ a smoothly bounded compact set with outer unit normal field ν. If f is hypermonogenic in Ω and $y \in K$ then*

$$f(y) = \frac{2^{n-1}}{\omega_{n+1}} \int_{\partial K} \frac{y_n^{n-1}\left((x-y)^{-1}\nu(x) f(x) - (\widehat{x}-y)^{-1}\widehat{\nu(x) f(x)}\right)}{|x-y|^{n-1}|y-\widehat{x}|^{n-1}} d\sigma(x).$$

This formula may be developed further as follows.

Theorem 3.10. *Let Ω be an open subset of $\mathbb{R}_+^{n+1}$ and $\overline{K} \subset \Omega$ a smoothly bounded compact set with outer unit normal field ν. If f is hypermonogenic in Ω and $y \in K$ then*

$$f(y) = \frac{2^{n-1}}{\omega_{n+1}} \int_{\partial K} \left(h_1(x,y) P(\nu(x) f(x)) + h_2(x,y) Q'(\nu(x) f(x))\right) d\sigma(x)$$

where the kernels h_1 and h_2 are hypermonogenic functions for any $y \in K$ given by

$$h_1(x,y) = -2x_n p(y,x)$$

and

$$h_2(x,y) = -y_n^{n-1} q(y,x) e_n.$$

Proof. Substituting

$$\nu(x) f(x) = P(\nu(x) f(x)) + Q(\nu(x) f(x)) e_n$$

and

$$\widehat{\nu(x) f(x)} = P(\nu(x) f(x)) - Q(\nu(x) f(x)) e_n$$

to the previous theorem we obtain

$$f(y) = \frac{2^{n-1}}{\omega_{n+1}} \int_{\partial K} \frac{y_n^{n-1}\left((x-y)^{-1} - (\widehat{x}-y)^{-1}\right)}{|x-y|^{n-1}|y-\widehat{x}|^{n-1}} P(\nu(x) f(x))\, d\sigma(x)$$

$$+ \frac{2^{n-1}}{\omega_{n+1}} \int_{\partial K} \frac{y_n^{n-1}\left((x-y)^{-1} + (\widehat{x}-y)^{-1}\right) e_n}{|x-y|^{n-1}|y-\widehat{x}|^{n-1}} Q'(\nu(x) f(x))\, d\sigma(x).$$

Since

$$y_n^{n-1} \frac{(x-y)^{-1} - (\widehat{x}-y)^{-1}}{|x-y|^{n-1}|y-\widehat{x}|^{n-1}} = -2x_n p(y,x)$$

and

$$\frac{y_n^{n-1}\left((x-y)^{-1}+(\widehat{x}-y)^{-1}\right)e_n}{|x-y|^{n-1}\,|y-\widehat{x}|^{n-1}} = -y_n^{n-1}q(y,x)\,e_n$$

completing the proof. □

Using the product rules (2.3) and (2.4) of P and Q we obtain an interesting decomposition.

Theorem 3.11. *Let Ω be an open subset of $\mathbb{R}_+^{n+1}$ and $\overline{K}\subset\Omega$ a smoothly bounded compact set with outer unit normal field ν. If f is hypermonogenic in Ω then there exists hypermonogenic functions f_1 and f_2 satisfying $f=f_1+f_2$ and f_1 is determined by Pf and f_2 by Qf as follows*

$$f_1(y)=\frac{2^{n-1}}{\omega_{n+1}}\int_{\partial K}\left(h_1(x,y)\,P(\nu(x))+h_2(x,y)\,\nu_n(x)\right)P(f(x))\,d\sigma(x),$$

$$f_2(y)=\frac{2^{n-1}}{\omega_{n+1}}\int_{\partial K}\left(h_2(x,y)\,P'(\nu(x))+h_1(x,y)\,\nu_n(x)\right)Q'(f(x))\,d\sigma(x).$$

Since hypermonogenic functions form the right $Cl_{0,n-1}$-module we obtain directly the result.

Theorem 3.12. *Let Ω be an open subset of $\mathbb{R}_+^{n+1}$ and $\overline{K}\subset\Omega$ a smoothly bounded compact set with outer unit normal field ν. If f is continuous function in $\partial\Omega$ then function*

$$g(y)=\frac{2^{n-1}}{\omega_{n+1}}\int_{\partial K}\left(h_1(x,y)\,P(\nu(x)\,f(x))+h_2(x,y)\,Q'(\nu(x)\,f(x))\right)d\sigma(x)$$

is hypermonogenic for all $y\in\Omega\backslash\partial K$.

4. Dual hypermonogenic functions

In geometric algebras a dual of an element u is an element uI, where I is a pseudo unit. Since a non zero k-hypermonogenic function multiplied from the right with e_n is not any more k-hypermonogenic we introduce the following new concept.

Definition 4.1. Let I be a pseudo unit in $C\ell_{0,n}$. A function f is called dual k-hypermonogenic in an open subset Ω of $\mathbb{R}^{n+1}$if fI is k-hypermonogenic on Ω . The set of dual k-hypermonogenic functions in Ω is denoted by $\mathcal{P}_k(\Omega)$.

Proposition 4.2. *Let I be a pseudo unit in $C\ell_{0,n}$. A function f is dual k-hypermonogenic in an open subset Ω of $\mathbb{R}^{n+1}$ if and only if f satisfy the equation*

$$x_nD_lf-ke_nP'f=0$$

in Ω.

Proof. Assume that f is dual k-hypermonogenic in an open subset Ω of $\mathbb{R}^{n+1}$. Since $Q(fI) = P(fe_1e_2\dots e_{n-1})$ we obtain

$$\begin{aligned} x_n M(fI) &= x_n\left(D_l\left(fe_1e_2\dots e_{n-1}\right)\right)e_n + kP'\left(fe_1e_2\dots e_{n-1}\right) \\ &= \left(x_n D_l\left(fe_1e_2\dots e_{n-1}\right) - ke_nP\left(fe_1e_2\dots e_{n-1}\right)\right)e_n = 0, \end{aligned}$$

which implies that fI is k-hypermonogenic. If fI is k-hypermonogenic then similarly we deduce that fI^2 is dual k-hypermonogenic, completing the proof. □

Corollary 4.3. *A function f is dual k-hypermonogenic in an open subset Ω of $\mathbb{R}^{n+1}$ if and only if fe_n or equivalently fI is k-hypermonogenic.*

The generalized Cauchy-Riemann equations (3.2) for dual k-hypermonogenic functions are easily computed.

Proposition 4.4. *A function f is dual k-hypermonogenic in an open subset Ω of $\mathbb{R}^{n+1}$ if and only if*

$$\begin{aligned} -D_{n-1}(Qf) - \frac{\partial(P'f)}{\partial x_n} + k\frac{P'f}{x_n} &= 0, \\ D_{n-1}(Pf) - \frac{\partial Q'(f)}{\partial x_n} &= 0. \end{aligned} \tag{4.1}$$

Dual k-hypermonogenic functions may be also characterized using M operators as follows.

Proposition 4.5. *A function f is dual k-hypermonogenic if and only if*

$$x_n M_k f = kf'e_n.$$

Proof. We just compute

$$\begin{aligned} -x_n M_k\left(fe_n\right)e_n &= x_n Df - kP'fe_n = x_n\left(M_k f\right) - kQ'f - kP'fe_n \\ &= x_n\left(M_k f\right) - kf'e_n, \end{aligned}$$

completing the proof. □

Using the preceding proposition we see that the function identically zero is the only function that is hypermonogenic and dual hypermonogenic.

Proposition 4.6. *If f is k-hypermonogenic and dual k-hypermonogenic in an open subset Ω of $\mathbb{R}^{n+1}$ then $f = 0$.*

Using the preceding result we obtain directly the result.

Theorem 4.7. *A right $C\ell_{0,n}$-module of k-hypermonogenic functions in an open subset Ω of $\mathbb{R}^{n+1}$ is $\mathcal{M}_k(\Omega) \oplus \mathcal{P}_k(\Omega)$.*

If f is k-hypermonogenic functions then $\frac{\partial f}{\partial x_n}$ is not generally hypermonogenic. However the following surprising result holds.

Proposition 4.8. *If f is dual k-hypermonogenic in an open subset Ω of $\mathbb{R}^{n+1}$ then*

$$g = \frac{\partial f}{\partial x_n} - k\frac{Pf}{x_n} = e_n D_{n-1} f$$

is hypermonogenic and $e_n D_{n-1} f e_n$ is dual k-hypermonogenic. Moreover if f is k-hypermonogenic then

$$h = \frac{\partial f}{\partial x_n} - k\frac{e_n Q' f}{x_n} = e_n D_{n-1} f$$

is dual k-hypermonogenic and $e_n D_{n-1} f e_n$ is k-hypermonogenic.

Proof. Assume that f is dual k-hypermonogenic in an open subset Ω of $\mathbb{R}^{n+1}$. Using $Df - k\frac{e_n Pf}{x_n} = 0$ we obtain

$$\begin{aligned} Mg = Dg + k\frac{Q'g}{x_n} &= \frac{\partial Df}{\partial x_n} - k\frac{DPf}{x_n} + k\frac{e_n Pf}{x_n^2} + \frac{k}{x_n}\frac{\partial Q' f}{\partial x_n} \\ &= \frac{\partial}{\partial x_n}\left(Df - k\frac{e_n Pf}{x_n}\right) - k\frac{D_{n-1} Pf}{x_n} + \frac{k}{x_n}\frac{\partial Q' f}{\partial x_n}. \end{aligned}$$

Applying (4.1) we deduce $Mg = 0$. Moreover, using $ke_n\frac{Pf}{x_n} = Df$ and the definition of g, we obtain

$$g = \frac{\partial f}{\partial x_n} + e_n Df = e_n D_{n-1} f.$$

If f is hypermonogenic then fe_n is dual hypermonogenic and the first part implies that

$$\begin{aligned} g = \frac{\partial f e_n}{\partial x_n} - k\frac{P(fe_n)}{x_n} &= \frac{\partial f}{\partial x_n} e_n + k\frac{Qf}{x_n} = \left(\frac{\partial f}{\partial x_n} - ke_n\frac{Q'f}{x_n}\right) e_n \\ &= e_n D_{n-1} f e_n \end{aligned}$$

is k-hypermonogenic. Hence $h = -ge_n$ is dual k-hypermonogenic. □

Corollary 4.9. *If h is $C\ell_{0,n-1}$-valued k-hyperbolic harmonic then*

$$g = \triangle_{n-1} h' + \overline{D}_{n-1}\left(\frac{\partial h}{\partial x_n}\right) e_n = \frac{k}{x_n}\frac{\partial h'}{\partial x_n} - \frac{\partial^2 h'}{\partial x_n^2} + \overline{D}_{n-1}\left(\frac{\partial h}{\partial x_n}\right) e_n$$

is k-hypermonogenic.

Proof. Assume that h is $C\ell_{0,n-1}$-valued k-hyperbolic harmonic. Then $\overline{D}h$ is hypermonogenic and the preceding theorem implies that

$$g = e_n D_{n-1} \overline{D} h e_n = \triangle_{n-1} h' + \overline{D}_{n-1}\left(\frac{\partial h}{\partial x_n}\right) e_n$$

is k-hypermonogenic. □

The kernel of the mapping $f \to e_n D_{n-1} f e_n$ is obtained from monogenic functions.

Proposition 4.10. *If f is k-hypermonogenic and $e_nD_{n-1}fe_n = 0$, then there exist monogenic functions w_0 and w_1 independent of x_n satisfying*

$$f = w_0 + w_1 x_n^k e_n.$$

Proof. Assume that f is k-hypermonogenic and $e_nD_{n-1}fe_n = 0$. Then $D_{n-1}Pf = PD_{n-1}f = 0$ and $D_{n-1}Qf = QD_{n-1}f = 0$. Since f is k-hypermonogenic. then applying the generalized Cauchy-Riemann equations we obtain

$$-\frac{\partial(Q'f)}{\partial x_n} + k\frac{Q'f}{x_n} = 0,$$
$$\frac{\partial P'(f)}{\partial x_n} = 0.$$

Hence $Pf = w_0$ is monogenic and $Qf = x_n^k w_1$ where w_1 is monogenic. □

Corollary 4.11. *If f and g are k-hypermonogenic in an open subset Ω of $\mathbb{R}^{n+1}_+$ and $D_{n-1}f = D_{n-1}g$ then there exist monogenic functions w_0 and w_1 independent of x_n satisfying*

$$f = g + w_0 + w_1 x_n^k e_n.$$

The integral formula for the dual hypermonogenic functions is obtained from 3.12.

Theorem 4.12. *Let Ω be an open subset of $\mathbb{R}^{n+1}_+$ and $\overline{K} \subset \Omega$ a smoothly bounded compact set with outer unit normal field ν. If f is dual hypermonogenic in Ω and $y \in K$*

$$f(y) = \frac{2^{n-1}}{\omega_{n+1}} \int_{\partial K} \left(h_1(x,y)\, e_n Q'(\nu(x) f(x)) - h_2(x,y)\, e_n P(\nu(x) f(x))\right) d\sigma(x).$$

where h_1 and h_2 are the same as in Theorem 3.12 and h_1e_n and h_2e_n are dual hypermonogenic functions.

Proof. Assume that f is dual hypermonogenic in Ω and $y \in K$. By Corollary 4.3 the function fe_n is hypermonogenic. Hence applying Theorem 3.12 we infer

$$f(x)\, e_n = \frac{2^{n-1}}{\omega_{n+1}} \int_{\partial K} \left(h_1(x,y)\, P(\nu(x) f(x)\, e_n) + h_2(x,y)\, Q'(\nu(x) f(x)\, e_n)\right) d\sigma.$$

Since $P(\nu(x) f(x)\, e_n) = -Q(\nu(x) f(x))$ and $Q(\nu(x) f(x)\, e_n) = P(\nu(x) f(x))$ we obtain

$$f(x)\, e_n = \frac{2^{n-1}}{\omega_{n+1}} \int_{\partial K} \left(-h_1(x,y)\, Q(\nu(x) f(x)) + h_2(x,y)\, P'(\nu(x) f(x))\right) d\sigma.$$

Multiplying both sides with $-e_n$ from the right we deduce

$$f(x) = \frac{2^{n-1}}{\omega_{n+1}} \int_{\partial K} \left(h_1(x,y)\, Q(\nu(x) f(x))\, e_n - h_2(x,y)\, P'(\nu(x) f(x))\right) e_n d\sigma.$$

Hence by virtue of (2.5) we conclude the result. □

Theorem 4.13. *Let Ω be an open subset of $\mathbf{R}_+^{n+1}$ and $\overline{K} \subset \Omega$ a smoothly bounded compact set with outer unit normal field ν. If f is continuous function in $\partial\Omega$, then function g defined by*

$$g(y) = \frac{2^{n-1}}{\omega_{n+1}} \int_{\partial K} \left(h_1(x,y)\, e_n Q'(\nu(x) f(x)) - h_2(x,y)\, e_n P(\nu(x) f(x)) \right) d\sigma(x) .$$

is dual hypermonogenic in $y \in \mathbb{R}_+^{n+1} \backslash \partial K$.

The integral formula for $(1-n)$-hypermonogenic functions is obtained from the following characterization.

Theorem 4.14. *Let Ω be an open subset of $\mathbb{R}^{n+1} \backslash \{x_n = 0\}$ and $f : \Omega \to C\ell_n$ be a $\mathcal{C}^1(\Omega, C\ell_n)$ function. A function $f : \Omega \to C\ell_n$ is $-k$-hypermonogenic if and only if the function $x_n^k f$ is dual k-hypermonogenic. Especially, a function $f : \Omega \to C\ell_n$ is $(1-n)$-hypermonogenic if and only if the function $x_n^{n-1} f$ is dual hypermonogenic.*

Proof. Applying Lemma 3.3 and Corollary 4.3 we obtain the result. □

Theorem 4.15. *Let Ω be an open subset of $\mathbb{R}_+^{n+1}$ and $\overline{K} \subset \Omega$ a smoothly bounded compact set with outer unit normal field ν. If f is $(1-n)$-hypermonogenic in Ω and $y \in K$ then*

$$f(y) = \frac{2^{n-1}}{\omega_{n+1}} \int_{\partial K} \left(s_1(x,y)\, Q'(\nu(x) f(x)) - s_2(x,y)\, P(\nu(x) f(x)) \right) d\sigma(x) ,$$

where

$$s_1(x,y) = y_n^{1-n} x_n^{n-1} h_1(x,y)\, e_n$$
$$s_2(x,y) = y_n^{1-n} x_n^{n-1} h_2(x,y)\, e_n$$

are $(1-n)$-hypermonogenic functions with respect to y.

References

[1] Brackx, F., Delanghe, R., and Sommen, F., *Clifford Analysis*, Pitman, Boston, London, Melbourne, 1982.

[2] Eriksson-Bique, S.-L., k-hypermonogenic functions. *In Progress in Analysis*, Vol I, World Scientific (2003), 337–348.

[3] Eriksson, S.-L., Integral formulas for hypermonogenic functions, *Bull. Bel. Math. Soc.* **11** (2004), 705–717.

[4] Eriksson, S.-L., *Cauchy-type integral formulas for k-hypermonogenic functions.* To appear in *Proceedings of the 5th ISAAC conference*, Catania, Italy 2005.

[5] Eriksson, S.-L., Overview to Hyperbolic Function Theory, submitted for publication.

[6] Eriksson-Bique, S.-L. and Leutwiler, H., On modified quaternionic analysis in $\mathbb{R}^3$, *Arch. Math.* **70** (1998), 228–234.

[7] Eriksson-Bique, S.-L. and Leutwiler, H., Hypermonogenic functions. In *Clifford Algebras and their Applications in Mathematical Physics*, Vol. 2, Birkhäuser, Boston, 2000, 287–302.

[8] Eriksson-Bique, S.-L. and Leutwiler, H., Hypermonogenic functions and Möbius transformations, *Advances in Applied Clifford algebras*, Vol **11** (S2), December (2001), 67–76.

[9] Eriksson, S.-L. and Leutwiler, H., Hypermonogenic functions and their Cauchy-type theorems. In *Trend in Mathematics: Advances in Analysis and Geometry*, Birkhäuser, Basel/Switzerland, 2004, 97–112.

[10] Eriksson, S.-L. and Leutwiler, H., Contributions to the theory of hypermonogenic functions, *Complex Variables and elliptic equations* **51**, Nos. 5-6 (2006), 547–561.

[11] Eriksson, S.-L. and Leutwiler, H., On hyperbolic function theory (to appear).

[12] Eriksson, S.-L. and Leutwiler, H., Hyperbolic Function Theory, Adv. appl. Clifford alg. 17 (2007), 437–450.

[13] Leutwiler, H., Modified Clifford analysis, *Complex Variables* **17** (1992), 153–171.

[14] Leutwiler, H., Modified quaternionic analysis in $\mathbb{R}^3$, *Complex Variables* **20** (1992), 19–51.

[15] Leutwiler, H., Quaternionic analysis in $\mathbb{R}^3$ versus its hyperbolic modification. In F. Brackx et al. (eds.) *Clifford Analysis and its Applications*, Kluwer, Dordrecht 2001, 193–211.

[16] Qiao, Y., Bernstein, S., Eriksson, S.-L. and Ryan, J., Function theory for Laplace and Dirac-Hodge operators in hyperbolic space, *Journal d'Analyse Mathématiques* **98** (2006), 43–64.

Sirkka-Liisa Eriksson
Department of Mathematics
Tampere University of Technology
P.O. Box 553
FI–33101 Tampere, Finland
e-mail: `sirkka-liisa.eriksson@tut.fi`

Quaternionic and Clifford Analysis
Trends in Mathematics, 151–164

Description of a Complex of Operators Acting Between Higher Spinor Modules

Peter Franek

Abstract. We construct a particular sequence of homomorphisms of generalized Verma modules and show that this sequence is a complex. The dual sequence can be identified with a complex of linear differential operators so that the first operator in this sequence is a generalization of the Dirac operator in many Clifford variables. Further, we use Zuckerman translation principle to show that a similar sequence exists for any higher spinor operator in a particular model of Cartan geometry, including, e.g., the Rarita-Schwinger operator in many variables. There are indications that this sequence may be exact, forming a resolvent of the first operator.

Mathematics Subject Classification (2000). 22E46, 32W99.

Keywords. Differential operator, complex, Dirac, Generalized Verma module.

1. Introduction

1.1. Motivation

Although this article deals with homomorphisms of algebraic objects, the motivation for it comes from differential geometry. The aim is to construct possible resolvent of well-known differential operators that would generalize the Dolbeault complex. It is shown in [5, 6] that in a particular type of Cartan geometry $G \to G/P$, where G is a Lie group and P its parabolic subgroup, there exists a G-invariant differential operator $\Gamma(G \times_P \mathbb{V}_1) \to \Gamma(G \times_P \mathbb{V}_2)$ between sections of associated vector bundles that can be locally identified with the Dirac operator in k variables $D : C^\infty(\mathbb{R}^n, \mathbb{S}) \to C^\infty(\mathbb{R}^n, \mathbb{C}^k \otimes \mathbb{S})$ described in [10].

On algebraic level, the differential operators correspond to homomorphisms of generalized Verma modules. In [5], we described the structure of the generalized Verma module homomorphisms dual to the differential operators that continue the Dirac operator in k variables in case n is odd. In this article, we prove that one can

The research were supported by the grant MSM 0021620839.

start with any higher spinor operator (including, e.g., Rarita-Schwinger operator in many variables, studied by [3]) and the structure of these sequences will remain similar. Further, we show that the generalized Verma module homomorphisms can be summed up to form a linear complex. We assume that this complex is exact, although the proof is not known yet.

1.2. Lie algebras and parabolic subalgebras

Let $\mathfrak{g}$ be a semisimple Lie algebra, $\mathfrak{h}$ a chosen Cartan subalgebra, Φ^+ a set of positive roots and Δ a set of simple roots. The Borel algebra $\mathfrak{b}$ is defined as $\mathfrak{h} \oplus \mathfrak{b}_+$, where $\mathfrak{b}_+$ is the span of all positive root spaces. A Lie subalgebra $\mathfrak{p} \subset \mathfrak{g}$ is called *parabolic*, if it contains $\mathfrak{b}$. Any such $\mathfrak{p}$ induces a grading $\mathfrak{g} = \oplus_{i=-k}^{k} \mathfrak{g}_i$ such that $\mathfrak{p} = \oplus_{k \geq 0} \mathfrak{g}_i$. Further, any $\mathfrak{p}$ is determined by a subset $\Sigma \subset \Delta$ such that $\mathfrak{p}$ is the span of $\mathfrak{h}$, $\mathfrak{b}$ and all negative root spaces $\mathfrak{g}_{-\phi}$ such that ϕ is a nonnegative integral combination of elements in $\Delta - \Sigma$. Let as denote by P^{++} the set of integral dominant weights, i.e., $P^{++} = \{\alpha \in \mathfrak{h}^*, \forall \phi \in \Phi^+ \ \alpha(H_\phi) \in \mathbb{N}_0\}$, where H_ϕ is the ϕ-coroot (defined by $H_\phi = 2\hat{\phi}/(\phi, \phi)$, where $(,)$ is the Killing form and $\hat{\phi} \in \mathfrak{h}$ is the dual of $\phi \in \mathfrak{h}^*$ via the Killing form). This is exactly the set of highest weights of finite-dimensional irreducible $\mathfrak{g}$-modules. Similarly, define $P_{\mathfrak{p}}^{++} := \{\alpha \in \mathfrak{h}^*; \forall \phi \in \Delta - \Sigma \ \alpha(H_\phi) \in \mathbb{N}_0\}$. This is the set of highest weights of irreducible finite-dimensional $\mathfrak{p}$-modules. Let $\delta := 1/2 \sum_{\phi \in \Phi^+} \phi$ be the "lowest form".

1.3. Homomorphisms of generalized Verma modules

Let $\mathcal{U}(\mathfrak{g})$, $\mathcal{U}(\mathfrak{p})$ and $\mathcal{U}(\mathfrak{b})$ be the universal enveloping algebras of $\mathfrak{g}$, $\mathfrak{p}$, $\mathfrak{b}$. $\mathcal{U}(\mathfrak{g})$ is a left $\mathcal{U}(\mathfrak{g})$-module and a right $\mathcal{U}(\mathfrak{b})$ module. For any $\alpha \in \mathfrak{h}^*$, we define the Verma module $M(\alpha) = \mathcal{U}(\mathfrak{g}) \otimes_{\mathcal{U}(\mathfrak{b})} \mathbb{C}_{\alpha-\delta}$, where $\mathbb{C}_{\alpha-\delta} = \mathbb{C}$ is the one-dimensional representation of $\mathfrak{b}$ defined by $h \cdot c = (\alpha - \delta)(h)c$ for $h \in \mathfrak{h}$, $c \in \mathbb{C}_{\alpha-\delta}$ and $b \cdot c = 0$ for $b \in \mathfrak{b}_+$. Each highest weight module is a factor of a Verma module with the same highest weight.

For each root ϕ, we define the reflection $s_\phi : \mathfrak{h}^* \to \mathfrak{h}^*$, $\alpha \mapsto \alpha - \alpha(H_\phi)$. The Weyl group W of $\mathfrak{g}$ is a finite group generated by all root reflections. The following facts were proved by Bernstein-Gelfand-Gelfand and Verma in [1, 2, 13]

Theorem 1.1. *Let $\mu, \lambda \in \mathfrak{h}^*$. Each homomorphism $M(\mu) \to M(\lambda)$ is injective and* $\dim(\mathrm{Hom}(M(\mu), M(\lambda))) \leq 1$. *Therefore, we can write $M(\mu) \subset M(\lambda)$ in such case.*

A nonzero homomorphism of Verma modules $M(\mu) \to M(\lambda)$ exists if and only if there exist weights $\lambda = \lambda_0, \lambda_1, \ldots, \lambda_k = \mu$ so that $\lambda_{i+1} = s_{\beta_i}\lambda_i$ for some positive roots β_i and $\lambda_i(H_{\beta_i}) \in \mathbb{N}$ for all i. Equivalently, $\lambda_i - \lambda_{i-1}$ is a positive integral multiple of some positive root for all i.

For a parabolic subgroup $\mathfrak{p} \subset \mathfrak{g}$ and a $\mathfrak{p}$-module $\mathbb{V}$ with highest weight $\mu - \delta$, we define the generalized Verma module (further GVM)

$$M_{\mathfrak{p}}(\mathbb{V}) := \mathcal{U}(\mathfrak{g}) \otimes_{\mathcal{U}(\mathfrak{p})} \mathbb{V}.$$

Usually, $\mathbb{V}$ will be irreducible finite dimensional, i.e., $\mu \in P_{\mathfrak{p}}^{++} + \delta$ and in this case, we write $M_{\mathfrak{p}}(\mu) := M_{\mathfrak{p}}(\mathbb{V})$. GVM's are highest weight modules and $M_{\mathfrak{p}}(\mu) \simeq M(\mu)/K$, where K is some submodule of $M(\mu)$. Any homomorphism of Verma modules $M(\nu) \to M(\mu)$ factors to a homomorphism $M_{\mathfrak{p}}(\nu) \to M_{\mathfrak{p}}(\mu)$ that is called standard homomorphism. The following theorem can be used to show that a standard homomorphism of GVM's is nonzero:

Theorem 1.2. *Let $M(\nu) \subset M(\mu)$ be an inclusion of Verma modules and let $h : M_{\mathfrak{p}}(\nu) \to M_{\mathfrak{p}}(\mu)$ be the corresponding standard homomorphism of GVM's. Then h is zero if and only if there exists a simple root $\alpha \in \Delta - \Sigma$ such that $M(\nu) \subset M(s_\alpha \mu) \subset M(\mu)$.*

A central character of $\mathcal{U}(\mathfrak{g})$ is a homomorphism from the center $Z(\mathcal{U}(\mathfrak{g}))$ of the universal enveloping algebra to $\mathbb{C}$. Each highest weight module admits a central character, i.e., for each highest weight $\mathfrak{g}$-module $\mathbb{V}$ there exists $\phi : Z(\mathcal{U}(\mathfrak{g})) \to \mathbb{C}$ such that $\forall v \in \mathbb{V}\ \forall u \in Z(\mathcal{U}(\mathfrak{g}))\ u \cdot v = \phi(u)v$. Any homomorphisms of highest weight modules preserves the central character and the Harris-Chandra theorem states that central characters of $M(\nu)$ and $M(\mu)$ are the same if and only if μ and ν are on the same orbit of the Weyl group ([9]).

2. Results on sequences of invariant operators starting with Dirac

2.1. Odd dimension

In this section, we will summarize the main results from [5, 6]. Let $\mathfrak{g} = \mathrm{so}(n+2k, \mathbb{C})$ for some odd n. Choosing the convention of [8], we can represent elements of $\mathfrak{g}$ as matrices antisymmetric with respect to the anti-diagonal. We can choose the Cartan subalgebra $\mathfrak{h}$ to be the subalgebra of diagonal matrices and the standard basis of $\mathfrak{h}^*$ to be $\{\epsilon_1, \ldots, \epsilon_{k+(n-1)/2}\}$, where

$$\epsilon_i(\mathrm{diag}(d_1, d_2, \ldots, d_{k+(n-1)/2}, 0, -d_{k+(n-1)/2}, \ldots, -d_1)) = d_i.$$

Let $\Delta = \{\alpha_1, \ldots, \alpha_{k+(n-1)/2}\}$ be the set of simple roots and $\mathfrak{p}$ be a parabolic subalgebra corresponding to $\Sigma = \{\alpha_k\}$. The subalgebra $\mathfrak{p}$ induces the grading

$$\mathfrak{g} = \left(\begin{array}{c|c|c} \mathfrak{g}_0 & \mathfrak{g}_1 & \mathfrak{g}_2 \\ \hline \mathfrak{g}_{-1} & \mathfrak{g}_0 & \mathfrak{g}_1 \\ \hline \mathfrak{g}_{-2} & \mathfrak{g}_{-1} & \mathfrak{g}_0 \end{array}\right),$$

the blocks having size k, n and k. The reductive subalgebra is $\mathfrak{g}_0 \simeq \mathrm{gl}(k, \mathbb{C}) \times \mathrm{so}(n, \mathbb{C})$. Let $G = \mathrm{Spin}(n+k, k)$ be the Lie group and P its parabolic subgroup so that the complexified Lie algebra of P is $\mathfrak{p}$. Any finite-dimensional irreducible P-module is a $\mathfrak{p}$-module (by the infinitesimal action) and any finite-dimensional irreducible $\mathfrak{p}$-module can be given the structure of a P-module. They are classified by highest weights and the set of (isomorphism classes of) irreducible finite-dimensional P-modules is isomorphic to $P_{\mathfrak{p}}^{++}$. The lowest form (in the ϵ_i-basis) is $\delta = [2((k+n)-1)/2, \ldots, 3/2, 1/2]$. In the ϵ_i-basis, the Weyl group is generated by transpositions, sign-transpositions $([\ldots, a_i, \ldots, a_j, \ldots] \mapsto$

$[\dots,-a_j,\dots,-a_i,\dots])$ and sign-flip $([\dots,a_i,\dots]\mapsto[\dots,-a_i,\dots])$. P-dominant weights are $[a_1,\dots a_k|b_1,\dots,b_n]$ such that $a_1\geq a_2\geq\cdots\geq a_k$, $b_1\geq\cdots\geq b_n\geq 0$, $a_i-a_j\in\mathbb{Z}$, $b_i-b_j\in\mathbb{Z}$ and the b_i's are either all integers or all half-integers. Consider the weight $\lambda=[(2k-1)/2,\dots,3/2,1/2|(n-1)/2,\dots,2,1]$. Then $\lambda-\delta\in P_{\mathfrak{p}}^{++}$ and, as a $\mathfrak{g}_0^{ss}$-module, the module with highest weight $\lambda-\delta$ $\mathbb{V}_{\lambda-\delta}\simeq\mathbb{C}\otimes\mathbb{S}$ is the product of the trivial representation of sl(k) and the spinor representation of so(n). The structure of GVM homomorphisms $M_{\mathfrak{p}}(\nu)\to M_{\mathfrak{p}}(\lambda)$, where $\nu,\mu\in P_{\mathfrak{p}}^{++}+\delta$ are on the Weyl group orbit of λ, is described in the following theorem (proved in [5]):

Theorem 2.1. *There are 2^k weights in $S_k=P_{\mathfrak{p}}^{++}+\delta\cap W\lambda$, each of them is of the form $[a_1,\dots,a_k|(n-1)/2,\dots,2,1]$, where $(a_1,\dots,a_k)$ is a decreasing sign-permutation of $((2k-1)/2,\dots,3/2,1/2)$. For $k=2$, there exist nonzero standard homomorphisms*

$$M_{\mathfrak{p}}([-\frac{1}{2},-\frac{3}{2}|\dots])\to M_{\mathfrak{p}}([\frac{1}{2},-\frac{3}{2}|\dots])\to M_{\mathfrak{p}}([\frac{3}{2},-\frac{1}{2}|\dots])\to M_{\mathfrak{p}}([\frac{3}{2},\frac{1}{2}|\dots]) \tag{2.1}$$

and the composition of any two is zero. For $k>2$, the set of weights can be split into two subsets: S^1 the set of weights $[(2k-1)/2,\dots|\dots]$ and S^2 the set of weights $[\dots,-(2k-1)/2|\dots]$. Let $i,j:\mathfrak{h}^*_{k+n-1}\to\mathfrak{h}^*_{k+n}$, $[a_1,\dots,a_{k-1}|\dots]\mapsto[(2k-1)/2,a_1\dots,a_{k-1}|\dots]$ and $j:[a_1,\dots,a_{k-1}|\dots]\mapsto[a_1\dots,a_{k-1},-(2k-1)/2|\dots]$. Then there exists a nonzero GVM homomorphism $M_{\mathfrak{p}}(i(\nu))\to M_{\mathfrak{p}}(i(\mu))$ (GVM's with highest weights $+\delta$ from S^1) $\Leftrightarrow$ there exists a nonzero GVM homomorphism $M_{\mathfrak{p}}(\nu)\to M_{\mathfrak{p}}(\mu)$ $\Leftrightarrow$ there exists a nonzero GVM homomorphism $M_{\mathfrak{p}}(j(\nu))\to M_{\mathfrak{p}}(j(\mu))$. So, the structure of GVM homomorphisms between GVM's from S^1 (i.e., with highest weights $+\delta$ from S^1) is similar then the structure of GVM homomorphisms between GVM's for $k-1$. The sets S^1,S^2 can be naturally subdivided into $S^{1,1}$, $S^{1,2}$, $S^{2,1}$ and $S^{2,2}$ ($S^{1,2}$, for example, is the set of weights $[(2k-1/2,\dots,-(2k-3)/2)|\dots])$. For any $\nu\in\mathfrak{h}^*_{n+k-2}$, there exists a nonzero GVM homomorphism $M_{\mathfrak{p}}(j_k i_{k-1}(\nu))\to M_{\mathfrak{p}}(i_k j_{k-1}(\nu))$ (i.e., a homomorphism connecting a GVM in S^2 with a GVM in S^1). Each GVM homomorphisms on this orbit is a composition of these.

The order of the dual differential operator is 1 for $k=1$ (the Dirac operator), 2 for any homomorphism connecting a GVM in S^1 with a GVM in S^2 and the others are determined inductively.

Graphically, the GVM homomorphisms have the structure show in Figure 1. Figure 2 show S_k for $k=3,4$ (the arrows go from down to up and from left to right):

It is proved in [6] that the "top" homomorphism connecting $M_{\mathfrak{p}}(\lambda)$ with the neighbor GVM is dual to a differential operator $\Gamma(G\times_P\mathbb{V}^*_{\mu-\delta})\to\Gamma(G\times_P\mathbb{V}^*_{\lambda-\delta})$ so that, after identifying the local section in $0=eP$ with vector-valued functions on $\mathfrak{g}_-$ and restricting to functions constant in $\mathfrak{g}_{-2}$, the operator can be identified with the Dirac operator in k variables (acting on $\mathbb{R}^n$, n odd), described in [10].

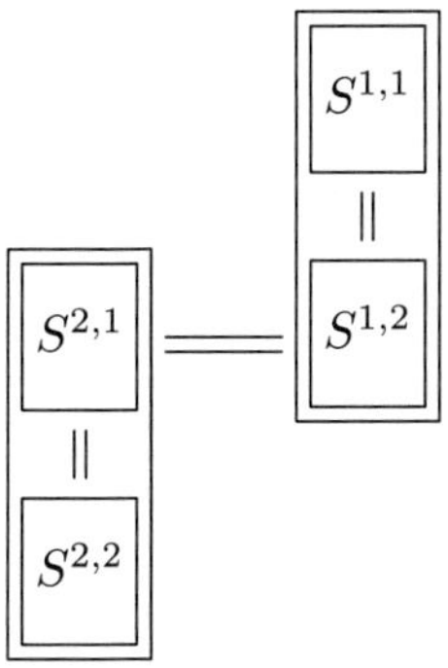

FIGURE 1. Structure of the GVM homomorphisms.

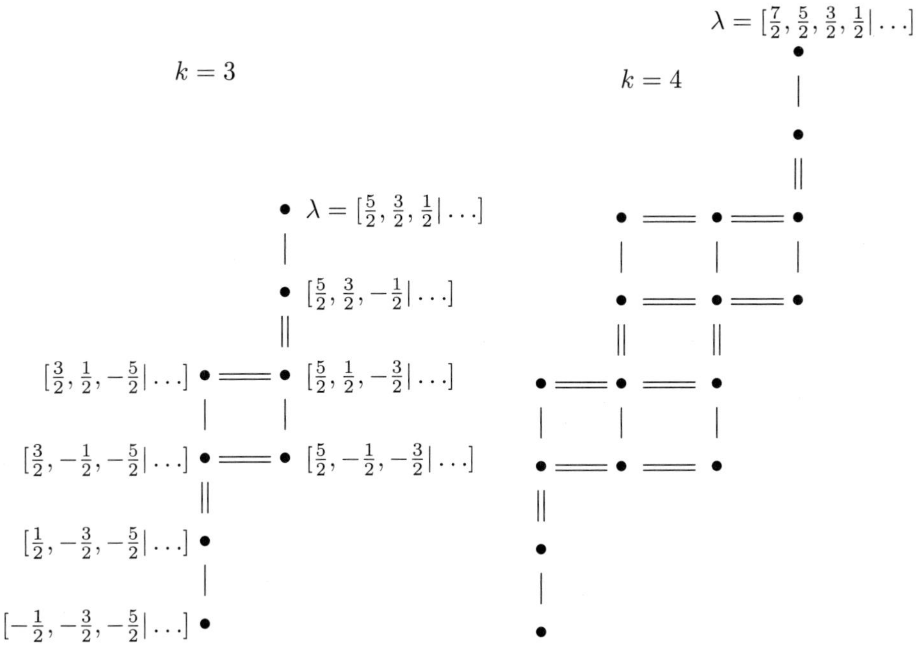

FIGURE 2. S_k for $k = 3, 4$ (the arrows goes from down to up and from left to right).

2.2. Even dimension

Let $\mathfrak{g} = \mathrm{so}(n+2k, \mathbb{C})$, n even. Similarly as before, we represent the elements of $\mathfrak{g}$ as matrices antisymmetric with respect to the antidiagonal, choose $\mathfrak{h}$ to be the algebra of diagonal matrices, $\mathfrak{p}$ be a parabolic subalgebra corresponding to $\Sigma = \{\alpha_k\}$, and

$\{\epsilon_i\}$ the basis of $\mathfrak{h}^*$, where

$$\epsilon_i(\mathrm{diag}(a_1, \ldots, a_{k+n/2}, -a_{k+n/2}, \ldots, -a_1)) = a_i.$$

Let $G = \mathrm{Spin}(n+k, k)$ be the Lie group and P its parabolic subgroup, as before. The lowest form is $\delta = [(k+n)-1, \ldots, 1, 0]$. In the ϵ_i-basis, the Weyl group is generated by transpositions and sign-transpositions. P-dominant weights are $[a_1, \ldots, a_k | b_1, \ldots, b_n]$ such that $a_1 \geq a_2 \geq \cdots \geq a_k$, $b_1 \geq \cdots \geq b_{n-1} \geq |b_n|$, $a_i - a_j \in \mathbb{Z}$, $b_i - b_j \in \mathbb{Z}$ and the b_i's are either all integers or all half-integers. Consider the weight $\lambda = [(2k-1)/2, \ldots, 3/2, 1/2|(2n-1)/2, \ldots, 3/2, 1/2]$. Then $\lambda - \delta \in P_{\mathfrak{p}}^{++}$ and the situation is similar to the odd case. There exists a homomorphism $M_{\mathfrak{p}}(\mu) \to M_{\mathfrak{p}}(\lambda)$ so that the dual differential operator may be identified with the Dirac operator in k variables.

To describe the orbit, we have to distinguish the case $n/2 \geq k$ (s.c. stable range) and $n/2 < k$ (non-stable range). If $n/2 \geq k$, then the Weyl orbit of the weight λ has a similar structure to that described in Theorem 2.1, only one must write $[\ldots, |(2n-1)/2, \ldots, 3/2, \pm 1/2]$ instead of $[\ldots|(n-1)/2, \ldots, 2, 1]$ in the expression of each weight. The structure of the GVM homomorphisms also seems to be the same, although we have no proof for that; we only proved in [6] that all the homomorphisms corresponding to first-order dual differential operators exist and the other homomorphisms are non-standard. For $k = 2$, the existence of a nonzero nonstandard homomorphism $M_{\mathfrak{p}}([1/2, -3/2| \ldots]) \to M_{\mathfrak{p}}(3/2, -1/2| \ldots)$ was proved in [7]. For higher k, however, it is only a conjecture that they all exist.

If $n/2 < k$, the situation becomes even more complicated, because the orbit $W\lambda \cap P_{\mathfrak{p}}^{++} + \delta$ contains more than 2^k elements. For example, the situation of the GVM homomorphisms (or dual differential operators) $k = 3, n = 2$ is pictured in Figure 3. The "new" weights in the diagram correspond to representations of P such that, as representation of $G_0^{ss} \simeq \mathrm{Sl}(k) \times \mathrm{Spin}(n)$ (its semisimple part), they have a form $\mathbb{V} \otimes \mathbb{S}$, where $\mathbb{S}$ are no more spinors, but some higher spinor representations.

3. Operators acting on higher spinor modules

3.1. Translation principle

Let G be a semisimple Lie group, P a parabolic subgroup, $\mathbb{E}$ an irreducible P-module and $\mathbb{W}$ an irreducible G-module. As P-module (by restriction), $\mathbb{W}$ has a filtration of P-modules

$$\mathbb{W} = W_l \supset W_{l-1} \supset \cdots \supset W_{-l}$$

with composition factors $\mathbb{W}_i = W_i/W_{i-1}$ that can be decomposed into irreducible P-modules. Let as suppose that $\mathbb{E} \otimes \mathbb{W}_i = \oplus_j \mathbb{E}_i^j$ is the decomposition into irreducible P-modules. This exists, because the action of $\mathfrak{p}^+$ on $\mathbb{W}_i$ and $\mathbb{E}$ is trivial, so it coincides with the G_0-module decomposition.

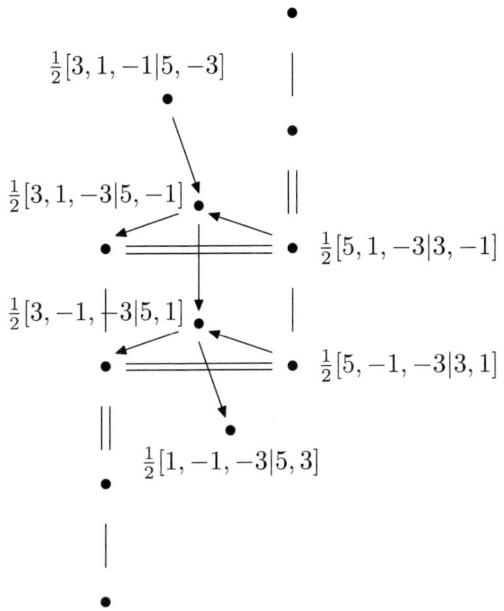

FIGURE 3. Situation of the GVM homomorphisms (or dual differential operators) $k = 3, n = 2$.

The following theorems are proved in [11]:

Theorem 3.1. *If $\mathbb{E}_i^j$ has a central character χ different from all the central characters of $\mathbb{E}_{i'}^{j'}$ for $i' \neq i, j' \neq j$, then the χ-eigenspace of $M_\mathfrak{p}(\mathbb{E} \otimes \mathbb{W})$ is its direct summand isomorphic to $M_\mathfrak{p}(\mathbb{E}_i^j)$.*

Theorem 3.2. *There exists a $\mathfrak{g}$-module isomorphismus $M_\mathfrak{p}(\mathbb{E} \otimes \mathbb{W}) \simeq M_\mathfrak{p}(\mathbb{E}) \otimes \mathbb{W}$ (the spaces are the same, but the actions of $\mathfrak{g}$ different).*

Let $\mathbb{F}$ be another P-module and $\mathbb{F} \otimes \mathbb{W}_i = \oplus_j \mathbb{F}_i^j$. It follows immediately that if $\mathbb{E}_i^j$ has a central character χ different from all other central characters of $\mathbb{E}_{i'}^{j'}$, $\mathbb{F}_k^l$ has the same central character χ and the central characters of all other $\mathbb{F}_{k'}^{l'}$ are different from χ, then the existence of a nonzero GVM homomorphism $M_\mathfrak{p}(\mathbb{E}) \to M_\mathfrak{p}(\mathbb{F})$ implies the existence of a nonzero GVM homomorphism $M_\mathfrak{p}(\mathbb{E}_i^j) \to M_\mathfrak{p}(\mathbb{F}_k^l)$.

3.2. Application

Let $G = \mathrm{Spin}(n+k, k)$ be the odd-dimensional spin group, P the parabolic subalgebra as in Section 2.1. Let $\mathbb{W}$ be the (complex) defining representation of G. As P-module, it has the filtration

$$W_{-1} \subset W_0 \subset W_1 = \mathbb{W}, \quad \text{where} \quad W_{-1} = \left\{ \begin{pmatrix} * \\ \hline 0 \\ \hline 0 \end{pmatrix} \right\}$$

is s subspace of $\mathbb{C}^{n+2k}$ having zero on $k+1,\dots,(2k+n)$th positions and

$$W_0 = \left\{ \begin{pmatrix} * \\ - \\ * \\ - \\ 0 \end{pmatrix} \right\}$$

has zeros on last k positions. As representations of G_0 (the reductive part of G), the quotients are $\mathbb{W}_{-1} = W_{-1} \simeq \mathbb{C}^k \otimes \mathbb{C}$ (tensor product of the defining representation of $\mathrm{GL}(k)$ and the trivial representation of $\mathrm{Spin}(n)$), $\mathbb{W}_0 = W_0/W_{-1} \simeq \mathbb{C} \otimes \mathbb{C}^n$ and $\mathbb{W}_1 = W_1/W_0 \simeq (\mathbb{C}^k)^* \otimes \mathbb{C}$. In terms of highest weights, we can write $\mathbb{W}_{-1} = [1,0,\dots,0|0,\dots,0]$, $\mathbb{W}_0 = [0,\dots,0|1,0,\dots,0]$ and $\mathbb{W}_1 = [0,\dots,0,-1|0,\dots,0]$.

Lemma 3.3. *Let Λ be the set of weights of $\mathbb{W}$, the defining representation of G (i.e., of weights $[0,\dots,0,\pm 1,0,\dots]$) and let V_ν be an irreducible representation of G_0 with highest weight ν. Then the tensor product $V_\nu \otimes \mathbb{W}$ of G_0-modules decomposes multiplicity-free as*

$$V_\nu \otimes \mathbb{W} \simeq \oplus_{\lambda \in \Theta} V_\lambda, \quad \text{where} \quad \Theta = (\nu + \Lambda) \cap P_{\mathfrak{p}}^{++}.$$

The proof follows from Klimyk formula, see [12], or [4] for details.

Theorem 3.4. *Let $(a_1,\dots,a_{(n-1)/2})$ be a decreasing sequence of positive integers, $(b_1,\dots,b_k)$ be a decreasing sequence of positive half-integers, let π_1,π_2 be permutations of $\{1,\dots,k\}$ and $s_1,\dots,s_k,\tilde{s}_1,\dots,\tilde{s}_k \in \{1,-1\}$. Let as suppose that*

$$(s_1 b_{\pi_1(1)}, s_2 b_{\pi_1(2)}, \dots, s_k b_{\pi_1(k)}) \quad \text{and} \quad (\tilde{s}_1 b_{\pi_2(1)}, \tilde{s}_2 b_{\pi_2(2)}, \dots, \tilde{s}_k b_{\pi_2(k)})$$

are decreasing sequences. Further, let $(c_1,\dots,c_k) = ((2k-1)/2,\dots,3/2,1/2)$.

Then there exists a nonzero GVM homomorphism

$$M_{\mathfrak{p}}([s_1 b_{\pi_1(1)},\dots,s_k b_{\pi_1(k)}|a_1,a_2\dots,a_{(n-1)/2}]) \\ \to M_{\mathfrak{p}}([\tilde{s}_1 b_{\pi_2(1)},\dots,\tilde{s}_k b_{\pi_2(k)}|a_1,\dots,a_{(n-1)/2}])$$

if and only if there exists a nonzero standard GVM homomorphism

$$M_{\mathfrak{p}}([s_1 c_{\pi_1(1)},\dots,s_k c_{\pi_1(k)}|\dots,2,1]) \to M_{\mathfrak{p}}([\tilde{s}_1 c_{\pi_2(1)},\dots,\tilde{s}_k c_{\pi_2(k)}|\dots,2,1])$$

Example. There exist a sequence of GVM homomorphisms

$$M_{\mathfrak{p}}\left(\frac{7}{2},\frac{1}{2}|6,3\right) \to M_{\mathfrak{p}}\left(\frac{7}{2},-\frac{1}{2}|6,3\right) \to M_{\mathfrak{p}}\left(\frac{1}{2},-\frac{7}{2}|6,3\right) \to M_{\mathfrak{p}}\left(-\frac{1}{2},-\frac{7}{2}|6,3\right),$$

analogously to (2.1).

Proof. Let $b_i + 1$ be different from all other b_j's, for some i and let $\mathbb{W}$ be the defining representation of G with composition factors $\mathbb{W}_{-1},\mathbb{W}_0,\mathbb{W}_1$ as before. Let $\nu = [s_1 b_{\pi_1(1)},\dots,s_k b_{\pi_1(k)}|c_1,\dots,c_{(n-1)/2}]$ and ν' be the weight having $\pm(b_i+1)$ instead of $\pm b_i$ on the particular position. The numbers $s_1 b_{\pi_1(1)},\dots,s_k b_{\pi_1(k)}$ are strictly decreasing, so interchanging b_i with b_i+1 does not change this property (if b_i+1 is different from all other b_j's). Therefore, $\nu' - \delta$ is P-dominant. It follows from Lemma 3.3 that, in the G_0-module decomposition of $\mathbb{V}_{\nu-\delta} \otimes (\mathbb{W}_{-1} \oplus \mathbb{W}_0 \oplus \mathbb{W}_1)$, the module $\mathbb{V}_{\nu'-\delta}$ occurs with multiplicity one. We state that $\mathbb{V}_{\nu-\delta}$ has different central character from all other composition factors of $V_\nu \otimes (\mathbb{W}_{-1} \oplus \mathbb{W}_0 \oplus \mathbb{W}_1)$.

To see this, assume, for contradiction, that $\alpha \in P_p^{++} + \delta$ is another weight so that $V_{\alpha-\delta}$ is in the decomposition and has the same central character as $V_{\nu'-\delta}$. This means that the coordinate expression of α is a sign-permutation of the coordinates of ν' and differs from ν by ± 1 in exactly one coordinate. The coordinate $\pm(b_i+1)$ is obtained either from $\pm b_i$ by adding ± 1, or from some $\pm b_k$. In the first case, $\alpha = \nu'$, the second case is impossible, because $\pm b_k$ would be missing in the expression of α. It follows from Theorems 3.1 and 3.2 that $M_{\mathfrak{p}}(\nu')$ is a direct summand of $M_{\mathfrak{p}}(\nu) \otimes \mathbb{W}$.

The same argument shows that, for $\mu = [\tilde{s}_1 b_{\pi_2(1)}, \ldots, \tilde{s}_k b_{\pi_2(k)} | c_1, \ldots, c_{(n-1)/2}]$ and μ' having $b_i + 1$ instead of b_i on the particular position, $M_{\mathfrak{p}}(\mu')$ is a direct summand of $M_{\mathfrak{p}}(\mu) \otimes \mathbb{W}$. Therefore, if there exist a nonzero homomorphism $M_{\mathfrak{p}}(\nu) \to M_{\mathfrak{p}}(\mu)$, there exists a nonzero homomorphism $M_{\mathfrak{p}}(\nu') \to M_{\mathfrak{p}}(\mu')$. On the other hand, an analogous argument then above with $b_i = (b_i + 1) - 1$ being different from all other b_j, shows that the existence of a nonzero homomorphism $M_{\mathfrak{p}}(\nu') \to M_{\mathfrak{p}}(\mu')$, implies the existence of a nonzero homomorphism $M_{\mathfrak{p}}(\nu) \to M_{\mathfrak{p}}(\mu)$.

Starting with $(d_1, \ldots, d_k) = ((2k-1)/2, \ldots, 1/2)$, and increasing the b_i's, we see by induction that a nonzero homomorphism

$$M_{\mathfrak{p}}(s_1 d_{\pi_1(1)}, \ldots, s_k d_{\pi_1(k)} | \ldots, 2, 1) \to M_{\mathfrak{p}}(\tilde{s}_1 d_{\pi_2(1)}, \ldots, \tilde{s}_k d_{\pi_2(k)} | \ldots, 2, 1)$$

exists if and only if there exists a nonzero homomorphism

$$M_{\mathfrak{p}}(s_1 b_{\pi_1(1)}, \ldots, s_k b_{\pi_1(k)} | \ldots, 2, 1) \to M_{\mathfrak{p}}(\tilde{s}_1 b_{\pi_2(1)}, \ldots, \tilde{s}_k b_{\pi_2(k)} | \ldots, 2, 1)$$

for any decreasing sequence of positive half-integers $(b_1, \ldots, b_k)$.

Fixing the b_i's, the same procedure can be done by increasing the integers $((n-1)/2, \ldots, 2, 1)$ to any decreasing sequence of positive integers $(a_1, \ldots, a_{(n-1)/2})$ and the theorem follows. $\square$

Choosing for example, the weight $\lambda = [(2k-1)/2, \ldots, 3/2, 1/2 | (n+1)/2, (n-3)/2, (n-5)/2, \ldots, 2, 1]$, $\mu = [(2k-1)/2, \ldots, 3/2, -1/2 | (n+1)/2, (n-3)/2, (n-5)/2, \ldots, 2, 1]$, the dual differential operator to $M_{\mathfrak{p}}(\mu) \to M_{\mathfrak{p}}(\lambda)$ maybe locally identified with the Rarita-Schwinger operator in several variables described, e.g., in [3].

4. The complex of operators

4.1. The main theorem

In this section, we will prove that the GVM homomorphisms described above can be chosen and the GVM's can be summed up in such a way, so that the sequence becomes a linear complex of homomorphisms. Because $\oplus_i M_{\mathfrak{p}}(\mathbb{V}_i) \simeq M_{\mathfrak{p}}(\oplus \mathbb{V}_i)$, we can dualize the homomorphisms and identify locally the dual differential operators with an operator acting on functions defined on the vector space $\mathfrak{g}_-$ with values in $\oplus \mathbb{V}_i^*$. Again, restricting to functions constant in $\mathfrak{g}_{-2}$, we expect this operators to form a resolvent of the Dirac, Rarita-Schwinger resp. other operators in k variables.

However, we will only prove that it forms a complex, the exactness leaving open so far.

Let $\lambda = [(2k-1)/2, \dots, 3/2, 1/2|(n-1)/2, \dots, 2, 1]$. The structure of GVM homomorphisms between GVM's $M_{\mathfrak{p}}(\mu)$, where $\mu \in W\lambda \cap P_{\mathfrak{p}}^{++} + \delta$ is described in Theorem 2.1 in detail. $M_{\mathfrak{p}}(\lambda)$ is on the "top" of the diagram (there exist $M_{\mathfrak{p}}(\mu) \to M_{\mathfrak{p}}(\lambda)$, but no $M_{\mathfrak{p}}(\lambda) \to M_{\mathfrak{p}}(\nu)$). Define the modules M_i so that $M_0 := M_{\mathfrak{p}}(\lambda)$ and $M_{i+1} = \oplus_{\nu \in \Gamma} M_{\mathfrak{p}}(\nu)$, where Γ is the set of weights ν having the following properties:

a) there exists a nonzero standard GVM homomorphism $M_{\mathfrak{p}}(\nu) \to M_{\mathfrak{p}}(\mu)$ for some μ so that $M_{\mathfrak{p}}(\mu)$ is a direct summand in M_i and
b) if there exists a sequence of nonzero GVM homomorphisms $M_{\mathfrak{p}}(\nu) \to M_{\mathfrak{p}}(\nu') \to M_{\mathfrak{p}}(\mu)$, where μ is like in a), then either $\nu' = \nu$ or $\nu' = \mu$.

Example. For $k = 3$, M_3 consists of the sum of two GVM's (the homomorphisms go from right to left, the dual differential operators from left to right):

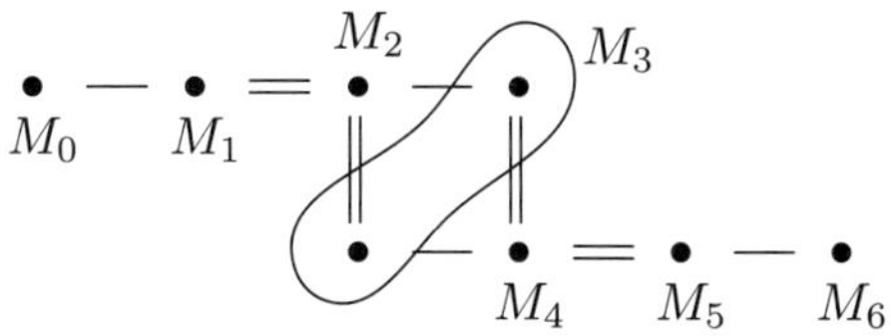

In this picture, the line from M_0 to M_1 represents a homomorphism $M_1 = M_{\mathfrak{p}}(\mu) \to M_{\mathfrak{p}}(\lambda) = M_0$ that is dual to the Dirac operator in several variables, and one line indicates that the dual differential operator is of first order.

Any choice of the standard GVM homomorphisms described in 2.1 (each of the homomorphisms is unique only up to scalar) determines a linear sequence of homomorphisms $\dots \stackrel{d_2}{\to} M_1 \stackrel{d_1}{\to} M_0$ (if M_i is a sum of more then one GVM's as M_3 in the picture, the homomorphisms leading in and out of the components of M_i sum up to d_i and d_{i-1}).

Theorem 4.1. *The homomorphisms of GVM's described in Theorem* 2.1 *can be chosen in such a way, so that the associated sequence of homomorphisms* $\{M_i, d_i\}$ *described above is a complex.*

Proof. The key ingredient is Lemma 1.2. For $\alpha \in \Delta - \Sigma$, the action of s_α on $\mathfrak{h}^*$ interchanges 2 neighbor coordinates (but not kth and $(k+1)$th) or the sign of the last one. In case $k = 2$, $s_{\alpha_1}[3/2, 1/2|\dots, 2, 1] = [1/2, 3/2|\dots]$, Theorem 1.1 implies $M([1/2, -3/2|\dots]) \subset M([1/2, 3/2|\dots])$ and it follows from Lemma 1.2 that the standard map $M_{\mathfrak{p}}([1/2, -3/2|\dots]) \to M_{\mathfrak{p}}([3/2, 1/2|\dots])$ is zero (which is the composition of the second and third homomorphism of (2.1)). Similarly, one shows that the composition of the first and second homomorphism in (2.1) is zero.

Now assume, by induction, that the theorem is true for some k. Let S_k, S^1, S^2, i, j be like in Theorem 2.1. Define the category $\mathcal{M}$ with objects $\{M_{\mathfrak{p}}(\nu), \nu \in S_k\}$, $\mathcal{M}^j$ with objects $\{M_{\mathfrak{p}}(\nu), \nu \in S^j_{k+1}\}$, $j = 1, 2$ and the morphisms to be standard

GVM homomorphisms. It follows from Theorem 2.1 that the functors $\varphi^j : \mathcal{M} \to \mathcal{M}^j$ defined by $M_\mathfrak{p}(\nu) \mapsto M_\mathfrak{p}(i(\nu))$ resp. $M_\mathfrak{p}(\nu) \mapsto M_\mathfrak{p}(j(\nu))$ are isomorphisms (they preserve the existence of a nonzero standard GVM homomorphism). Define

$$M_i^1 := M_i \cap \oplus_{M_\mathfrak{p}(\nu)\in\mathcal{M}^1} M_\mathfrak{p}(\nu) \text{ and } M_i^2 := M_i \cap \oplus_{M_\mathfrak{p}(\nu)\in\mathcal{M}^2} M_\mathfrak{p}(\nu).$$

Figure 4 shows the situation for $k = 4$.

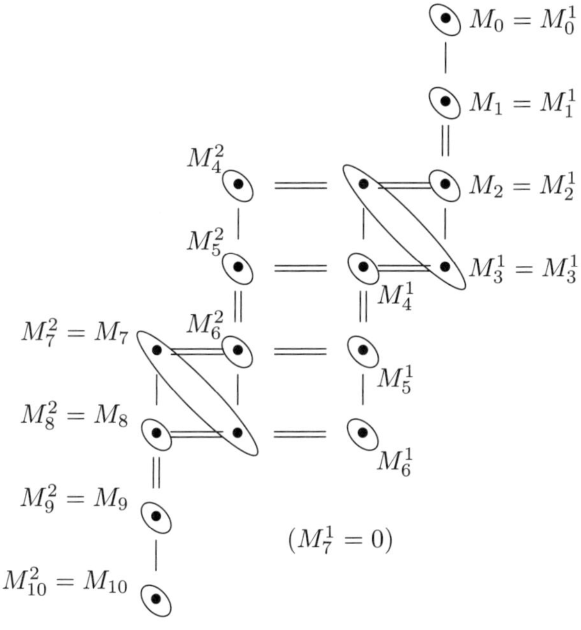

FIGURE 4.

The maps $M_\mathfrak{p}(\nu) \mapsto M_\mathfrak{p}(i(\nu))$ and $M_\mathfrak{p}(\nu) \mapsto M_\mathfrak{p}(j(\nu))$ can be extended to $M_i \to M_i^1$ and $M_i \to M_i^2$ (although some M_i^j may be zero). By induction, the homomorphisms between GVM's from $\mathcal{M}$ can be chosen so that the associated homomorphisms $d_i d_{i+1} : M_{i+1} \to M_{i-1}$ are zero. Because φ^j are isomorphisms of categories, the homomorphisms between GVM's in $\mathcal{M}^j$ can be chosen in such a way, so that $d_i^j d_{i+1}^j : M_{i+1}^j \to M_{i-1}^j$ is zero, for $j = 1, 2$. Assume that these homomorphisms have been fixed. So, $\{M_i^j, d_i^j\}$ are subcomplexes of $\{M_i, d_i\}$, $j = 1, 2$.

Now, if $M_i = M_i^1$, then M_i is a sum of GVM's with highest weights $+\delta$ from S^1. So, M_{i-1} and M_{i-2} have the same property, because there is no GVM homomorphism $M_\mathfrak{p}(\nu) \to M_\mathfrak{p}(\mu), \nu \in S^1, \mu \in S^2$. So, in this case, $d_{i-1}d_i$ is zero. Similarly, if $M_i = M_i^2$, then M_{i+1} and M_{i+2} have the same property and $d_{i+1}d_{i+2} = d_{i+1}^2 d_{i+2}^2 = 0$.

It remains to fix the "connecting homomorphisms", i.e., homomorphisms $M_\mathfrak{p}(\nu) \to M_\mathfrak{p}(\mu)$ with $\nu \in S^2, \mu \in S^1$ (in the above picture, these are the 4 middle-homomorphisms). Let $\nu \in S^{2,1}$, $M_{\mathfrak{p}}(\nu) \in M_i$. We know that there exists a nonzero homomorphisms $M_{\mathfrak{p}}(\nu) \to M_{\mathfrak{p}}(\mu)$, where $\mu \in S^{1,2}$. Consider a homomorphism $M_{\mathfrak{p}}(\mu) \to M_{\mathfrak{p}}(\beta)$, where $M_{\mathfrak{p}}(\beta) \in M_{i-2}$. There are two possibilities: either $\beta \in S^{1,1}$, or $\beta \in S^{1,2}$.

Case A. First let $\beta \in S^{1,1}$. This implies that there exists a homomorphisms from $M_{\mathfrak{p}}(\mu)$, $\mu \in S^{1,2}$ to $M_{\mathfrak{p}}(\beta)$, $\beta \in S^{1,1}$ and hence $\mu \in S^{1,2,1}$, $\beta \in S^{1,1,2}$. This means that

$$\nu = [(2k-3)/2, (2k-5)/2, a_1, \dots, a_{k-3}, -(2k-1)/2|\dots],$$
$$\mu = [(2k-1)/2, (2k-5)/2, a_1, \dots, a_{k-3}, -(2k-3)/2|\dots],$$
$$\beta = [(2k-1)/2, (2k-3)/2, a_1, \dots, a_{k-3}, -(2k-5)/2|\dots].$$

In such case, choosing any nonzero standard homomorphism $l : M_{\mathfrak{p}}(\nu) \to M_{\mathfrak{p}}(\mu)$, the composition $M_{\mathfrak{p}}(\nu) \to M_{\mathfrak{p}}(\mu) \to M_{\mathfrak{p}}(\beta)$ is zero, because $s_{\alpha_1}\beta = [(2k-3)/2, (2k-1)/2, \dots, -(2k-5)/2|\dots]$ and $M(\nu) \subset M(s_{\alpha_1}\beta)$ by Theorem 1.1 ($\nu = s_\gamma s_{\alpha_1}\beta$, where s_γ sign-interchanges the $(2k-5)/2$ and $(2k-1)/2$). So, Lemma 1.2 implies that the composition is zero.

Case B. The second case is $\beta \in S^{1,2}$, as well as μ. We have already fixed a standard homomorphisms $k : M_{\mathfrak{p}}(\mu) \to M_{\mathfrak{p}}(\beta)$. The condicion $\beta \in S^{1,2}$ implies that β is of the form

$$\beta = [(2k-1)/2, a_1, \dots, a_{k-2}, -(2k-3)/2|\dots],$$

and

$$\mu = [(2k-1)/2, b_1, \dots, b_{k-2}, -(2k-3)/2|\dots],$$
$$\nu = [(2k-3)/2, b_1, \dots, b_{k-2}, -(2k-1)/2|\dots],$$

where $(b_1, \dots, b_{k-2})$ differs from $(a_1, \dots, a_{k-2})$ by a (sign)-inversion $(a_i, a_j) \to ((-)a_j, (-)a_i)$. Let s_γ be the root reflection so that $s_\gamma\beta = \mu$. Then for the weight

$$\nu' = s_\gamma\nu = [(2k-3), a_1, \dots, a_{k-2}, -(2k-1)/2|\dots]$$

we see $\nu' \in S^{2,1}$ and Theorem 2.1 implies that there exist a nonzero standard homomorphism $n : M_{\mathfrak{p}}(\nu) \to M_{\mathfrak{p}}(\nu')$ and a nonzero standard homomorphism $m : M_{\mathfrak{p}}(\nu') \to M_{\mathfrak{p}}(\beta)$. So, we have the following square of standard GVM homomorphisms:

$$\begin{array}{ccc} M_p(\nu') & \xrightarrow{m} & M_p(\beta) \\ {\scriptstyle n}\uparrow & & \uparrow{\scriptstyle k} \\ M_p(\nu) & \xrightarrow[l]{} & M_p(\mu) \end{array}$$

Now, we want to prove that the homomorphisms l, n can be chosen so that $\{M_i, d_i\}$ is a complex. The necessary and sufficient condition is that the restriction of $d_{i-1}d_i$ to any direct summand $M_{\mathfrak{p}}(\nu)$ of M_i is zero. This happens if and only if the projection of the image of the restriction of $d_{i-1}d_i|_{M_{\mathfrak{p}}(\nu)}$ to each $M_{\mathfrak{p}}(\beta)$ (direct summand of M_{i-2}) is zero.

Let i_0 be the smallest integer so that $M_i \neq M_i^1$ and let $M_\mathfrak{p}(\nu) \in M_i$, $\nu \in S^2$. This ν is unique, namely

$$\nu = [(2k-3)/2, \ldots, 3/2, 1/2, -(2k-1)/2)| \ldots]$$

and there is only one standard homomorphism $M_\mathfrak{p}(\nu) \to M_\mathfrak{p}(\mu)$ for $\mu \in S^1$, namely for

$$\mu = [(2k-1)/2, \ldots, 3/2, 1/2, -(2k-3)/2| \ldots].$$

For any standard homomorphism $M_\mathfrak{p}(\mu) \to M_\mathfrak{p}(\beta)$, where $M_\mathfrak{p}(\beta) \in \mathcal{M}^1_{i_0-2}$, β has to be in $S^{1,1}$ (case A) and the composition $M_\mathfrak{p}(\nu) \to M_\mathfrak{p}(\beta)$ is zero. So, we may choose the homomorphism $M_\mathfrak{p}(\nu) \to M_\mathfrak{p}(\mu)$ to be arbitrary nonzero and $d_{i_0-1}d_{i_0} = 0$.

Let as suppose, by further induction, that for $i = i_0, i_0+1, \ldots, i_0+s-1$, all the "connecting homomorphisms" $M_\mathfrak{p}(\nu) \to M_\mathfrak{p}(\mu)$ have been fixed for any $\nu \in S^2$, $\mu \in S^1$, $M_\mathfrak{p}(\nu)$ being a direct summand of M_i and $M_\mathfrak{p}(\mu)$ a direct summand of M_{i-1}, so that $d_{i-1}d_i = 0$. Now, let $j = i_0 + s$, $\nu \in S^2$, $\mu \in S^1$ $M_\mathfrak{p}(\nu)$ be a direct summand of M_j and $M_\mathfrak{p}(\mu)$ a direct summand of M_{j-1} so that there exist a nonzero standard homomorphism $l : M_\mathfrak{p}(\nu) \to M_\mathfrak{p}(\mu)$. Such a μ is unique ($\mu = s_\gamma \nu$, where s_γ sign-interchanges the $(2k-1)/2$ and $(2k-3)/2$ on the 1st and kth positions). Let $k : M_\mathfrak{p}(\mu) \to M_\mathfrak{p}(\beta)$ be a standard homomorphism, $\beta \in S^{1,2}$, $M_\mathfrak{p}(\beta)$ being a direct summand of M_{j-2}. This is the situation of "case B" and we know that there exists $n : M_\mathfrak{p}(\nu) \to M_\mathfrak{p}(\nu')$ and $m : M_\mathfrak{p}(\nu') \to M_\mathfrak{p}(\beta)$, $M_\mathfrak{p}(\nu')$ being a direct summand of M_{j-1}. The homomorphism m have been already fixed, by induction. Because a standard homomorphism $M_\mathfrak{p}(\nu) \to M_\mathfrak{p}(\beta)$ is determined uniquely up to a scalar, it follows that we can choose the homomorphisms $l : M_\mathfrak{p}(\nu) \to M_\mathfrak{p}(\mu)$ so that the compositions $k \circ l$ and $m \circ n$ cancel. In other words, the $M_\mathfrak{p}(\beta)$-component of the image of the restriction of $d_{i-1}d_i$ to $M_\mathfrak{p}(\nu)$ is zero. Similarly, if $\beta \in S^{1,1}$, then the composition $M_\mathfrak{p}(\nu) \to M_\mathfrak{p}(\beta)$ vanishes (Case A). So, the restriction of $d_{i-1}d_i$ to any such $M_\mathfrak{p}(\nu)$ is zero. The restriction of $d_{i-1}d_i$ on $M_\mathfrak{p}(\gamma)$ for $\gamma \in S^1$ is zero, as $\{M_i^1, d_i^1\}$ is a subcomplex of $\{M_i, d_i\}$. So, $d_{i-1}d_i = 0$.

In this way, we fix all the homomorphisms $M_\mathfrak{p}(\nu) \to M_\mathfrak{p}(\mu)$, $\nu \in S^2$, $\mu \in S^1$.

Let further $\xi \in S^{2,2}$ so that there exist nonzero $M_\mathfrak{p}(\xi) \to M_\mathfrak{p}(\nu)$ and $M_\mathfrak{p}(\nu) \to M_\mathfrak{p}(\mu)$, $\nu \in S^{2,1}$, $\mu \in S^{1,2}$. In this case, Lemma 1.2 implies that the composition $M_\mathfrak{p}(\xi) \to M_\mathfrak{p}(\mu)$ is zero, similarly as in "case A". So, the restriction of $d_{i-1}d_i$ to any $M_\mathfrak{p}(\nu)$ is zero and the result follows. □

Again, the translation principle described in Section 3 can be applied to the complex just constructed. Dualizing this, we obtain the complex of G-invariant differential operators $\Gamma(G \times_P \mathbb{V}_1) \to \Gamma(G \times_P \mathbb{V}_2) \to \Gamma(G \times_P \mathbb{V}_3) \to \cdots$ such that, as a $G_0^{ss} \simeq \mathrm{Sl}(k) \times \mathrm{Spin}(n)$-module, $\mathbb{V}_1 = T_1 \otimes U$ where T_1 may be any irreducible finite-dimensional representation of $\mathrm{Sl}(k)$ and U may be any higher spinor representation of $\mathrm{Spin}(n)$, i.e., a representation with highest weight $[a_1, \ldots, a_{(n-1)/2}]$ such that all the a_i's are half-integers. It follows from the construction that all the modules $\mathbb{V}_i$ are of the form $T_i \otimes U$ where T_i is a representation of $\mathrm{Sl}(k)$ and the "spinor part" U does not change.

References

[1] J.H. Bernstein, I.M. Gelfand, S.I. Gelfand, *Differential Operators on the Base Affine Space and a Study of g-Modules*, Lie Groups and Their Representations, I.M. Gelfand, Ed., Adam Hilger, London, 1975.

[2] J.H. Bernstein, I.M. Gelfand, S.I. Gelfand, *Structure of Representations generated by vectors of highest weight*, Functional. Anal. Appl. 5 (1971).

[3] A. Damiano, *Algebraic analysis of the Rarita-Schwinger operator in dimension three*, Arch. Math. 42 (2007) suppl. 197–211.

[4] H.D. Fegan, *Conformally invariant first order differential operators*, Q J Math.1976; 27: 371–378.

[5] P. Franek, *Generalized Dolbeault sequences in Parabolic geometry*, to appear in Journal of Lie Theory.

[6] P. Franek, *Several Dirac operators in parabolic geometry*, Dissertation thesis.

[7] P. Franek, *Dirac Operator in Two Variables from the Viewpoint of Parabolic Geometry*, Advances in Applied Clifford Algebras, Birkhäuser Basel, vol. 17, 2007.

[8] R. Goodmann, N.R. Wallach, *Representations and Invariants of the Classical Groups*, Cambridge University Press, Cambridge 1998.

[9] J. Humphreys, *Introduction to Lie Algebras and Representation Theory*, Springer Verlag, 1980.

[10] F. Colombo, I. Sabadini, F. Sommen, D.C. Struppa, *Analysis of Dirac Systems and Computational Algebra*, Birkhäuser, 2004.

[11] J. Slovak, M. Eastwood, *Semi-holonomic Verma Modules*, Journal of Algebra, 197 (1997), 424–448.

[12] J. Slovak, V. Soucek, *Invariant operators of the first order on manifolds with a given parabolic structure*, Global analysis and harmonic analysis (Marseille-Luminy, 1999), 251–276, Sémin. Congr., 4, Soc. Math. France, Paris, 2000.

[13] N. Verma, *Structure of certain induced representations of complex semisimple Lie algebras*, Bull. Amer. Math. Soc. 74 (1968).

Peter Franek
Sokolovska 83
18675 Praha
Czech Republic
e-mail: `franp9am@artax.karlin.mff.cuni.cz`

Quaternionic and Clifford Analysis
Trends in Mathematics, 165–185

Recent Developments for Regular Functions of a Hypercomplex Variable

Graziano Gentili, Caterina Stoppato, Daniele C. Struppa and Fabio Vlacci

Abstract. In this paper we survey a series of recent developments in the theory of functions of a hypercomplex variable. The central idea underlying these developments consists in requiring a function to be holomorphic on suitable slices of the space on which the function itself is defined. Specifically, we apply this approach to functions defined on the space $\mathbb{H}$ of quaternions, on the space $\mathbb{O}$ of octonions, and finally on the Clifford algebra of type $(0, 3)$, denoted $Cl(0, 3)$. The properties of these functions resemble those of holomorphic functions, and yet the different nature of the three algebras on which we work introduces new and exciting phenomena.

Mathematics Subject Classification (2000). Primary: 30G35, 32W05.

Keywords. Functions of hypercomplex variables, $\overline{\partial}$-type operators.

1. Introduction

In the last couple of years, the authors developed a new theory of regularity for functions defined on the algebras of quaternions and octonions, as well as on an eight-dimensional Clifford algebra, [17, 18, 19, 20]. The new definition of regularity is quite simple: it only requires holomorphicity on complex slices of the algebra under consideration. Despite its simplicity, the theory ends up being quite powerful in replicating the fundamental results of the theory of holomorphic functions of a complex variable. More importantly, polynomial functions as well as convergent power series can be treated in this setting, unlike what happens in the usual theories of monogenic and Fueter-regular functions. This aspect of the theory is quite relevant because it can be applied to the study of a functional calculus in a non-commutative setting, [3, 4, 5, 6]. In this survey paper, we highlight in a unified

This work was partially supported by GNSAGA of the INdAM and by PRIN "Proprietà geometriche delle varietà reali e complesse" and "Geometria Differenziale e Analisi Globale" of the MIUR.

fashion the fundamental ideas of the theory, and we anticipate a few results which have not been published yet. We want to thank the anonymous referee, whose comments have significantly improved the final version of this paper.

2. Hamilton, Cayley and Clifford algebras

It is very rare, in mathematics, to have a clear and explicit description of the birth and the origin of a new theory: in most cases, in fact, it is even difficult to identify a date which could be taken as its birthday. Luckily, the situation is different for the birth of the theory of quaternions as well as for the theory of octonions[1].

Sir William R. Hamilton was looking for ways of extending complex numbers (which can be viewed as points on the 2-dimensional Gauss plane) to higher spatial dimensions. He could not do so for dimension 3, and in fact it was later shown by Frobenius that this task is actually impossible: the only associative division algebras which are finite dimensional over the real numbers are the real numbers $\mathbb{R}$, the complex numbers $\mathbb{C}$ and the non-commutative algebra $\mathbb{H}$ of quaternions or Hamilton numbers. Furthermore, the only non-associative division algebra which is finite dimensional over the real numbers is the algebra $\mathbb{O}$ of octonions or Cayley numbers (see, e.g., [1, 22]). Eventually Hamilton tried dimension 4 and created quaternions[2].

1) *Extract from a letter from Sir W.R. Hamilton to Professor P.G. Tait.*
Letter dated October 15, 1858.

... P.S. – To-morrow will be the 15*th birthday of the Quaternions. They started into life, or light, full grown, on [Monday] the* 16*th of October,* 1843, *as I was walking with Lady Hamilton to Dublin, and came up to Brougham Bridge, which my boys have since called the Quaternion Bridge. That is to say, I then and there felt the galvanic circuit of thought close; and the sparks which fell from it were the fundamental equations between i, j, k; exactly such as I have used them ever since. I pulled out on the spot a pocket-book, which still exists, and made an entry, on which, at the very moment, I felt that it might be worth my while to expend the labour of at least ten (or it might be fifteen) years to come. But then it is fair to say that this was because I felt a problem to have been at that moment solved – an intellectual want relieved – which had haunted me for at least fifteen years before.*

Less than an hour elapsed before I had asked and obtained leave of the Council of the Royal Irish Academy, of which Society I was, at that time, the President – to read at the next General Meeting a Paper on Quaternions; which I accordingly did, on November 13, 1843.

Some of those early communications of mine to the Academy may still have some interest for a person like you, who has since so well studied my volume, which was not published for ten years afterwards.

In the meantime, will you not do honour to the birthday to-morrow, in an extra cup of – ink? for it may be obsolete now to propose XXX, or even XYZ.

Note: Robert P. Graves notes in his biography of Hamilton that 'Brougham Bridge', referred to by Hamilton in his letters regarding the discovery of quaternions, is properly referred to as Broome Bridge: so called from the name of a family residing near.

Source: [23]

2) According to Hamilton, on October 16 he was out walking along the Royal Canal in Dublin with his wife when the solution in the form of the equation

$$i^2 = j^2 = k^2 = ijk = -1 \tag{1}$$

suddenly occurred to him; Hamilton then promptly carved this equation into the side of the nearby Broom Bridge (which Hamilton called Brougham Bridge). Since 1989, the National University of Ireland, Maynooth has organized a pilgrimage, where mathematicians take a walk from Dunsink observatory to the bridge where, unfortunately, no trace of the carving remains, though a stone plaque does commemorate the discovery.

Letter from Sir W.R. Hamilton to Rev. Archibald H. Hamilton.

Letter dated August 5, 1865.

My dear Archibald – (1) *I had been wishing for an occasion of corresponding a little with you on QUATERNIONS: and such now presents itself, by your mentioning in your note of yesterday, received this morning, that you "have been reflecting on several points connected with them"* (*the quaternions*), *"particularly on the Multiplication of Vectors."* (2) *No more important, or indeed fundamental question, in the whole Theory of Quaternions, can be proposed than that which thus inquires What is such MULTIPLICATION? What are its Rules, its Objects, its Results? What Analogies exist between it and other Operations, which have received the same general Name? And finally, what is* (*if any*) *its Utility?* (3) *If I may be allowed to speak of myself in connection with the subject, I might do so in a way which would bring you in, by referring to an ante-quaternionic time, when you were a mere child, but had caught from me the conception of a Vector, as represented by a Triplet: and indeed I happen to be able to put the finger of memory upon the year and month – October,* 1843 – *when having recently returned from visits to Cork and Parsonstown, connected with a meeting of the British Association, the desire to discover the laws of the multiplication referred to regained with me a certain strength and earnestness, which had for years been dormant, but was then on the point of being gratified, and was occasionally talked of with you. Every morning in the early part of the above-cited month, on my coming down to breakfast, your* (*then*) *little brother William Edwin, and yourself, used to ask me, "Well, Papa, can you multiply triplets"? Whereto I was always obliged to reply, with a sad shake of the head: "No, I can only add and subtract them."* (4) *But on the* 16*th day of the same month – which happened to be a Monday, and a Council day of the Royal Irish Academy – I was walking in to attend and preside, and your mother was walking with me, along the Royal Canal, to which she had perhaps driven; and although she talked with me now and then, yet an under-current of thought was going on in my mind, which gave at last a result, whereof it is not too much to say that I felt at once the importance. An electric circuit seemed to close; and a spark flashed forth, the herald* (*as I foresaw, immediately*) *of many long years to come of definitely directed thought and work, by myself if spared, and at all events on the part of others, if I should even be allowed to live long enough distinctly to communicate the discovery. Nor could I resist the impulse – unphilosophical as it may have been – to cut with a knife on a stone of Brougham Bridge, as we passed it, the fundamental formula with the symbols, i, j, k; namely, $i^2 = j^2 = k^2 = ijk = -1$ which contains the Solution of the Problem, but of course, as an inscription, has long since mouldered away. A more durable notice remains, however, on the Council Books of the Academy for that day* (*October* 16*th*, 1843), *which records the fact, that I then asked for and obtained leave to read a Paper on Quaternions, at the First General Meeting of the session: which reading took place accordingly, on Monday the* 13*th of the November following. With this quaternion of paragraphs I close this letter I.; but I hope to follow it up very shortly with another.*

Your affectionate father, William Rowan HAMILTON.

Source: [23]

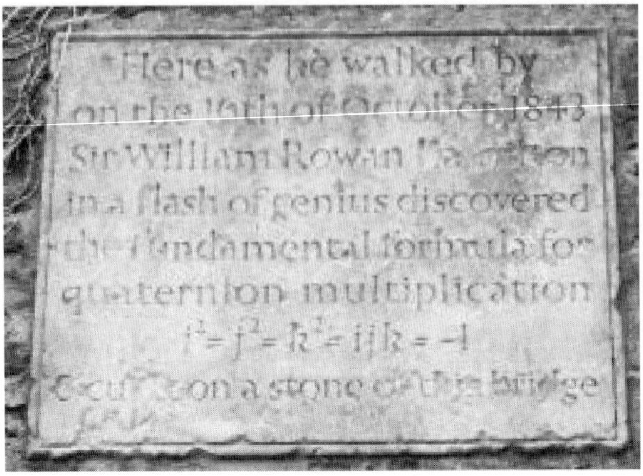

Commemorating stone plaque on Broom Bridge

The quaternions are of the form $q = x_0 + ix_1 + jx_2 + kx_3$ where the x_l are real ($l = 0, \ldots, 3$), and i, j, k, are imaginary units (i.e., their square equals -1) such that, according to (1), $ij = -ji = k$, $jk = -kj = i$, and $ki = -ik = j$. In this way, $\mathbb{H}$ can be considered as a vector space over the real numbers of dimension 4. Notice that a generic element q of $\mathbb{H}$ can be written as

$$q = (x_0 + x_1 i) + (x_2 + x_3 i)j, \tag{2}$$

a fact that allows an identification of $\mathbb{H}$ with pairs of complex numbers, each in $\mathbb{R} + \mathbb{R}i$; this split $\mathbb{H} = \mathbb{C} + \mathbb{C}j$ will play a crucial role in the next section.

We quoted explicitly from Hamilton's letters to P.G. Tait (dated October 15, 1858) and to his son A.H.Hamilton (dated August 5, 1865), but an even more important correspondence took place on October 17, 1843, when Hamilton described his discovery of quaternions to his good friend John T. Graves [24]. As Baez points out in [1], it was Graves' interest in algebra that stimulated some of Hamilton's own work in this direction. Soon after Hamilton discovered the quaternions, Graves developed his own theory of octonions (which he called *octaves* and described in a letter to Hamilton, dated December 26, 1843). Interestingly enough, octonions are now associated to the name of Cayley, because Graves did not publish his discovery, while Cayley did in March, 1845.

The algebra $\mathbb{O}$ of octonions can be described in a similar way to what we have done for quaternions. Consider a basis $\mathcal{E} = \{e_0 = 1, e_1, \ldots, e_6, e_7\}$ of $\mathbb{R}^8$ and relations

$$e_\alpha e_\beta = -\delta_{\alpha\beta} + \psi_{\alpha\beta\gamma} e_\gamma, \qquad \alpha, \beta, \gamma = 1, 2, \ldots, 7$$

where $\delta_{\alpha\beta}$ is the Kronecker delta and $\psi_{\alpha\beta\gamma}$ equals 1 when (α, β, γ) is one of the following combinations

$$(1,2,3), (1,4,5), (2,4,6), (3,4,7), (2,5,7), (1,6,7), (5,3,6),$$

it is totally antisymmetric in α, β, γ, and it equals 0 in the remaining cases. The generic element of $\mathbb{O}$ can be written as

$$w = x_0 + x_1 e_1 + \cdots + x_7 e_7.$$

but one can show that $(1, e_1, e_2, e_1 e_2)$ form a basis for a subalgebra of $\mathbb{O}$ isomorphic to the algebra $\mathbb{H}$ of quaternions. In fact we have the decomposition

$$\begin{aligned} \mathbb{O} &= (\mathbb{R} + \mathbb{R}e_1) + (\mathbb{R} + \mathbb{R}e_1)e_2 + [(\mathbb{R} + \mathbb{R}e_1) + (\mathbb{R} + \mathbb{R}e_1)e_2]e_4 \\ &= \mathbb{C} + \mathbb{C}e_2 + (\mathbb{C} + \mathbb{C}e_2)e_4 = \mathbb{H} + \mathbb{H}e_4. \end{aligned} \tag{3}$$

Given a generic element w of $\mathbb{H}$ or $\mathbb{O}$ we define in a natural fashion its conjugate $\overline{w} = x_0 - \sum_{k\geq 1} x_k e_k$, and its square norm $|w|^2 = w\overline{w} = \sum_{k\geq 0} x_k^2$.

The following is immediate and yet important:

Proposition 2.1. *For any non-real quaternion or octonion w, there exist, and are unique, $x, y \in \mathbb{R}$ with $y > 0$, and an imaginary unit I_w such that $w = x + yI_w$.*

Definition 2.2. *Given any imaginary unit I, the set $\mathbb{R} + \mathbb{R}I$ will be denoted by L_I.*

Notice that after identifying the imaginary unit I_w in $\mathbb{H}$ or in $\mathbb{O}$ with the imaginary unit i of $\mathbb{C}$, the set L_{I_w} may be considered as a complex plane in $\mathbb{H}$ or $\mathbb{O}$ passing through $0, 1$ and w. In this way both $\mathbb{H}$ and $\mathbb{O}$ can be obtained as an infinite union of complex planes (which will be also called *slices*). Now, let $Cl(0,3)$ denote the real Clifford algebra of signature $(0,3)$. This algebra can be defined as follows (see [8] for this and other related definitions): let $E = \{e_1, e_2, e_3\}$ be the canonical orthonormal basis for $\mathbb{R}^3$ with defining relations $e_i e_j + e_j e_i = -2\delta_{ij}$. An element of the Clifford algebra $Cl(0,3)$ can be written in a unique way as

$$w = x_0 + x_1 e_1 + x_2 e_2 + x_3 e_3 + x_{12} e_1 e_2 + x_{13} e_1 e_3 + x_{23} e_2 e_3 + x_{123} e_1 e_2 e_3$$

where the coefficients x_i, x_{ij}, x_{ijk} are real numbers. Thus, we see that $Cl(0,3)$ is an eight-dimensional real space, endowed with a natural multiplicative structure. Notice that the square of each unit e_i, $e_i e_j$ is -1, while the square of $e_1 e_2 e_3$ is 1. For this reason, the element $e_1 e_2 e_3$ is often referred to as a pseudoscalar. Therefore it is appropriate to define the set of Clifford real numbers as $\mathcal{R} = \{w = x_0 + x_{123} e_1 e_2 e_3\}$ and the set of Clifford imaginary numbers as $\mathcal{I} = \{w = x_1 e_1 + x_2 e_2 + x_3 e_3 + x_{12} e_1 e_2 + x_{13} e_1 e_3 + x_{23} e_2 e_3\}$. With these definitions, we can decompose any element w of $Cl(0,3)$ as the sum of an element in $\mathcal{R}$, its real part $\mathrm{Re}(w)$, and an element in $\mathcal{I}$, its imaginary part $\mathrm{Im}(w)$.

Let $\mathbb{A}$ denote any of the algebras $\mathbb{C}, \mathbb{H}, \mathbb{O}, Cl(0,3)$. For each of these algebras, we can define the set of roots of -1 as $\mathbb{S}_{\mathbb{A}} = \{w \in \mathbb{A} : w^2 = -1\}$. This set will be referred to as the *sphere of imaginary units of* $\mathbb{A}$. A second set of interest is what we call the *unit imaginary sphere*, namely $\mathbb{U}_{\mathbb{A}} = \{w \in \mathbb{A} : \mathrm{Re}(w) = 0, |\mathrm{Im}(w)| = 1\}$. The identity between $\mathbb{S}_{\mathbb{A}}$ and $\mathbb{U}_{\mathbb{A}}$ which holds for $\mathbb{A} = \mathbb{C}, \mathbb{H}, \mathbb{O}$ can be easily verified (see [17, 18, 19]) and implies that $\mathbb{S}_{\mathbb{C}}$, $\mathbb{S}_{\mathbb{H}}$ and $\mathbb{S}_{\mathbb{O}}$ are, respectively, 0-dimensional, 2-dimensional and 6-dimensional spheres.

On the other hand (see [20]), an element $w = x_0 + x_1e_1 + x_2e_2 + x_3e_3 + x_{12}e_1e_2 + x_{13}e_1e_3 + x_{23}e_2e_3 + x_{123}e_1e_2e_3$ in $Cl(0,3)$ belongs to $\mathbb{S}_{Cl(0,3)}$ if and only if it belongs to $\mathbb{U}_{Cl(0,3)}$ and its coordinates satisfy

$$x_1x_{23} - x_2x_{13} + x_3x_{12} = 0.$$

Moreover, it turns out that the set of $w \in Cl(0,3)$ such that $w^2 = 1$ reduces to $w = \pm 1$ and $w = \pm e_1e_2e_3$.

If one were to attempt to replicate proposition 2.1 in the Clifford algebra $Cl(0,3)$, one would need to be able to write every element w of $Cl(0,3)$ as the sum of a Clifford real number and the product of an element of $\mathbb{S}_{Cl(0,3)}$ by a Clifford real number, namely

$$w = (\alpha + \beta e_1e_2e_3) + I(\gamma + \delta e_1e_2e_3)$$

where $\alpha, \beta, \gamma, \delta$ are real numbers and $I \in \mathbb{S}_{Cl(0,3)}$. This is not always possible. Indeed, denote by $\mathcal{U}$ the set of all Clifford numbers $w = x_0 + x_1e_1 + x_2e_2 + x_3e_3 + x_{12}e_1e_2 + x_{13}e_1e_3 + x_{23}e_2e_3 + x_{123}e_1e_2e_3$ in $Cl(0,3)$ such that $w \in \mathcal{R}$ or $(x_1, x_2, x_3) \neq \pm(x_{23}, -x_{13}, x_{12})$. In [20] we proved

Proposition 2.3. *Let $w \in Cl(0,3)$. There exist $I \in \mathbb{S}_{Cl(0,3)}$ and $\alpha, \beta, \gamma, \delta \in \mathbb{R}$ such that $w = (\alpha + \beta e_1e_2e_3) + I(\gamma + \delta e_1e_2e_3)$ if and only if $w \in \mathcal{U}$. The above representation is unique up to substituting (I, γ, δ) by $-(I, \gamma, \delta)$ or $\pm(Ie_1e_2e_3, \delta, \gamma)$.*

3. Regular functions

Since the beginning of last century, mathematicians have been interested in creating a theory of quaternion-valued functions of a quaternionic variable, which would somehow resemble the classical theory of holomorphic functions of one complex variable. The simplest extension, namely the request for a quaternionic function to have a quaternionic derivative, fails to be of any interest since (see [41])

Proposition 3.1. *Let Ω be a simply-connected domain in $\mathbb{H}$ and let $f : \Omega \to \mathbb{H}$. If for any $q_0 \in \Omega$*

$$\lim_{q \to q_0} (q - q_0)^{-1}(f(q) - f(q_0))$$

exists in $\mathbb{H}$, then necessarily $f(q) = qa + b$ for some $a, b \in \mathbb{H}$. If for any $q_0 \in \Omega$

$$\lim_{q \to q_0} (f(q) - f(q_0))(q - q_0)^{-1}$$

exists in $\mathbb{H}$, then necessarily $f(q) = aq + b$ for some $a, b \in \mathbb{H}$.

Another quite natural approach could be to consider the class of functions which admit (local) series expansions of the form

$$\sum_s m_s(q - q_0),$$

with $m_s(q)$ finite sum of monomials of the type $a_0 \cdot q \cdot a_1 \dots a_{s-1} \cdot q \cdot a_s$.

But if $q = x_0 + ix_1 + jx_2 + kx_3$, then it is very easy to verify that $x_0 = \frac{1}{4}(q - iqi - jqj - kqk)$, $x_1 = \frac{1}{4i}(q - iqi + jqj + kqk)$, $x_2 = \frac{1}{4j}(q + iqi - jqj + kqk)$, $x_3 = \frac{1}{4k}(q + iqi + jqj - kqk)$, so that the class of maps considered coincides with the class of real analytic maps of $\mathbb{R}^4$ in $\mathbb{R}^4$.

In the 1930s, Fueter (see [12, 13]) introduced the differential operator

$$\frac{\partial}{\partial \bar{q}} = \frac{1}{4}\left(\frac{\partial}{\partial x_0} + i\frac{\partial}{\partial x_1} + j\frac{\partial}{\partial x_2} + k\frac{\partial}{\partial x_3}\right)$$

now known as the *Cauchy–Fueter operator* and defined the space of regular functions as the space of solutions of the equation associated to this operator. This theory of regular functions is very well developed, in many different directions, and we refer the reader to [41] for the basic features of these functions.

For more recent work in this area we refer the reader to [8, 27, 32, 33] and references therein. Furthermore, an analogue to the Cauchy–Fueter operator, called the *Dirac operator*, has been defined in Clifford algebras (see [8]); in this environment, any solution of the Dirac operator is known as a *monogenic function*. Fueter's Theorem [12] gives a method for constructing a regular function of a quaternion variable from an analytic function of one complex variable. Fueter's Theorem and its generalizations to higher dimensions and to the case of monogenic functions are deducible from the results contained in [36].

While the theory of Fueter-regular functions is extremely successful in replicating many important properties of holomorphic functions (and not only in one variable, see [8]), even the simplest quaternionic polynomials fail to be regular in the sense of Fueter. Another unexpected consequence in the definition of regular functions in the sense of Fueter is that their zero-sets can vary a lot and have real dimension zero, one, two or four (see [41]).

A first step towards a different definition was taken by Cullen in [11] on the basis of the notion of intrinsic functions as developed in [37]. This definition has the advantage that polynomials and even power series of the form $\sum_{n=0}^{+\infty} w^n a_n$ are regular in this sense.

Polynomials and power series of a quaternionic variable also belong to the interesting class of holomorphic functions over quaternions, which was defined by Fueter [12] and more recently generalized and developed by Laville and Ramadanoff (see [29], [30]), who studied the theory of holomorphic Cliffordian functions. It turns out that the set of Cullen regular functions and the set of Fueter regular functions, strictly contained in the set of holomorphic functions over quaternions, do not coincide.

Cullen regular functions are also closely related to a class of functions of the reduced quaternionic variable $x_0 + ix_1 + jx_2$, studied by Leutwiler in [31]. This class consists of all the solutions of a generalized Cauchy–Riemann system of equations, it contains the natural polynomials, and supports the series expansion of its elements as well.

Following the idea of Cullen [11] two of us were inspired to give the following definition (see [17, 18, 19]):

Definition 3.2. *Let $\mathbb{K}$ be either $\mathbb{H}$ or $\mathbb{O}$. If Ω is a domain in $\mathbb{K}$, a real differentiable function $f : \Omega \to \mathbb{K}$ is said to be* Cullen-regular *if, for every $I \in \mathbb{S}_\mathbb{K}$, its restriction f_I to the complex line $L_I = \mathbb{R} + \mathbb{R}I$ passing through the origin and containing* 1 *and I is holomorphic on $\Omega \cap L_I$.*

Throughout the paper, since no confusion can arise, we will refer to Cullen-regular functions as regular functions *tout court.* Note that in our most recent work we have used the expression *slice regular* instead.

In the spirit of Gateaux, a notion of I-derivative is defined as follows:

Definition 3.3. *Let Ω be a domain in $\mathbb{K}$ and let $f : \Omega \to \mathbb{K}$ be a real differentiable function. For any $I \in \mathbb{S}_\mathbb{K}$ and any point $q = x + yI$ in Ω (x and y are real numbers here) we define the I-derivative of f at q as*

$$\partial_I f(x + yI) := \frac{1}{2}\left(\frac{\partial}{\partial x} - I\frac{\partial}{\partial y}\right) f_I(x + yI)$$

The definition of regularity extends to the case of $Cl(0,3)$ as follows (see [20]). Recall that $\mathcal{U} = \{w = x_0 + x_1e_1 + x_2e_2 + x_3e_3 + x_{12}e_1e_2 + x_{13}e_1e_3 + x_{23}e_2e_3 + x_{123}e_1e_2e_3$ in $Cl(0,3)$ such that $w \in \mathcal{R}$ or $(x_1, x_2, x_3) \neq \pm(x_{23}, -x_{13}, x_{12})\}$.

Definition 3.4. *Let Ω be a domain in $\mathcal{U}$. A real differentiable function $f : \Omega \to Cl(0,3)$ is said to be regular if, for every $I \in \mathbb{S}_{Cl(0,3)}$, its restriction f_I to the four-dimensional plane $L_I = \mathcal{R} + I\mathcal{R} = \{(t_1 + t_2e_1e_2e_3) + I(t_3 + t_4e_1e_2e_3)\}$ passing through the origin and containing* 1 *and I satisfies, on $\Omega \cap L_I$, the system*

$$2D_If_I := (d_{12} + Id_{34})f_I = 0, \tag{4}$$

where $d_{ij} = d_{t_it_j}$ indicates the (real) differential with respect to the variables t_i and t_j.

Fix $I \in \mathbb{S}_{Cl(0,3)}$ and let $K = Ie_1e_2e_3$. Each $w \in L_I = L_K$ can be represented in the following two ways

$$\begin{aligned} w &= (t_1 + t_2e_1e_2e_3) + I(t_3 + t_4e_1e_2e_3) \;= (t_1 + It_3) + (t_2 + It_4)e_1e_2e_3 \\ w &= (t_1 + t_2e_1e_2e_3) + K(t_4 + t_3e_1e_2e_3) = (t_1 + Kt_4) + (t_2 + Kt_3)e_1e_2e_3. \end{aligned} \tag{5}$$

Chosen $J \in \mathbb{S}_{Cl(0,3)}$, we can represent $f_I = f_K$ (see [20]) either as

$$\begin{aligned} f_I &= (f_{00} + f_{01}e_1e_2e_3) + I(f_{10} + f_{11}e_1e_2e_3) \\ &\qquad + [(g_{00} + g_{01}e_1e_2e_3) + I(g_{10} + g_{11}e_1e_2e_3)]J \\ &= (f_{00} + If_{10}) + (f_{01} + If_{11})e_1e_2e_3 + [(g_{00} + Ig_{10}) + (g_{01} + Ig_{11})e_1e_2e_3]J \\ &= F_0 + F_1e_1e_2e_3 + (G_0 + G_1e_1e_2e_3)J \end{aligned} \tag{6}$$

or as

$$\begin{aligned} f_K &= (f_{00} + f_{01}e_1e_2e_3) + K(f_{11} + f_{10}e_1e_2e_3) \\ &\qquad + [(g_{00} + g_{01}e_1e_2e_3) + K(g_{11} + g_{10}e_1e_2e_3)]J \\ &= (f_{00} + Kf_{11}) + (f_{01} + Kf_{10})e_1e_2e_3 + [(g_{00} + Kg_{11}) + (g_{01} + Kg_{10})e_1e_2e_3]J \\ &= M_0 + M_1e_1e_2e_3 + (N_0 + N_1e_1e_2e_3)J. \end{aligned} \tag{7}$$

The previous equations imply that $2D_I f_I = (d_{12} + Id_{34})f_I = 0$ if, and only if,

$$\begin{array}{ll} \begin{cases} d_{12}f_{00} = d_{34}f_{10} \\ d_{12}f_{10} = -d_{34}f_{00} \end{cases} & \begin{cases} d_{12}f_{01} = d_{34}f_{11} \\ d_{12}f_{11} = -d_{34}f_{01} \end{cases} \\ \begin{cases} d_{12}g_{00} = d_{34}g_{10} \\ d_{12}g_{10} = -d_{34}g_{00} \end{cases} & \begin{cases} d_{12}g_{01} = d_{34}g_{11} \\ d_{12}g_{11} = -d_{34}g_{01} \end{cases} \end{array} \tag{8}$$

and $2D_K f_K = (d_{12} + Kd_{43})f_K = 0$ if, and only if,

$$\begin{array}{ll} \begin{cases} d_{12}f_{00} = d_{43}f_{11} \\ d_{12}f_{11} = -d_{43}f_{00} \end{cases} & \begin{cases} d_{12}f_{01} = d_{43}f_{10} \\ d_{12}f_{10} = -d_{43}f_{01} \end{cases} \\ \begin{cases} d_{12}g_{00} = d_{43}g_{11} \\ d_{12}g_{11} = -d_{43}g_{00} \end{cases} & \begin{cases} d_{12}g_{01} = d_{43}g_{10} \\ d_{12}g_{10} = -d_{43}g_{01}. \end{cases} \end{array} \tag{9}$$

The two-dimensional Cauchy–Riemann systems (8) and (9) are compatible (i.e., they share solutions) and it turns out that the set of regular functions is not empty (see [20]).

We conclude this section recalling the following representations of f_I, which derive from the geometric properties of imaginary units shown in the first section. We will often refer to these results as the *Splitting Lemmas* (see [17, 18, 19]).

Lemma 3.5. *If f is a regular function on $\Omega \subset \mathbb{K}$, then for every $I_1 \in \mathbb{S}_\mathbb{K}$*

1. *if $\mathbb{K} = \mathbb{H}$, for any $J \in \mathbb{S}_\mathbb{H}$, $J \perp I_1$, there exist two holomorphic functions $F, G : \Omega \cap L_{I_1} \to L_{I_1}$ such that*
$$f_{I_1}(z) = F(z) + G(z)J;$$
2. *if $\mathbb{K} = \mathbb{O}$ and $I_2, I_4 \in \mathbb{S}_\mathbb{O}$ are properly chosen, (see [18]), then there exist four holomorphic functions $F_1, F_2, G_1, G_2 : \Omega \cap L_{I_1} \to L_{I_1}$ such that*
$$f_{I_1}(z) = F_1(z) + F_2(z)I_2 + (G_1(z) + G_2(z)I_2)I_4.$$

In the case of $Cl(0,3)$ we have (see [20])

Lemma 3.6. *Let Ω be a domain in $\mathcal{U} \subseteq Cl(0,3)$ and let f be regular in Ω. Fix $I \in \mathbb{S}_{Cl(0,3)}$, set $K = Ie_1e_2e_3$. According to equations (5), each $w \in L_I = L_K$ can be represented as $w = z_1 + z_2e_1e_2e_3$ with $z_1, z_2 \in \mathbb{R} + I\mathbb{R} \simeq \mathbb{C}$ and as $w = \zeta_1 + \zeta_2e_1e_2e_3$ with $\zeta_1, \zeta_2 \in \mathbb{R} + K\mathbb{R} \simeq \mathbb{C}$.*

Chosen $J \in \mathbb{S}_{Cl(0,3)}$ there exist F_0, F_1, G_0, G_1 holomorphic in (z_1, z_2) and M_0, M_1, N_0, N_1 holomorphic in (ζ_1, ζ_2) (see [20]) such that

$$f_I = F_0 + F_1e_1e_2e_3 + (G_0 + G_1e_1e_2e_3)J = M_0 + M_1e_1e_2e_3 + (N_0 + N_1e_1e_2e_3)J = f_K$$

as in equations (6) and (7).

4. Power series and series expansions for regular functions

Given $a \in \mathbb{K}$, the basic monomial $w^n a$ defines a regular function in $\mathbb{K}$ according to Definition 3.2. Since the sum of regular functions is regular, we immediately have that polynomials with coefficients in $\mathbb{K}$ on the right side are regular. Moreover, since convergence of power series is uniform on compact sets, it turns out that power series $\sum_{n=0}^{+\infty} w^n a_n$ with coefficients in $\mathbb{K}$ are also regular in their domain of convergence, which proves to be a ball $B(0,R) = \{w \in \mathbb{K} : |w| < R\}$. The following result is proven in [19]:

Theorem 4.1. *A function $f : B = B(0,R) \to \mathbb{K}$ is regular if, and only if, it has a series expansion of the form*

$$f(q) = \sum_{n=0}^{+\infty} q^n \frac{1}{n!} \frac{\partial^n f}{\partial x^n}(0)$$

converging in B. In particular if f is regular then it is C^∞ in B.

Recall that (see [20]) for an arbitrary point w in $\mathcal{U}$, there exist $I \in \mathbb{S}_{Cl(0,3)}$ and $z_1, z_2 \in \mathbb{R} + I\mathbb{R}$ such that $w = z_1 + z_2 e_1 e_2 e_3$. With this in mind, we can state the analogue of Theorem 4.1 in $Cl(0,3)$.

Theorem 4.2. *Let f be a regular function in $\mathcal{U}$. For each choice of $I \in \mathbb{S}_{Cl(0,3)}$ and for all $z_1, z_2 \in \mathbb{R} + I\mathbb{R}$ we have*

$$f(z_1 + z_2 e_1 e_2 e_3) = \sum_{m,n \in \mathbb{N}} z_1^m z_2^n \frac{\partial^{m+n} f(0)}{\partial z_1^m \partial z_2^n}.$$

The power series expansion which we have obtained for regular functions is the key ingredient in proving the analogues of many well-known results of holomorphic functions in one complex variable. For this reason, in what follows we will always restrict our attention to functions which are regular in a ball $B(0,R)$ centered at the origin of $\mathbb{K}$ ($\mathbb{K} = \mathbb{H}$ or $\mathbb{K} = \mathbb{O}$).

Remark 4.3. *The forthcoming book [7] will study regular quaternionic functions on the larger class of* slice domains, *namely those domains $\Omega \subseteq \mathbb{H}$ such that $\Omega \cap \mathbb{R}$ is non-empty and, for any $I \in \mathbb{S}$, $L_I \cap \Omega$ is a domain in L_I.*

5. Cauchy formula and related results

We will now list some interesting results for regular functions. The proofs, which are contained in [18, 19], mainly involve the Splitting Lemmas and techniques which are imported from the theory of holomorphic functions of one complex variable. The notations are the same as in the previous sections. We begin with this version of the identity principle.

Theorem 5.1. *Let f and g be regular functions on $B = B(0, R)$. If there exists $I \in \mathbb{S}_{\mathbb{K}}$ such that $f \equiv g$ on a subset of $L_I \bigcap B$ having an accumulation point in $L_I \bigcap B$, then $f \equiv g$ in B.*

The following versions of the mean value property and the maximum modulus principle hold for regular functions.

Proposition 5.2. *If $f : B = B(0, R) \to \mathbb{K}$ is a regular function, $I \in \mathbb{S}_{\mathbb{K}}$ and $a \in B \cap L_I$, then, for any $r > 0$ small enough,*

$$f(a) = \frac{1}{2\pi} \int_0^{2\pi} f_I(a + re^{I\vartheta})\, d\vartheta.$$

Theorem 5.3. *Let $f : B = B(0, R) \to \mathbb{K}$ be a regular function. If $|f|$ has a maximum at $a \in B$, then f is constant in B.*

Perhaps the most important consequence of the Splitting Lemma 3.5 is the analogue, for regular functions, of the Cauchy representation formula. In order to state it appropriately, we will adopt the following notation. If $w \in \mathbb{K}$, we set

$$I_w = \begin{cases} \dfrac{\mathrm{Im}(w)}{|\mathrm{Im}(w)|} \in \mathbb{S}_{\mathbb{K}} & \text{if } \mathrm{Im}(w) \neq 0 \\ \text{any element of } \mathbb{S}_{\mathbb{K}} \text{ otherwise.} \end{cases}$$

Notice that for any $\zeta \in L_{I_w}$, $\zeta \neq w$ the equality $(\zeta - w)^{-1} d\zeta = d\zeta(\zeta - w)^{-1}$ holds. We can now state this integral representation formula for regular functions.

Theorem 5.4. *Let $f : B = B(0, R) \to \mathbb{K}$ be a regular function, and let $w \in B$. Then*

$$f(w) = \frac{1}{2\pi I_w} \int_{\partial\Delta_{I_w}(0,r)} \frac{d\zeta}{(\zeta - w)} f(\zeta)$$

where $\zeta \in L_{I_w} \bigcap B$, and where $r > 0$ is such that

$$\overline{\Delta_{I_w}(0, r)} = \{x + yI_w : x^2 + y^2 \leq r^2\}$$

is contained in B and $w \in \Delta_{I_w}(0, r)$.

As a consequence we obtain:

Theorem 5.5 (Cauchy estimates). *Let $f : B = B(0, R) \to \mathbb{K}$ be a regular function, let $r < R$, $I \in \mathbb{S}_{\mathbb{K}}$, and $\partial\Delta_I(0, r) = \{(x+yI) : x^2+y^2 = r^2\}$. If $M_I = \max\{|f(w)| : w \in \partial\Delta_I(0, r)\}$ and if $M = \inf\{M_I : I \in \mathbb{S}_{\mathbb{K}}\}$ then, for all $n \geq 0$,*

$$\frac{1}{n!} \left| \frac{\partial^n f}{\partial x^n}(0) \right| \leq \frac{M}{r^n}.$$

We now state the analogue of the Liouville theorem.

Theorem 5.6. *Let $f : \mathbb{K} \to \mathbb{K}$ be an entire regular map (i.e., a regular map defined and regular everywhere on $\mathbb{K}$). If f is bounded, i.e., there exists a positive number M such that $|f(w)| \leq M$ for all $w \in \mathbb{K}$, then f is constant.*

We close this section with this version of Morera's theorem.

Theorem 5.7. *Let $f : B = B(0,R) \to \mathbb{K}$ be a differentiable function. If, for every $I \in \mathbb{S}_{\mathbb{K}}$, the differential form $f(z)dz, z = x + yI, x, y \in \mathbb{R}$, defined on $L_I \bigcap B$ is closed, then the function f is regular.*

6. The Schwarz lemma of the unit ball and its boundary generalizations

Before focussing our attention on other peculiar properties of regular functions, we want to devote this section to some rigidity properties shared by regular and holomorphic functions (see – for the holomorphic case –, e.g., [2, 26, 34, 42]; for the regular case we refer to [21]).

We begin by stating an analogue of the classical Schwarz lemma for regular functions: its proof (see [18, 19]) is not a consequence of the Splitting Lemmas. The argument relies upon the fact that a regular function has a power series expansion.

Theorem 6.1. *Let $B(0,1) = \{w \in \mathbb{K} : |w| < 1\}$. If $f : B(0,1) \to B(0,1)$ is a regular function such that $f(0) = 0$. Then for every $w \in B$,*

$$|f(w)| \leq |w| \tag{10}$$

and

$$|f'(0)| \leq 1. \tag{11}$$

The statement of Theorem 6.1 with equality in (11) can be considered as a Cartan-type result (see [43]) for regular functions in the ball $B(0,1)$. Generalizing it to the case of a generic fixed point $w_0 \in B(0,1)$ is not easy. First of all, Moebius transformations (in the sense of [25]) are not regular in general: in [18, 19] we proved that

Proposition 6.2. *For any $w_0 \in B(0,1) \setminus \mathbb{R}$, the map η defined by*

$$\eta(w) = (w - w_0)(1 - \overline{w_0}w)^{-1}$$

is a diffeomorphism of $B(0,1)$ onto itself but it is not regular.

This problem is addressed, in the case of quaternions, in the forthcoming paper [16], where the author studies the new class of *regular Moebius transformations.* It is also possible to transform the unit ball biregularly into the the right half-space $\mathbb{K}^+ = \{w \in \mathbb{K} : \mathrm{Re}(w) > 0\}$: if we define the *Cayley transformation* as $\psi(w) = (1 - w)^{-1}(1 + w)$, then (see [18, 19])

Lemma 6.3. *The Cayley transformation $\psi : B(0,1) \to \mathbb{K}^+$ is a one-to-one regular function of w with (regular) inverse the function $\phi(w) = (w - 1)(w + 1)^{-1}$.*

Nevertheless, since the regularity of functions is not preserved under composition, the above-mentioned Cayley transformation and regular Moebius transformations do not help in generalizing Theorem 6.1.

The following is indeed proven by using a different technique (see [21]):

Theorem 6.4. *Assume that* $f : B(0,1) \to B(0,1)$ *is a regular function and there exists* $w_0 \in B(0,1)$ *such that* $f(w_0) = w_0$ *and that* $f'(w_0) = 1$. *Then* $f(w) = w$ *for every* $w \in B(0,1)$.

An interesting result, which can be regarded as transposition of a result due to Rudin (see [39]) is the following

Theorem 6.5. *Let* f *be a regular self-map of the unit ball* $B(0,1)$ *of* $\mathbb{K}$ *and suppose there exists* $r \in \mathbb{R}$ *such that* $f(r) = r$. *Then either* f *has no other fixed points in* $B(0,1)$ *or* f *is the Identity of* $B(0,1)$.

We now want to consider "boundary" generalizations of Schwarz–Cartan type Theorems in the setting of regular functions. For this purpose, we will start by recalling (see [2])

Theorem 6.6. *Let* Δ *be the open unit disc of* $\mathbb{C}$ *and let* $f : \Delta \to \Delta$ *be holomorphic and such that*

$$f(z) = 1 + (z - 1) + o(|z - 1|^3)$$

as $z \to 1$. *Then* $f \equiv Id_\Delta$.

This (sharp) result, which has been extensively used to obtain rigidity results for holomorphic self-maps in some pseudoconvex domains of $\mathbb{C}^n$ (see [2, 26]), is achieved by applying Herglotz representation formula and Hopf's Lemma. In [42] the authors use a more direct approach to prove a generalization of Theorem 6.6 which gives some insight for other rigidity results. For the regular case in [21] the following analogue of Theorem 6.6 is obtained

Theorem 6.7. *Let* f *be a regular self-map of* $B(0,1) \subset \mathbb{K}$. *Assume there exists* $w_0 \in \partial B(0,1)$ *such that*

$$f(w) = w_0 + (w - w_0) + o(|w - w_0|^3)$$

as $w \to w_0$. *Then* $f \equiv Id_{B(0,1)}$.

7. Zeroes of regular functions

In this section, we recall the main results which give a complete description of the zero-sets for regular quaternionic functions; quite surprisingly, these zero-sets have a structure which turns out to be significantly different from that of Fueter-regular functions or holomorphic functions of two complex variables.

Furthermore, the approach used for regular power series offers a shorter proof of the results stated for polynomials in [35]. From now on we will denote $\mathbb{S}_{\mathbb{H}}$ by $\mathbb{S}$. The crucial and key point in the description of the zero-set of a regular function is the following (see [17, 18, 19]).

Theorem 7.1. *Let* $\sum_{n=0}^{+\infty} q^n a_n$ *be a given quaternionic power series with radius of convergence* R. *Suppose that there exist* $x_0, y_0 \in \mathbb{R}$ *and* $I, J \in \mathbb{S}$ *with* $I \neq J$ *such that*

$$\sum_{n=0}^{+\infty} (x_0 + y_0 I)^n a_n = 0 \tag{12}$$

and

$$\sum_{n=0}^{+\infty} (x_0 + y_0 J)^n a_n = 0. \tag{13}$$

Then for all $L \in \mathbb{S}$ *we have*

$$\sum_{n=0}^{+\infty} (x_0 + y_0 L)^n a_n = 0.$$

The main result which describes the geometric properties of the zero-set is the following (see [15])

Theorem 7.2 (Structure of the zero-set). *Let* f *be a regular function on an open ball* $B(0, R)$ *centered in the origin of* $\mathbb{H}$. *If* f *is not identically zero then its zero-set* Z_f *consists of isolated points or isolated spheres of the form* $x + y\mathbb{S}$, *for* $x, y \in \mathbb{R}$, $y \neq 0$.

This result has a curious consequence concerning the zeroes of holomorphic functions. Since (the power series expansion of) any holomorphic function f can be uniquely extended to (the power series expansion of) a regular function $\widetilde{f}$ over quaternions, the question of distinguishing which zeroes of f will remain isolated after the extension, and which will become "spherical", naturally arises. It turns out that each pair of conjugate zeroes of f contributes a spherical zero of $\widetilde{f}$, while the other zeroes of f correspond to isolated zeroes of $\widetilde{f}$. The techniques employed to prove Theorem 7.2 suggest the use of the following multiplication between regular power series.

Definition 7.3. *Let*

$$f(q) = \sum_{n=0}^{+\infty} q^n a_n \quad \textit{and} \quad g(q) = \sum_{n=0}^{+\infty} q^n b_n$$

be given quaternionic power series with radii of convergence greater than R. *We define the* regular product *of* f *and* g *as the series*

$$f * g(q) = \sum_{n=0}^{+\infty} q^n c_n,$$

where $c_n = \sum_{k=0}^{n} a_k b_{n-k}$ *for all* n.

We point out that the sequence of the coefficients of the regular product $f*g$ is the discrete convolution of the sequences of the coefficients of f and g. In the polynomial case, the regular multiplication coincides with the classical multiplication of the polynomial ring over the quaternions, $\mathbb{H}[X]$. In the sequel, $h^{*n} = h * \cdots * h$ will denote the nth power of a regular function h with respect to $*$-multiplication.

We have the following result.

Theorem 7.4. *Let $f : B(0,R) \to \mathbb{H}$ be a regular function and let p belong to $B(0,R)$. Then $f(p) = 0$ if and only if there exists a regular function $g : B(0,R) \to \mathbb{H}$ such that $f(q) = (q-p) * g(q)$.*

Furthermore we can describe the zero-set of a regular product in terms of the zero-sets of the factors:

Theorem 7.5 (Zeros of a regular product). *Let $f, g : B(0,R) \to \mathbb{H}$ be regular and $p \in B(0,R)$. Then $f * g(p) = 0$ if and only if $f(p) = 0$ or $f(p) \neq 0$ and $g(f(p)^{-1}pf(p)) = 0$.*

This theorem extends to quaternionic power series the theory presented in [28] for polynomials.

8. Poles of regular functions

Given the peculiar properties of the zeros which are summarized in the previous section, a new question arises. Do regular functions have singularities resembling the poles of holomorphic complex functions? This question receives a positive answer in [40]:

Proposition 8.1. *Consider a quaternionic Laurent series $f(q) = \sum_{n\in\mathbb{Z}} q^n a_n$ with quaternionic coefficients $a_n \in \mathbb{H}$. There exists a spherical shell*

$$A = A(0, R_1, R_2) = \{q \in \mathbb{H} : R_1 < |q| < R_2\}$$

such that:

(i) *the series*

$$f^+(q) = \sum_{n=0}^{+\infty} q^n a_n \quad \text{and} \quad f^-(q) = \sum_{n=1}^{+\infty} q^{-n} a_{-n}$$

both converge absolutely and uniformly on the compact subsets of A;

(ii) *$f^+(q)$ diverges for $|q| > R_2$;*

(iii) *$f^-(q)$ diverges for $|q| < R_1$.*

If A is not empty (i.e., $0 \leq R_1 < R_2$) then the function $f : A \to \mathbb{H}$ defined by

$$f(q) = \sum_{n\in\mathbb{Z}} q^n a_n = f^+(q) + f^-(q)$$

is regular.

This allows the construction of functions which are regular on a punctured ball $B(0,R)\setminus\{0\}$ and have a singularity at 0. Moreover, any function which is regular on a spherical shell $A(0,R_1,R_2)$ admits a Laurent series expansion centered at 0. The latter is a special case of the following.

Theorem 8.2. *Let f be a regular function on a domain Ω, let $p\in\mathbb{H}$ and let L_I be a complex line through p. If Ω contains an annulus $A_I = A(p,R_1,R_2)\cap L_I$ then there exists $\{a_n\}_{n\in\mathbb{Z}}\subseteq\mathbb{H}$ such that*

$$f_I(z)=\sum_{n\in\mathbb{Z}}(z-p)^n a_n \quad \textit{for all } z\in A_I.$$

If, moreover, $p\in\mathbb{R}$ then

$$f(q)=\sum_{n\in\mathbb{Z}}(q-p)^n a_n \quad \textit{for all } q\in A(p,R_1,R_2)\cap\Omega.$$

Definition 8.3. *Let f, p and $\{a_n\}_{n\in\mathbb{Z}}$ be as in Theorem 8.2. The point p is called a* pole *if there exists an $n\in\mathbb{N}$ such that $a_{-m}=0$ for all $m>n$; the minimum of such $n\in\mathbb{N}$ is called the* order *of the pole and denoted as $ord_f(p)$. If p is not a pole for f then we call it an* essential singularity *for f.*

Notice that, by the final statement of Theorem 8.2, real singularities are completely analogous to singularities of holomorphic functions of one complex variable and there is no resemblance to the case of several complex variables. As for non-real singularities, Theorem 8.2 only provides information on the complex line L_I through the point p; we apparently cannot predict the behavior of the function in a (four-dimensional) neighborhood of p. In order to overcome this difficulty, we need to introduce a couple of new definitions.

Definition 8.4. *For a regular function $f: B(0,R)\to\mathbb{H}$ having power series expansion $\sum_{n=0}^{+\infty} q^n a_n$ we define the* regular conjugate *f^c and the* symmetrization *f^s of f as*

$$f^c(q)=\sum_{n=0}^{+\infty} q^n\bar{a}_n, \qquad f^s(q)=f*f^c(q)=f^c*f(q)=\sum_{n=0}^{+\infty} q^n r_n$$

with

$$r_n=\sum_{k=0}^{n} a_k\bar{a}_{n-k}\in\mathbb{R}.$$

Definition 8.5. *Let $f,g: B=B(0,R)\to\mathbb{H}$ be regular functions and suppose $f\not\equiv 0$. The* (left) regular quotient *of f and g is the function $f^{-*}*g$ defined on $B\setminus Z_{f^s}$ by $f^{-*}*g(q)=\frac{1}{f^s(q)}f^c*g(q)$. Moreover, the* regular reciprocal *of f is the function $f^{-*}=f^{-*}*1$.*

Regular quotients are regular on their domains of definition, and the algebraic meaning of this construction is explained by the following result.

Proposition 8.6. *Fix R and consider the associative real algebra $(\mathcal{D}_R, +, *)$ of regular functions on $B(0, R)$. If we endow the set of left regular quotients $\mathcal{L}_R = \{f^{-*}*g : f, g \in \mathcal{D}_R, f \not\equiv 0\}$ with the multiplication $*$ defined by $(f^{-*}*g)*(h^{-*}*k) = \frac{1}{f^s h^s} f^c * g * h^c * k$ then $(\mathcal{L}_R, +, *)$ is a division algebra over $\mathbb{R}$ and it is the classical ring of quotients of $(\mathcal{D}_R, +, *)$.*

For the definition of the classical ring of quotients, see [38]. Regular quotients allow a detailed study of the poles. Define a function f *semiregular* if it does not have essential singularities or, equivalently, if the restriction f_I is meromorphic for all $I \in \mathbb{S}$. As proven in [40], f is semiregular on $B(0, R_0)$ if and only if $f_{|_{B(0,R)}}$ is a left regular quotient for all $R < R_0$. This allows the definition of a multiplication operation $*$ on the set of semiregular functions on a ball and the proof of the following result.

Theorem 8.7. *Let f be a semiregular function on $B = B(0, R)$, choose $p \in B$ and let $m = ord_f(p), n = ord_f(\bar{p})$. There exists a unique semiregular function g on B such that*

$$f(q) = [(q-p)^{*m} * (q-\bar{p})^{*n}]^{-*} * g(q) \tag{14}$$

The function g is regular near p and $\bar{p}$ and $g(p) \neq 0, g(\bar{p}) \neq 0$, provided $m > 0$ or $n > 0$.

The previous result allows the study of the structure of the poles.

Theorem 8.8 (Structure of the poles). *If f is a semiregular function on $B = B(0, R)$ then f extends to a regular function on B minus a union of isolated real points or isolated 2-spheres of the type $x + y\mathbb{S} = \{x + yI : I \in \mathbb{S}\}$ with $x, y \in \mathbb{R}, y \neq 0$. All the poles on each 2-sphere $x + y\mathbb{S}$ have the same order with the possible exception of one, which must have lesser order.*

9. The open mapping theorem

One of the most celebrated properties of holomorphic functions of one complex variable is the fact that they are open. In this section we will present a version of the open mapping theorem for regular quaternionic functions, which states that each such function is open when restricted to an appropriate open subset of its domain of definition.

The regular quotients introduced in the previous section allow the proof of the minimum modulus principle for regular quaternionic functions, which leads to the open mapping theorem. The first step in this direction is taken by proving the following relation between the regular quotient $f^{-*} * g(q)$ and the quotient $f(q)^{-1}g(q) = \frac{1}{f(q)}g(g)$ (see [14, 40]).

Theorem 9.1. *Let f, g be regular functions on $B = B(0, R)$. Setting $T_f(q) = f^c(q)^{-1}qf^c(q)$ defines a diffeomorphism of $B \setminus Z_{f^s}$ onto itself, with inverse function*

$T_f^{-1} = T_{f^c}$. *Moreover,*

$$f^{-*} * g(q) = \frac{1}{f(T_f(q))} g(T_f(q)) \tag{15}$$

for all $q \in B \setminus Z_{f^s}$.

Let $f : B = B(0,R) \to \mathbb{H}$ be a regular function. Applying the maximum modulus principle to the regular reciprocal $f^{-*}(q) = \frac{1}{f(T_f(q))}$ proves that, if $|f|$ has a local minimum point $p = x + yI \in B$, then either f is constant or f has a zero in $x + y\mathbb{S}$. This is not enough for our proof of the open mapping theorem, which requires the following (stronger) property.

Theorem 9.2 (Minimum modulus principle). *Let* $f : B = B(0,R) \to \mathbb{H}$ *be a regular function. If* $|f|$ *has a local minimum point* $p \in B$ *then either* $f(p) = 0$ *or* f *is constant.*

The proof is given in [14] and depends on the following peculiar property of regular functions.

Theorem 9.3. *Let* $f : B = B(0,R) \to \mathbb{H}$ *be a regular function. For all* $x, y \in \mathbb{R}$ *such that* $x + y\mathbb{S} \subseteq B$ *there exist* $b, c \in \mathbb{H}$ *such that* $f(x+yI) = b + Ic$ *for all* $I \in \mathbb{S}$.

In other words, when restricted to a 2-sphere $x + y\mathbb{S}$, a regular function is either constant or an affine map of $x + y\mathbb{S}$ onto a 2-sphere $b + \mathbb{S}c$ with $b, c \in \mathbb{H}$.

Example 9.4. *Consider the polynomial function* $f(q) = q^2$. *For* $x, y \in \mathbb{R}$ *we have* $f(x + yI) = x^2 - y^2 + I2xy$ *for all* $I \in \mathbb{S}$. *Thus* f *maps each 2-sphere* $x + y\mathbb{S}$ *with* $y \neq 0$ *onto* $(x^2 - y^2) + \mathbb{S}(2xy)$. *In particular, if* $x = 0$ *then* f *is constant on* $x + y\mathbb{S} = y\mathbb{S}$.

Proposition 9.5. *If* $f : B = B(0,R) \to \mathbb{H}$ *is a regular function, denote by* D_f *the union of all the 2-spheres* $x + y\mathbb{S}$ *(for* $x, y \in \mathbb{R}, y \neq 0$*) on which* f *is constant. Then* D_f*, which we call the* degenerate set *of* f*, is closed in* $B \setminus \mathbb{R}$. *Moreover, if* f *is not constant then the interior of* D_f *is empty.*

We can now state the main result of [14].

Theorem 9.6 (Open mapping theorem). *Let* $f : B(0,R) \to \mathbb{H}$ *be a non-constant regular function and let* D_f *be its degenerate set. Then* $f : B(0,R) \setminus \overline{D_f} \to \mathbb{H}$ *is open.*

As already mentioned, the proof is based on the minimum modulus principle. Theorem 9.6 is a sharp result: the function $f(q) = q^2$ in example 9.4 proves to be open on $\mathbb{H} \setminus \overline{D_f} = \mathbb{H} \setminus \{q \in \mathbb{H} : \mathrm{Re}(q) = 0\}$ but not on $\mathbb{H}$. Nevertheless, Theorem 9.6 extends as follows. We say that $U \subseteq \mathbb{H}$ is *circular* if, for all $x + yI \in U$ with $y \neq 0$, the whole 2-sphere $x + y\mathbb{S}$ is contained in U.

Theorem 9.7. *Let* $f : B(0,R) \to \mathbb{H}$ *be a regular function. If* U *is a circular open subset of* $B(0,R)$*, then* $f(U)$ *is open. In particular* $f(B(0,R))$ *is an open set.*

10. Conclusions

Every time mathematicians define a new concept, two questions arise naturally. The first is whether the new concept can lead to an interesting theory by itself. We believe that the previous sections have demonstrated that the notion of slice-regularity is indeed a very stable notion, that suitably generalizes the fundamental properties of holomorphicity to the hypercomplex setting. At the same time, we identified some crucial differences between the complex and the hypercomplex case. The second question is whether the new theory, in addition to its intrinsic value, can also contribute to the solution of some outstanding problem. Two of us, with different coauthors, prove that this is the case in [6]. This paper offers a survey of how the ideas discussed in this paper have been applied to develop two new functional calculi in non-commutative settings. Those calculi are based on the notion of slice regularity and a similar notion of slice monogenicity. In addition to [6], the reader interested in the details is referred to [3, 4, 5, 7, 9], and [10].

References

[1] J. Baez, The octonions. *Bull. Amer. Math. Soc.* **39** (2002), 145–205.

[2] D.M. Burns, S.G. Krantz, Rigidity of holomorphic mappings and a new Schwarz lemma at the boundary. *J. Amer. Math. Soc.* **7** (1994), 661–676.

[3] F. Colombo, G. Gentili, I. Sabadini, D.C. Struppa, A functional calculus in a non commutative setting. *Electron. Res. Announc. Math. Sci.*, **14** (2007), 60-68.

[4] F. Colombo, G. Gentili, I. Sabadini, D.C. Struppa, Non commutative functional calculus: bounded operators, E-print. arXiv:0708.3591v1 [math.SP]

[5] F. Colombo, G. Gentili, I. Sabadini, D.C. Struppa, Non commutative functional calculus: unbounded operators, E-print. arXiv:0708.3592v2 [math.SP]

[6] F. Colombo, G. Gentili, I. Sabadini, D.C. Struppa, An overview of functional calculus in different settings, this volume.

[7] F. Colombo, G. Gentili, I. Sabadini, D.C. Struppa, Functional Calculus in a Non-commutative Setting, in preparation.

[8] F. Colombo. I. Sabadini, F. Sommen, D.C. Struppa, *Analysis of Dirac systems and computational algebra.* Birkhäuser, Boston, 2004.

[9] F. Colombo, I. Sabadini, D.C. Struppa, A new functional calculus for non commuting operators. *J. Funct. Anal.*, **254** (2008), 2255–2274.

[10] F. Colombo, I. Sabadini, D.C. Struppa, Slice monogenic functions. To appear in *Isr. J. Math.*

[11] C.G. Cullen, An integral theorem for analytic intrinsic functions on quaternions. *Duke Math. J.* **32** (1965), 139–148.

[12] R. Fueter, Die Funktionentheorie der Differentialgleichungen $\triangle u = 0$ und $\triangle\triangle u = 0$ mit vier reellen Variablen. *Comm. Math. Helv.* **7** (1934), 307–330.

[13] R. Fueter, Über einen Hartogs'schen Satz. *Comment. Math. Helv.* **12** (1939/40), 75–80.

[14] G. Gentili, C. Stoppato, The open mapping theorem for quaternionic regular functions, E-print. arXiv:0802.3861v1 [math.CV]

[15] G. Gentili, C. Stoppato, Zeros of regular functions and polynomials of a quaternionic variable. To appear in *Michigan Math. J.*

[16] C. Stoppato, Regular Moebius transformations of the space of quaternions, in preparation.

[17] G. Gentili, D.C. Struppa, A new approach to Cullen-regular functions of a quaternionic variable. *C. R. Acad. Sci. Paris*, Ser. I **342** (2006), 741–744.

[18] G. Gentili, D.C. Struppa, Regular functions on the space of Cayley numbers, Preprint. Dipartimento di Matematica "U. Dini", Università di Firenze, n. 13, (2006).

[19] G. Gentili, D.C. Struppa, A new theory of regular functions of a quaternionic variable. *Adv. Math.* **216** (2007), 279–301.

[20] G. Gentili, D.C. Struppa, Regular functions on a Clifford Algebra. *Complex Var. Elliptic Equ.* **53** (2008), no. 5, 475–483.

[21] G. Gentili, F. Vlacci, Rigidity for regular functions over Hamilton and Cayley numbers and a boundary Schwarz' Lemma. To appear in *Indag. Mathem.*

[22] R. Godement, *Cours d'algèbre.* Herman, Paris, 1978.

[23] R.P. Graves, *Life of Sir William Rowan Hamilton*, Arno Press, 1975. Volume II, Chapter XXVIII.

[24] W.R. Hamilton, Copy of a letter from Sir William R. Hamilton to John T. Graves, Esq. *Philosophical Magazine*, 3rd series, **25** 1844, 489–495.

[25] R. Heidrich, G. Jank, On the Iteration of Quaternionic Moebius Transformations. *Complex Var. Theory Appl.* **29** (1996), no. 4, 313–318.

[26] X. Huang, A boundary rigidity problem for holomorphic mappings on some weakly pseudoconvex domains. *Can. J. Math.* **47** (2) (1995), 405–420.

[27] V.V. Kravchenko, M. Shapiro, *Integral Representations for Spatial Models of Mathematical Physics.* Pitman Res. Notes in Math., 351. Longman, Harlow, 1996.

[28] T.Y. Lam, *A first course in noncommutative rings.* Graduate Texts in Mathematics, 123. Springer-Verlag, New York, 1991. 261–263.

[29] G. Laville, I. Ramadanoff, Holomorphic Cliffordian functions. *Adv. Appl. Clifford Algebras* **8** (1998), 323–340.

[30] G. Laville, I. Ramadanoff, Elliptic Cliffordian functions. *Complex Var. Theory Appl.* **45** (2001), no. 4, 297–318.

[31] H. Leutwiler, Modified quaternionic analysis in $\mathbb{R}^3$. *Complex Var. Theory Appl.* **20** (1992), no. 3-4, 19–51.

[32] M.E. Luna-Elizarrarás, M.A. Macías-Cedeño, M. V. Shapiro, On some relations between the derivative and the two-dimensional directional derivatives of a quaternionic function. AIP Conference Proceedings, **936**, 2007, 758–760.

[33] M.E. Luna-Elizarrarás, M.A. Macías-Cedeño, M. V. Shapiro, On the derivatives of quaternionic functions along two-dimensional planes. To appear in *Adv. Appl. Clifford Algebr.*

[34] S. Migliorini, F. Vlacci, A new rigidity result for holomorphic maps. *Indag. Mathem. (N.S.)* **13** (2002), no. 4, 537–549.

[35] A. Pogorui, M.V. Shapiro, On the structure of the set of zeros of quaternionic polynomials. *Complex Var. Theory Appl.* **49** (2004), no. 6, 379–389.

[36] T. Qian, F. Sommen, Deriving harmonic functions in higher dimensional spaces, *Z. Anal. Anwendungen* **22** (2003), no. 2, 275–288.

[37] R.F. Rinehart, Elements of a theory of intrinsic functions on algebras. *Duke Math. J.* **27** (1960), 1–19.

[38] L.H. Rowen, *Ring theory. Student edition.* Academic press, San Diego, 1991. 272–279.

[39] W. Rudin, *Function theory in the unit ball of* $\mathbb{C}^n$. Springer, Berlin, 1980.

[40] C. Stoppato, Poles of regular quaternionic functions. To appear in Complex Var. Elliptic Equ.

[41] A. Sudbery, Quaternionic analysis. *Math. Proc. Camb. Phil. Soc.* **85** (1979), 199–225.

[42] R. Tauraso, F. Vlacci, Rigidity at the boundary for self-maps of the unit disc. *Complex Var. Theory Appl.* **45** (2001), no. 2, 151–165.

[43] E. Vesentini, *Capitoli scelti della teoria delle funzioni olomorfe.* Unione Matematica Italiana, Bologna, 1984.

Graziano Gentili, Caterina Stoppato and Fabio Vlacci
Dipartimento di Matematica "U. Dini" – Università di Firenze
Viale Morgagni, 67/A
I-50134 Firenze, Italy
e-mail: gentili@math.unifi.it
stoppato@math.unifi.it
vlacci@math.unifi.it

Daniele C. Struppa
Department of Mathematics and Computer Science
Schmid College of Science – Chapman University
One University Drive
Orange, CA 92866 USA
e-mail: struppa@chapman.edu

Quaternionic and Clifford Analysis
Trends in Mathematics, 187–205

Differential Equations in Algebras

Yakov Krasnov

Abstract. The aim of this work is to investigate how topological and dynamical properties of differential equations (in the sequel DE) are reflected in the associated algebras, as well as to show how basic algebraic concepts provide valuable insights in DE.

Mathematics Subject Classification (2000). Primary 17A, 34G20, 35E20, 47H60.

Keywords. Non-associative algebra, Riccati equation, Dirac equation, idempotents, Peirce number, symmetry operator.

1. Preliminaries

Historically, mathematics has been divided into three distinct areas: Algebra, Geometry and Analysis. It is our belief that nowadays, the theory of differential equations performs as a common root unifying these (at first glance) different mathematical branches.

To clarify the point, recall

Definition 1.1. A differential equation in algebra is any equation written using algebraic operation and containing derivatives, either ordinary or partial.

From now on, $\mathbb{A} = (\mathbb{R}^n, \circ)$ stands for a real n-dimensional algebra. Consider two typical examples of DEs in $\mathbb{A}$:

Example 1. The Initial Value Problem (or IVP) for the Riccati equation in $\mathbb{A}$:

$$\frac{dx(t)}{dt} = x(t) \circ x(t), \quad x(0) = a. \tag{1.1}$$

Example 2. Take the standard basis $(e_1, \dots, e_n)$ in $\mathbb{A}$ and denote by $x_1, \dots, x_n$ the corresponding coordinates. Consider the Dirac equation $D \circ f(x) = 0$ in $\mathbb{A}$, where $f(x) = u_1(x)e_1 + \cdots + u_n(x)e_n$ and

$$D := e_1 \frac{\partial}{\partial x_1} + \cdots + e_n \frac{\partial}{\partial x_n}. \tag{1.2}$$

In the differential equations listed above the usage of the $\mathbb{A}$-multiplication in their definitions is the common point. At first glance, both examples seem to be two very special particular examples of DEs. In fact, as we will see later on, many DEs can be associated with certain algebraic structures. Namely (see Section 2), all polynomial ODEs can be imbedded into the Riccati equation, while (see Section 7) any linear PDE is the Dirac equation in an appropriate algebra.

The following fact shows a connection between the type of PDE and some properties of the associated algebra.

Proposition 1.2. *(see Subsection 7.5) The (well-defined) Dirac equation in $\mathbb{A}$ is elliptic iff $\mathbb{A}$ is divisible.*

Let us return to ODEs. Recall that, two vector fields in $\mathbb{R}^n$ are said to be locally *equivalent* at singular points x and y if there is a one-to-one mapping $H : \mathbb{R}^n \to \mathbb{R}^n$ taking x to y and conjugating the local phase flows of the equations at x and y.

Under the above notations, take two locally equivalent vector fields and call the associated local phase flows to be *linearly equivalent* (resp. *topologically equivalent*, *differentially equivalent*) if H is a *linear automorphism* (resp. *homeomorphism*, *diffeomorphism*).

A difference between the above equivalences rests on an answer to the following

Question 1.3. *Which equivalences of algebras $\mathbb{A}$ and $\mathbb{B}$ do provide the corresponding local phase flows to be:*

1. *linearly equivalent,*
2. *differentially equivalent,*
3. *topologically equivalent?*

In this paper, most of our efforts will be concentrated on answering the third question for a sufficiently large class of ODEs. In Section 5, we will give a partial answer to Question 1.3 as well as on the following one:

Question 1.4. *Under which conditions on the algebra $\mathbb{A}$ does the Riccati equation admit:*

(i) *a planar solution;*
(ii) *a planar bounded solution;*
(iii) *a planar first integral being polynomial (cf., for instance, [35])?*

Meanwhile, we resume the answers to Question 1.4 in purely algebraic terms as follows:

(i)′ iff there exists a two-dimensional subalgebra $\mathbb{B} \subset \mathbb{A}$;
(ii)′ iff there exist *complex structures* in $\mathbb{B}$ (cf. [5]);
(iii)′ iff **(i)** is satisfied and, in addition, $\mathbb{B}$ is diagonalizable and all Peirce numbers are negative and rational (cf. Theorem 6.2 below).

2. Polynomial ODEs

The following result (cf. [17]) shows that the applications of the algebraic approach developed in this paper are not exhausted by the Riccati equation – in principal, using this approach one can treat *any* polynomial system.

Proposition 2.1. *Any n-dimensional polynomial differential system $\dot{x} = P(x)$ with $\deg P = m$ can be embedded into a Riccati equation considered in an algebra A of dimension $\geq n$.*

Proof. First, we homogenize the system by adding the variable $x_{n+1} = 1$, and therefore $\dot{x}_{n+1} = 0$.

Next, if $m > 2$, define new variables y_i by

$$y_i = x_1^{i_1} x_2^{i_2} \cdots x_{n+1}^{i_{n+1}}, \quad i_1 + \cdots + i_{n+1} = m - 1.$$

with $i_1, \ldots, i_{n+1}$ positive integers. By direct verification, $\dot{y}_i$ is a quasi-homogeneous polynomial (of degree $2(m-1)$ in x_k-variables and quadratic in y_i-variable). The obtained Riccati equation can be essentially simplified by cancelling certain "fictive" y_i-variables (thus, passing to an algebra of smaller dimension). □

Example 3. The system $\dot{x} = x^2 y$, $\dot{y} = -x^3$ after the change of variables $z = xy + ix^2$ leads to the Riccati equation $\dot{z} = z^2$, $z \in \mathbb{C}$, in algebra of complex numbers.

3. Solutions to Riccati equation

Assume A is a power associative unital *division algebra* with unit e (see, for instance, [17, 35, 4, 5] and references therein). Then, a solution of the Riccati equation (1.1) may be written explicitly:

$$x(t) = (a^{-1} - te)^{-1}. \tag{3.1}$$

Although (3.1) requires the inversion, one can avoid it by using instead of (3.1) the power series solution representation (see, for instance, [17, 35, 4, 5]):

$$x(t) = a + a^2 t + \cdots + a^k t^{k-1} + \cdots . \tag{3.2}$$

Formula (3.2) can be extended (at least formally) to the case of an arbitrary (binary) algebra $\mathbb{A}$ (without additional assumption on its power associativity), using the recurrently defined *symmetric powers*

$$a^{[1]} = a, \quad a^{[k+1]} = \frac{1}{k} \sum_{i=0}^{k-1} a^{[i+1]} \cdot a^{[k-i]}, \tag{3.3}$$

as follows:

$$x(t) = a + a^2 t + a^{[3]} t^2 + \cdots + a^{[k]} t^{k-1} + \cdots \tag{3.4}$$

Since any Riccati equation is analytic, the convergence of series (3.4) is provided by the standard existence theorem from [14].

Remark 3.1. General formula (3.4) can be essentially simplified if certain specific properties of the underlying algebras are taken into account. It would be useful to study the properties of algebras providing reasonable simplifications of formula (3.4).

4. Multi-linear algebra behind idempotents

Let $\mathbb{A}$ be an n-dimensional real (in general, non-associative) commutative algebra, $u \in \mathbb{A}$ an idempotent and $L_u : \mathbb{A} \to \mathbb{A}$ a (linear) operator defined by $x \to u \cdot x$. Clearly, $\mu_0 = 1$ is always an eigenvalue of L_u.

Definition 4.1. The remaining (possibly, complex) eigenvalues of L_u (denoted by $\mu_1, \dots, \mu_k$, $k \leq n-1$), are called *Peirce numbers* associated with u.

The importance of the above notion is provided by the following result (cf. [27]).

Proposition 4.2. *Peirce numbers are invariant under linear transformations.*

4.1. Kronecker tensor product

Definition 4.3. Given a vector $X = \{x_1, \dots, x_n\}$, define its *Kronecker tensor square* $Y = X \times X \in \mathbb{R}^m$ by

$$Y := \{x_1^2, 2x_1x_2, \dots, 2x_1x_n, x_2^2, 2x_2x_3, \dots, 2x_{n-1}x_n, x_n^2\} \tag{4.1}$$

where $m = \frac{1}{2}n(n+1)$.

Definition 4.4. Vectors $X_1, \dots, X_k \in \mathbb{R}^n$ are said to be *quadratically independent*, if their Kronecker tensor squares $Y_1, \dots, Y_k \in \mathbb{R}^m$ (cf. (4.1)) are linearly independent in $\mathbb{R}^m$.

Remark 4.5. Clearly, formula (4.1) allows in certain cases to convert a nonlinear problem to the linear one considered in a linear space of higher dimension.

4.2. Diagonalizable algebras

Given an algebra $\mathbb{A}$, denote by $\mathcal{P}(\mathbb{A})$ (resp. $\mathcal{N}(\mathbb{A})$) the set of its non-zero idempotents (resp. nilpotents).

Definition 4.6. Let $\mathbb{A}$ be an n-dimensional commutative (in general, non-associative) real algebra. If there exist at least $m = \frac{1}{2}n(n+1)$ quadratically independent elements in $\mathcal{P}(\mathbb{A}) \cup \mathcal{N}(\mathbb{A})$, then $\mathbb{A}$ is called *diagonalizable.*

Remark 4.7. This term is avowed in a unary algebra if the multiplication is given by a diagonalizable matrix.

Theorem 4.8. *Suppose $\mathbb{A}$ is a diagonalizable commutative n-dimensional algebra. Then, the set $\mathcal{P}(\mathbb{A}) \cup \mathcal{N}(\mathbb{A})$ completely determines the multiplication table in $\mathbb{A}$.*

Proof. Let $\{e_1, e_2, \ldots, e_n\}$ be a basis in $\mathbb{A}$, $u = \sum_{i=1}^n u_i e_i \in \mathcal{P}(\mathbb{A}) \cup \mathcal{N}(\mathbb{A})$ and $\gamma_{i,j}^k$ structure constants in this basis. Then, u^2 may be written as follows:

$$u^2 = \sum_{i,j,k} \gamma_{i,j}^k u_i u_j e_k. \tag{4.2}$$

One can rewrite the equation $u^2 = \lambda u$ in the matrix form as follows:

$$\begin{pmatrix} \gamma_{1,1}^1 & \gamma_{1,2}^1 & \cdots & \gamma_{2,2}^1 & \gamma_{n,n}^1 \\ \gamma_{1,1}^2 & \gamma_{1,2}^2 & \cdots & \gamma_{2,2}^2 & \gamma_{n,n}^2 \\ \cdots & \cdots & & \cdots & \\ \gamma_{1,1}^n & \gamma_{1,2}^n & \cdots & \gamma_{2,2}^n & \gamma_{n,n}^n \end{pmatrix} \begin{pmatrix} u_1 u_1 \\ u_1 u_2 \\ \cdots \\ u_n u_n \end{pmatrix} = \lambda \begin{pmatrix} u_1 \\ u_2 \\ \cdots \\ u_n \end{pmatrix}, \tag{4.3}$$

By assumption, there exists a quadratically independent system $u^1, u^2, \ldots, u^m \in \mathcal{P}(\mathbb{A}) \cup \mathcal{N}(\mathbb{A})$.

Hence, the following $(m \times m)$ matrix S is non-singular:

$$S = \begin{pmatrix} u_1^1 u_1^1 & u_1^2 u_1^2 & \cdots & u_1^m u_1^m \\ u_1^1 u_2^1 & u_1^2 u_2^2 & \cdots & u_1^m u_2^m \\ \cdots & \cdots & \cdots & \cdots \\ u_n^1 u_n^1 & u_n^2 u_n^2 & \cdots & u_n^m u_n^m \end{pmatrix} \tag{4.4}$$

Combining this with (4.3) shows that the structure constants are completely determined by $\mathcal{P}(\mathbb{A}) \cup \mathcal{N}$. □

Example 4. In the two-dimensional case, one has $m = 3$. Let $u^i = x^i e_1 + y^i e_2$, $i = 1, 2, 3$. Then,

$$\det(S) = \prod_{1 \le i < j \le 3} (x^i y^j - x^j y^i) \tag{4.5}$$

is the generalized Vandermond determinant and the quadratic independence of vectors u^1, u^2, u^3 is equivalent to their pairwise linear independence.

Observe that if $n > 2$, simple examples show that quadratic independence does not imply the linear one and vice versa.

Remark 4.9. As is well known, there are several possibilities to define a multiplication in an algebra, for instance, one can use the canonical passage from a quadratic map to the corresponding symmetric bilinear one. The main advantage of the method provided by Theorem 4.8 rests on the fact that so-defined multiplication is nicely compatible with the topology of the (local) flow induced by the Riccati equation considered in the underlying algebra.

4.3. Main syzygies among idempotents and Peirce numbers

4.3.1. Preliminary Combinatorial Estimates. Denote by $p = \mathrm{card}(\mathcal{P}(\mathbb{A}))$ the cardinality of $\mathcal{P}(\mathbb{A})$. According to the Bézout theorem, if p is finite, then $p \le 2^n - 1$. Assume there exist $2^n - 1$ distinct simple idempotents $\mathrm{u}_1, \ldots, \mathrm{u}_{2^n-1}$ such that $\{\mathrm{u}_1, \ldots, \mathrm{u}_n\}$ is a basis of $\mathbb{A}$. Recall,

$$\mathrm{u}_i \cdot \mathrm{u}_j = \sum_{k=1}^n \gamma_{ij}^k \mathrm{u}_k, \quad 1 \le i < j \le n.$$

Thus, there exist

$$n\binom{n}{2} = \frac{n^2(n-1)}{2}$$

structure constants $\gamma_{ij}^k \in \mathbb{R}$ which completely determine the multiplication in $\mathbb{A}$. On the other hand,

$$\mathrm{u}_i = \sum_{k=1}^{n} c_i^k \mathrm{u}_k, \quad n+1 \leq i \leq 2^n - 1,$$

i.e., u_i are determined by $n(2^n - n - 1)$ coefficients $c_i^k \in \mathbb{R}$ for $n+1 \leq i \leq 2^n - 1$. It is easy to show that $n(2^n - n - 1) > \frac{1}{2}n^2(n-1)$ for $n \geq 3$. Thus, some combinations of the $2^n - 1 - n$ idempotents u_i are not allowed for $n+1 \leq i \leq 2^n - 1$, $n \geq 3$. Similar considerations can be applied to the Peirce numbers.

Remark 4.10. The idempotents and their Peirce numbers follow a pattern consistent with a specific rule. In the invariant theory this fact is known as the existence of *syzygy*.

4.3.2. Two-dimensional case.

Theorem 4.11. *Given a complex two-dimensional m-ary algebra $\mathbb{A}$, suppose there exist three distinct simple (i.e., pairwise linearly independent) idempotents u_1, u_2, u_3. Let μ_1, μ_2 and μ_3 be the Peirce numbers associated with u_1, u_2 and u_3, respectively (obviously, by condition, each idempotent admits precisely one Peirce number). Then, there exist the following syzygies:*

$$\sum_{k=1}^{3} \frac{1}{1-2\mu_k} = 1, \qquad \sum_{k=1}^{3} \frac{u_k}{1-2\mu_k} = 0. \tag{4.6}$$

Proof. To begin with, recall the classical residue theorem.

Theorem 4.12. *Let $z - f(z)$ be a meromorphic function with simple zeros $z_1, \ldots, z_k$ in the complex plane. Suppose that for sufficiently large z, $|f(z)|$ is bounded. Then, the following formula is true:*

$$\sum_{i=1}^{k} \frac{1}{1-f'(z_i)} = 1. \tag{4.7}$$

Let (P, Q) be the homogeneous vector polynomial mapping in $\mathbb{R}^2$, where $P(x, y)$ and $Q(x, y)$ are homogeneous polynomials of degree m, associated with the multiplication in $\mathbb{A}$. Choose in $\mathbb{R}^2$ coordinates (x, y) in such a way that $P(1, 0) = 1$ and $Q(1, 0) = 0$. Then:

$$P(x, y) = x^m + \text{lower terms w.r.t } x, \quad Q(x, y) = m\mu x^{m-1}y + \text{lower terms w.r.t. } x$$

with some parameter μ. Next, define $f(z) := \frac{Q(1,z)}{P(1,z)}$ (considered as a function of complex variable). Then, $f(z)$ is holomorphic, $f(0) = 0$ and $f'(0) = m\mu$. By the same token, μ is the Peirce number, associated to the idempotent $e = \{1, 0\}$.

Applying the same procedure to other idempotents and using Theorem 4.12 yield the required syzygies. □

4.3.3. The case of arbitrary dimensions. There exist several syzygies among idempotents and their Peirce numbers. All of them follow from the Lefschetz fixed point formula for elliptic complexes obtained by M.F. Atiyah and R. Bott [3].

Let E_i be vector bundles over a manifold M and assume that a differential complex

$$\mathcal{E} : 0 \to \Gamma(E_0) \xrightarrow{D} \Gamma(E_1) \xrightarrow{D} \cdots, \quad D^2 = 0$$

is elliptic. The following theorem was proved in [3]:

Theorem 4.13. *Given an elliptic complex $\mathcal{E}$ on a compact manifold M, suppose $f : M \to M$ has a lifting $\varphi_i : f^{-1}E_i \to E_i$ for each i such that the induced maps $T_i : \Gamma(E_i) \to \Gamma(E_i)$ give an endomorphism of the elliptic complex. Then, the Lefschetz number of T is given by formula*

$$L(T) = \sum_{f(x)=x} \frac{\sum(-1)^i \operatorname{tr} \varphi_{i,x}}{|\det(1 - f_{*,x})|}.$$

The following result is a direct consequence of Theorem 4.13.

Theorem 4.14. *Suppose there exist exactly $2^n - 1$ distinct nonzero idempotents u_i in an n-dimensional commutative real algebra $\mathbb{A}$ and μ_{ij}, $1 \leq j \leq n-1$, are the Peirce numbers associated with u_i $(1 \leq i \leq 2^n - 1)$. Then, the following syzygies hold:*

$$\sum_{i=1}^{2^n-1} \sigma_k(u_i; \mu_{i,1}, \ldots, \mu_{i,n-1}) \prod_{j=1}^{n-1} \frac{1}{1 - 2\mu_{i,j}} = \sigma_k(0), \quad k = 0, 1, \ldots, n-1. \quad (4.8)$$

Here $\sigma_k(x_1, x_2, \ldots, x_n)$ is any symmetric homogeneous function of order k.

Remark 4.15. Suppose there exist three distinct simple (i.e., pairwise linearly independent) possibly complex idempotents $\mathrm{u}_1, \mathrm{u}_2, \mathrm{u}_3$ in a two-dimensional complex algebra $\mathbb{A}$. Obviously, with each idempotent, there is associated exactly one Peirce number μ_1, μ_2 and μ_3 respectively. By Theorem 4.14, there exist syzygies (4.6) among them.

5. Classification of ODEs

In this section, we discuss the phase portrait classification of polynomial homogeneous differential systems up to the orbital topological equivalence.

Definition 5.1. Two (local) flows Φ and Ψ are said to be locally orbitally topologically equivalent (in short, OTE-equivalent) if there exists an orientation preserving homeomorphism taking trajectories (resp. singular points) of Φ onto trajectories (resp. singular points) of Ψ and preserving the move directions.

The problem of studying OTE-invariants is attracting a big deal of attention for a long time. There are at least two approaches to this problem: (i) singularity theory [2] and (ii) algebraic approach [21]. Our approach follows the algebraic framework and can be summarized as follows:

- The first step in studying OTE is understanding a dynamics in all subalgebras of an algebra in question.
- Next step is to study the adjustment of phase flow near these subalgebras. Here we generalize Shilov's [29] method elaborated for one-dimensional subalgebras (= ray solution to ODEs).
- Finally, one should glue up the the local phase portraits into a global one.

This approach requires the hierarchial studying of OTE (with dimension increasing). In this section, we mainly restrict ourselves with the two-dimensional case and consider isolated origins only. The 3D results will be published in the subsequent papers.

Without loss of generality, we assume that the field in question has at least one (but finitely many) fixed directions. Then, the two-dimensional phase portrait splits into finitely many sectors. As is well known, in the considered case, topologically there exist precisely three types of sectors: elliptic, parabolic and hyperbolic (see Subsection 5.2). The number of sectors together with the order of their appearance, mainly determine the 2D OTE (see, for instance, [14]). This translates to the algebraic language as a coding the OTE-classes by words written in the alphabet generated by types of sectors. Observe that not any word is allowed meaning that the alphabet is subordinated by a certain grammar. In fact, many authors involved in studying OTE were creating the appropriate grammar. The goal of this section is two-fold: (i) to completely describe the rules of the grammar in the two-dimensional case and (ii) to outline possible extensions to higher dimensions.

As a matter of fact, the OTE-classification in higher dimensions is not semialgebraic in nature, therefore, the "sector schemes" are not applied. Moreover, in contrast to the two-dimensional case, known examples show that the set of OTE-classes is not discrete in general. To fill out the gap, one needs to change the paradigm - we follow [29].

5.1. Dynamics near the one-dimensional subalgebras

To study the OTE-equivalence of homogeneous quadratic ODEs (with isolated origin), we start with investigation of a phase portrait near a fixed direction. In his study of the planar dynamics, Markus (see [21]) assigned to any fixed direction of a field Φ either $+$ or $-$ depending on whether the radial velocity along the ray is positive or negative. This way, he associated to Φ a *cyclic combinatorial scheme* and proved that two planar quadratic systems (with isolated origin) are topologically equivalent iff they have the same cyclic combinatorial schemes.

As it was indicated above, this approach cannot be extended to the case of higher dimensions. To overcome this obstacle (at least, in part), in the statement following below, we translate Shilov's approach to the algebraic language.

Theorem 5.2. *Let u be a real idempotent in $\mathbb{A}$ and μ a real Peirce number associated to u. Let ξ be an eigenvector of the multiplication by u operator: $u \circ \xi = \mu\xi$. Denote by $\mathbb{R}[u, \xi]$ the plane spanned by u and ξ. Then, the qualitative behavior of solutions to* (1.1) *in $\mathbb{R}[u, \xi]$ (near the fixed direction $\mathbb{R}[u]$) may be of five types depending on a type of $\mathbb{R}[u]$, namely, three generic cases (cf.* [29]*):*

1. *generic ray of type p_1, if $\mu > \frac{1}{2}$,*
2. *generic ray of type p_2, if $0 < \mu < \frac{1}{2}$,*
3. *generic ray of type h , if $\mu < 0$;*

and two exceptional cases:

4. *exceptional ray of type e_0, if $\mu = 0$,*
5. *exceptional ray of type e_1, if $\mu = \frac{1}{2}$.*

Proof. Let a solution near $\mathbb{R}[\mathrm{u}]$ be chosen in a form

$$x(t)\mathrm{u} + y(t)\xi + o(t), \tag{5.1}$$

where $o(t)$ stands for the higher terms w.r.t. t. Take a point $(x_0, y_0) \in \mathbb{R}[\mathrm{u}, \xi]$ near $\mathbb{R}[\mathrm{u}]$ and substitute (5.1) into the original equation. Then, for t small enough,

$$y = y_0 \left(\frac{x}{x_0}\right)^{2\mu} + o(t). \tag{5.2}$$

It is easy to see that Figure 1 represents all the topologically different phase portraits near the non-exceptional rays.

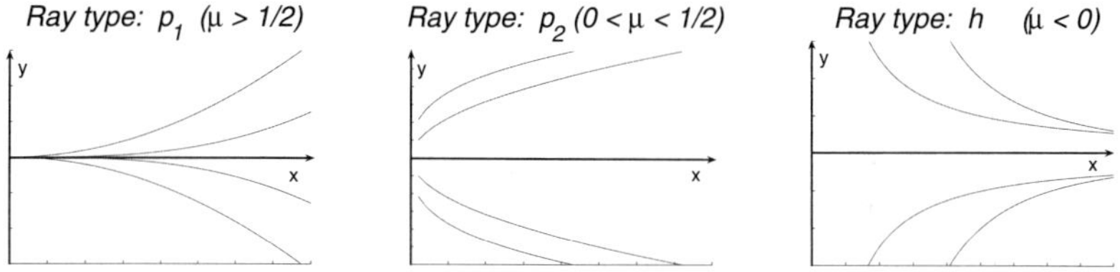

FIGURE 1. Phase trajectories near ray solution ($y = 0$).

Also, in contrast to the above "stable" cases, where the values of μ completely determine the local topological dynamics, the values $\mu = 0, \frac{1}{2}$ correspond to the "non-stable" ones. Therefore, in these cases, to establish the local dynamics, one needs to additionally analyze the term $o(t)$ in (5.2). □

5.2. Sector types

Definition 5.3. Given a planar homogeneous ODE, define a **sector** as an open domain bounded by two subsequent rays.

By Theorem 5.2, there exist precisely three non-exceptional types of rays, therefore, there exist precisely six types of regular sectors presented in Table 1 and Figure 1.

Type of l_1	Type of l_2	Sector Type
p_1	p_1	E
h	h	H_1
h	p_2	H_2
p_2	p_2	H_3
p_1	p_2	P_1
h	p_1	P_2

TABLE 1. Regular sector types

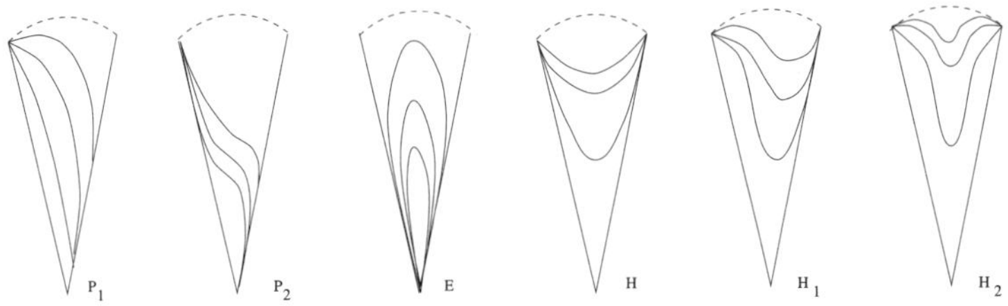

FIGURE 2. Non exceptional types of sectors

Definition 5.4. A sector composed by two subsequent rays such that at least one of them is exceptional, is called *exceptional.*

In fact, no rigorous topological behavior may be established in exceptional types of sectors and they require a special investigation (cf. [7], [30]).

Some of the authors (see [30][7]) used the above six types of the sectors while the other authors (see, for instance, [21]) used only three of them. Namely:

1. *Parabolic sector* (*type* ***P*** *sector*), where all trajectories start (resp. end) at the origin and end (resp. start) at infinity.
2. *Hyperbolic sector* (*type* ***H*** *sector*), where all trajectories start and end at infinity.
3. *Elliptic sector* (*type* ***E*** *sector*), where all trajectories start and end at the origin.

5.3. Multi-sectorial matching

The local phase portrait in a neighborhood of an isolated equilibrium point of a planar dynamical system (1.1) allows a multi-sectorial chain recurrence matching constructed as an ordered set of a different type sectors.

Definition 5.5. Such an ordered multi-sectorial matching near the origin will be called a **sector scheme**.

The presence of syzygies (4.8) yields some restrictions on possible sector schemes.

5.4. Formal language for OTE

In mathematics, a formal language is defined by an alphabet and rules. The alphabet is a set of symbols on which this language is built. The rules specify which strings of symbols are allowed. The well-posed strings of symbols are called words. The rules postulate which word belongs to the language or how to construct the words belonging to the language. Below is presented a language allowing to classify homogeneous planar ODEs up to OTE.

Definition 5.6. Denote by Σ_r an *alphabet* with small letters $\Sigma_r = \{h, p_1, p_2\}$ and by Σ_s an alphabet with capital letters $\Sigma_s = \{E, H_1, H_2, H_3, P_1, P_2\}$ (the subscript r (resp. s) stands for the ray (resp. sector) scheme).

The set of all possible finite length strings formed from the alphabet Σ_r (resp. Σ_s) is denoted by Σ_r^* (resp. Σ_s^*).

Definition 5.7. A *formal language* over Σ_r (resp. Σ_s) is a subset of Σ_r^* (resp. Σ_s^*) compatible with a grammatical and semantical structure and is denoted by $\mathbb{L}(\Sigma_r)$ (resp. $\mathbb{L}(\Sigma_s)$.

Definition 5.8. A *determinative grammar* for $\mathbb{L}(\Sigma_r)$ (resp. $\mathbb{L}(\Sigma_s)$ is a set of rules through which any element of Σ_r^* (resp. Σ_s^*) can be determined to be an element of $\mathbb{L}(\Sigma_r)$ (resp. $\mathbb{L}(\Sigma_s)$. We denote by $\mathbb{G}(\mathbb{L}(\Sigma_r))$ (resp. $\mathbb{G}(\mathbb{L}(\Sigma_s)))$ a grammar that recognizes the language $\mathbb{L}(\Sigma_r)$ (resp. $\mathbb{L}(\Sigma_s)$.

The presence of the main syzygy (4.6) is the key to constructing $\mathbb{G}(\mathbb{L}(\Sigma_r))$, i.e., it is responsible for composing strings correctly in $\mathbb{L}(\Sigma_r)$ and $\mathbb{G}(\mathbb{L}(\Sigma_s))$ (constructed according to valid pairs of two subsequent sector types).

It is much more easy to deal with Σ_r^* rather than with Σ_s^*, since the only restriction on the grammar in Σ_r^* is syzygy (4.6). This means that one should study the solubility of equation (4.6) in the interval arithmetics provided by Theorem 5.2, namely $I_1 = \{-\infty, 0\}$, $I_2 = \{0, \frac{1}{2}\}$ and $I_3 = \{\frac{1}{2}, \infty\}$.

We summarize in Table 2 all possible sector schemas allowed in formal language $\mathbb{L}(\Sigma_r)$ (resp. $\mathbb{L}(\Sigma_s)$ for the plane quadratic vector fields.

As an application of above stated results, we consider in the next section polynomial integrability of the Riccati equation (1.1).

6. Polynomial first integrals

As it was mentioned in [13], theory of integrability of differential equations (existence of polynomial, rational and analytic first integrals) was played an important role in the development of ODEs in the nineteenth and early twentieth centuries. Most of an effort was done by Kovalevskaya, Fuchs and other mathematicians. The usual technique was based on the linearization of ODEs near separatrices and study its (linearized) spectral properties. Such type of analysis leads to the defini-

	Ray scheme	Sector scheme
I	p_1hh or p_1p_1h	HPP
II	p_1p_1h	EPP
III	hhh or hhp_2	HHH
IV	e_1h or e_1p_2	HP
V	p_1e_1	EP
VI	p_2 or h	H
VII	p_1	E

TABLE 2. Markus' classification of planar quadratic ODEs with an isolated singular point

tion of the Kovalevskaya exponents, the Fuchs' indices of the linearized differential system that play an important role for integrability theory and in stability theory.

In this chapter, we first set the properties of the Kovalevskaya exponents become united with the Peirce numbers in underlying algebra. Namely: let u be an idempotent in algebra $\mathbb{A}$ and $\varrho_1 = -1, \varrho_2, \dots, \varrho_n$ are Kovalevskaya's exponents. Denote by $\mu_2, \dots, \mu_n$ the Peirce numbers associated with u, then

$$\varrho_i = 1 - 2\mu_i, \quad i = 2, \dots, n. \tag{6.1}$$

In 1996, Furta [10] and Goriely [13] established the existence of a link between the properties of the Kovalevskaya's exponents of quasi-homogeneous polynomial differential systems and their quasi-homogeneous polynomial first integrals.

In the particular case when Kovalevskaya's matrix associated with a quadratic homogeneous polynomial differential system (1.1) is diagonalizable, an improvement of this result was obtained by Tsygvintsev [33].

Theorem 6.1. (*cf.* [10], [13], [33]) *Suppose that the differential system* (1.1) *in* $\mathbb{R}^n$ *possesses a homogeneous first integral of degree* M. *Then there exists a set of non-negative integers* $k_2, \dots, k_n$ *such that*

$$k_2\varrho_2 + \cdots + k_n\varrho_n = M, \quad k_2 + \cdots + k_n \leq M, \tag{6.2}$$

where $\varrho_2, \dots, \varrho_n$ *are Kovalevskaya's exponents.*

In fact, Kovalevskaya's exponents $\varrho_2, \dots, \varrho_n$ in Theorem 6.1 one can substitute with the Peirce numbers $\mu_2, \dots, \mu_n$ respectively. Moreover in $\mathbb{R}^2$ the following interesting observation is true:

Theorem 6.2. *Let the Riccati equation* (1.1) *in two-dimensional* m*-ary algebra* $\mathbb{A}$ *possesses a homogeneous polynomial first integral. Then*

(i) *all Peirce numbers in* $\mathbb{A}$ *are non-positive rational, meaning that*
(ii) *ray scheme of* (1.1) *is* $hhh\dots h$ *and*
(iii) *the sector scheme of* (1.1) *is* $H_1H_1\dots H_1$ *only.*

Proof. By Theorem 6.1 in $\mathbb{R}^2$

$$1 - 2\mu_2 = \varrho_2 = \frac{M}{k_2} \geq 1, \quad \text{since} \quad k_2 \leq M.$$

Therefore μ_2 is rational and non-positive. (i) is proven. The same property is true among all idempotents in $\mathbb{A}$.

The property $\mu_2 \leq 0$, by Theorem 5.2, means that all rays are of type h or equivalently, all sectors must be type H_1 hyperbolic sectors (see Table 1). □

7. PDEs in algebra

As it was established in [6], [19], any homogeneous first-order system of PDEs $\mathcal{L}(D)f = 0$,

$$\{\mathcal{L}f\}_k := \sum_{i,j=1}^{n} \gamma_{ij}^k \partial_i u_j(x) = 0, \tag{7.1}$$

with the constant coefficients γ_{ij}^k is the Dirac equation $D \circ f = 0$ in the underlying algebra $\mathbb{A} = A_{\mathcal{L}}$ with multiplication (4.2).

Similarly to the classical case of *Complex Analysis*, any function of the form

$$\begin{aligned} f(x) := e_1 u_1(x_1, \ldots, x_n) + e_2 u_2(x_1, \ldots, x_n) \\ \cdots + e_n u_n(x_1, \ldots, x_n) \end{aligned}$$

where $u_i(x)$ are real analytic functions, is called an $\mathbb{A}$-*valued function.*

Definition 7.1. An $\mathbb{A}$-valued function $f(x)$ is called $\mathbb{A}$-**analytic** (denoted $f \in \mathrm{Hol}(\mathbb{A})$), if $f(x)$ is a solution to the Dirac equation

$$D \circ f(x) := \sum_{i,j=1}^{n} e_i \circ e_j \partial_i u_j(x_1, \ldots, x_n) = 0. \tag{7.2}$$

If $\mathbb{A}$ in Definition 7.1 is an algebra of complex numbers $\mathbb{C}$, then (7.2) coincides with the Cauchy-Riemann equations. Thus, our notation $\mathrm{Hol}(\mathbb{A})$ is in a complete agreement with the definition of holomorphic functions in complex analysis (denoted by $\mathrm{Hol}(\mathbb{C})$).

Remark 7.2. By $\mathbb{A}$-analysis we mean the systematic study of $\mathrm{Hol}(\mathbb{A})$.

One of the goals of this article is to show that many properties of PDEs $\mathcal{L}f = 0$ may be described in terms of the underlying algebra $\mathbb{A}_{\mathcal{L}} = \mathbb{A}$.

7.1. Algebraic approach to function theories

Let $\mathbb{A}$ be a real algebra (not necessarily commutative and/or associative). An analytic (holomorphic) function theory over such algebras has been developed by many authors (see [11], [6], [32], [34]). There are three distinct approaches in these investigations.

- The first one (Weierstrass approach) regards functions on $\mathbb{A}$ as their convergent (in some sense) power series expansions (cf. [16]).
- The second one (Cauchy-Riemann) approach is concentrated on solutions to the Dirac equation in algebra $\mathbb{A}$ (cf. [18], [19]).
- The third one is based on the function-theoretic properties known for complex analytic functions, such as Cauchy theorem, residues theory, Cauchy integral formula etc. (cf. [23], [12]).

All these methods look like generalization of the *analytic function theory* of complex variables (cf. [32], [34]). We use the term $\mathbb{A}$-*analysis* for such cases (cf. Clifford or quaternionic analysis [6], [32] if $\mathbb{A}$ is embedded into a Clifford algebra).

Remark 7.3. The totality of functions on a regular algebra (in general non-commutative and/or non-associative) splits into non-equivalent classes. These classes are uniquely characterized by their **unital hearts**. If such a heart is, in addition, an associative algebra, then an $\mathbb{A}$-analytic function may be expanded into the commutative operator-valued power series.

7.2. Isotopy classes

In this subsection, we will be concerned with *Albert isotopies* of algebras.

Definition 7.4. Two n-dimensional algebras $(\mathbb{A}_1, \circ)$ and $(\mathbb{A}_2, \star)$ are said to be **isotopic** (denoted $\mathbb{A}_1 \sim \mathbb{A}_2$) if there exist nonsingular linear operators K, L, M such that

$$x \circ y = M(Kx \star Ly). \tag{7.3}$$

Obviously, if, in addition, $K \equiv L \equiv M^{-1}$, then two algebras $\mathbb{A}_1$ and $\mathbb{A}_2$ are isomorphic (denoted $\mathbb{A}_1 \simeq \mathbb{A}_2$).

Definition 7.5. If for given two algebras $\mathbb{A}_1$ and $\mathbb{A}_2$ there exist nonsingular linear operators P, Q such that for every $g(\mathrm{x}) \in \mathrm{Hol}(\mathbb{A}_2)$, the function $f(\mathrm{x}) = Pg(Q\mathrm{x})$ belongs to $\mathrm{Hol}(\mathbb{A}_1)$ and vice versa, we say that the two function theories are **equivalent** and write $\mathrm{Hol}(\mathbb{A}_1) \simeq \mathrm{Hol}(\mathbb{A}_2)$.

With these definitions in hands, we have the important

Theorem 7.6. *Two function theories are equivalent iff the corresponding algebras are isotopic. Any properties of* (7.1) *must be isotopically invariant.*

Definition 7.7. If, for all (fixed) $x_0 \in \Omega \subset \mathbb{R}^n$, $\gamma_{ij}^k(x_0)$ in (7.1) constitute a set of isotopic algebras, then we say that $\mathcal{L}$ is of uniquely defined PDE type in Ω, otherwise, of *mixed* PDE type. An algebra $\mathbb{A}_0$ with structure constants $\gamma_{ij}^k(x_0)$ is called the algebra correspondingly to $\mathcal{L}$ in $\mathrm{x}_0 \in \Omega$.

Obviously, if the coefficients $\gamma_{ij}^k(x)$ are thought of as constants and b_i^k vanish identically, then operator $\mathcal{L}$ in (7.1) coincides with the Dirac operator D defined by (1.2), and a solution to homogeneous equation $\mathcal{L}f = 0$ is a (left) $\mathbb{A}$-analytic function for the operator D.

7.3. Classification of the first-order PDE

Here we study many important notions from PDE theory using the algebraic language. We begin by examining the conditions providing the Dirac equation in $\mathbb{A}$ (7.2) to be a well-defined system of PDEs.

7.4. Under- and over-determined system

Let $P(D)u(x) = f(x)$ be a system of partial differential equations, where $P(D)$ is a given $k \times l$ matrix of differential operators with constant coefficients, $f(x)$ (resp. $u(x)$) be a given (resp. unknown) k- (resp. l-) tuples of functions or distributions in $x \in \mathbb{R}^m$. Many authors (cf. [24]) usually assume that the system is under- (over-) determined, if the rank of $P(\xi)$ (resp. of its transpose $P'(\xi)$) is less than l for all (resp. for some) nonzero $\xi \in \mathbb{R}^m$.

The algebraic formulation of the fact that PDE (7.2) with constant coefficients is under- (over-) determined, can be given as follows.

Definition 7.8. A real n-dimensional algebra $\mathbb{A}$ is called left (resp. right) **regular** if there exists $v \in \mathbb{A}$ such that the linear operators $L_v, R_v : \mathbb{R}^n \to \mathbb{R}^n$ defined by $x \to v \circ x$ (resp. $x \to x \circ v$) are both invertible. Otherwise, $\mathbb{A}$ is called a *left (resp. right) singular* algebra. In other words, $\mathbb{A}$ is regular iff $\mathbb{A} \subset \mathbb{A}^2$

Recall an element $u \in \mathbb{A}$ (resp $v \in \mathbb{A}$) is called a left (resp. right) **annihilator** if $u \circ x = 0$, (resp. $x \circ v = 0$) for all $x \in \mathbb{A}$.

Theorem 7.9. *The Dirac operator D in algebra $\mathbb{A}$ is under determined iff $\mathbb{A}$ is singular and is over determined iff $\mathbb{A}$ is regular and contains an annihilator.*

Proof. For a given Dirac operator D in the corresponding algebra $\mathbb{A}$, take the left (resp. right) multiplication operators $L_v, R_v : \mathbb{R}^n \to \mathbb{R}^n$ as in Definition 7.8. If L_ξ, R_ξ are both invertible for some ξ, then D is well determined. Conversely, let L_v (respectively, R_v) be $k_1 \times l_1$ (resp. $k_2 \times l_2$) matrices of differential operators. Then, D is under determined if $k_1 < l_1$ and/or $k_2 > l_2$ and is over-determined if $k_1 > l_1$ and/or $k_2 < l_2$. The only case $k_1 = l_1 = k_2 = l_2 = n$ stands for the regular algebras $\mathbb{A}$ without annihilators and for a well-determined Dirac operator D in $\mathbb{R}^n$. □

In the next sections we deal with PDE's of different type.

7.5. PDEs of elliptic type

The ellipticity is one of the basic concepts in PDE. Actually, by Theorem 7.6, the *ellipticity* of the Dirac operator in a regular algebra $\mathbb{A}$ is a property invariant with respect to isotopy. In fact, one can reformulate it as a property of a singular algebra as well.

Definition 7.10. An algebra $\mathbb{A}$ is called a **division algebra** if both operators of left and right multiplications by any nonzero element are invertible.

Proposition 7.11. *The well-determined Dirac operator D in a (necessarily regular by Theorem 7.9) algebra $\mathbb{A}$ is elliptic iff $\mathbb{A}$ is a division algebra.*

Proof. The matrix symbol of a well-determined elliptic partial differential operator $\sigma(D)(\xi)$ is invertible for all $\xi \neq 0$ (cf. [11], [26]). Conversely, it follows immediately from the definition of ellipticity that the symbol of partial differential operator D is an invertible matrix for all $\xi \neq 0$. □

Example 5. (*D. Rusin*) Let $\mathbb{Q}_\varepsilon$ be constructed from the algebra of quaternions $\mathbb{Q}$ leaving the multiplication table unchanged except for the square of i: $i^2 = -1 + \varepsilon j$.

Two algebras $\mathbb{Q}_\varepsilon$ and $\mathbb{Q}$ are non isotopic 4-dimensional division algebras if $|\varepsilon| < 2$.

The Example 5 shows that in the four-dimensional space, there exist non equivalent elliptic function theories in the sense of Definition 7.5.

Question 7.12. *How many non-equivalent (well-defined elliptic) function theories do exist in* $\mathbb{R}^n$, $n = 4, 8$*?*

Clearly, Question 7.12 may be answered in terms of the existence of non isotopic classes of division algebras. In particular, from Example 5 follows, that not any solution to the linear first-order elliptic partial differential system in $\mathbb{R}^4$ may be obtained using quaternion analysis. Below, we study conditions providing an algebra to be an underlying one for an elliptic system.

Let $\mathcal{A}$ stand for an ideal in $\mathbb{A}$ of all left annihilators (it turns out to be maximal). Of course, the multiplication in $\mathcal{A}$ is trivial.

Definition 7.13. An algebra $\mathbb{A}$ is said to be **quasi-divisible** if the factor algebra $\mathbb{A}/\mathcal{A}$ is a division algebra.

In turn, in a quasi-division algebra, the equation $a \circ x = b$ is soluble for all a, b except for those a being the left annihilators of $\mathbb{A}$. We are now in position to formulate the algebraic analogue of ellipticity [1] for under- and over-determined system.

Proposition 7.14. *The Dirac operator D defined in* (1.2) *is elliptic iff the corresponding to D algebra $\mathbb{A}$* (*see* (4.2)) *is quasi-divisible.*

Proof immediately follows (cf. [11], [26]) from the definition of ellipticity [1] of the symbol of a partial differential operator D. □

Of course, if there are no other left/right annihilators except 0, every quasi division (in the sense of Definition 7.13) algebra is a *division* one.

For regular algebras, A. Albert (see [25]) obtained

Theorem 7.15. *Every regular algebra is isotopic to a unital algebra* (*with unit e*). *Every n-dimensional unital division algebra* (*with $n \geq 2$*) *contains an "imaginary unit" i being a square root of $-e$* ($i^2 = -e$).

Example 6. Consider two systems of PDEs in $\mathbb{R}^4$: the spherical and usual Dirac equation (see [23]) for scalar and vector functions $u(t,x), \overline{v}(t,x)$:

$$\mathrm{div}\overline{v} = 0; \quad \mathrm{grad}u + \mathrm{curl}\overline{v} = 0. \tag{7.4}$$

$$\partial_t u - \mathrm{div}\overline{v} = 0; \quad \mathrm{grad}u + \partial_t\overline{v} + \mathrm{curl}\overline{v} = 0. \tag{7.5}$$

Both systems are elliptic, the first one is over-determined. To see that, recall that the quaternion algebra $\mathbb{Q}$ stands behind (7.5), while an algebra $\mathbb{Q}'$ with the multiplication $x \circ y = 1/2(\overline{x} - x) \star y$, where "$\star$" is the quaternion multiplication, stands behind (7.4).

In turn, both algebras are quasi-divisible, but only $\mathbb{Q}'$ has a nontrivial ideal of (left) annihilators.

7.6. PDE of parabolic and hyperbolic type

Comparing the definitions of elliptic [1], hyperbolic [26] and parabolic [8] PDE with Theorem 7.9, we can see that parabolic and hyperbolic type Dirac operators correspond to algebras with **zero divisors**. Moreover, the property of algebras to be regular and quasi-divisible, the number and location of zero divisors and annihilators are invariant with respect to the isotopy relation. This gives rise to the following

Proposition 7.16. *The Dirac operator in a regular algebra* $\mathbb{A}$ *is:*

(i) *parabolic* [8] *iff* $\mathbb{A}/\mathcal{A}$ *contains exactly one (up to scalar factor) left zero divisor. In this case* $\mathbb{A}/\mathcal{A}$ *is isotopic to an algebra with one nilpotent and no more other zero divisors;*

(ii) *hyperbolic* [26] *iff* $\mathbb{A}/\mathcal{A}$ *contains at least two (left) zero divisors.*

Conclusions

The Peirce numbers and idempotents allocation are fundamental properties that characterize also the behavior of solutions to polynomial ODEs around their singularities. After being defined in algebra more than hundred years ago, they look more than ubiquitous in nonlinear analysis, and it is our belief that they will definitely take a firm position in applied mathematics. In fact:

- any homogeneous polynomial ODE may be thought of as a Riccati equation in an algebra;
- the equivalence of Riccati equations in algebras strongly depends on idempotent locations and arrangement of their Peirce numbers;
- every first-order PDO with constant coefficients is the *Dirac operator* in the corresponding algebra;
- solutions to Dirac equations in isotopic algebras yield equivalent function theories,
- the $\mathbb{A}$-analysis in regular algebras is equivalent to the canonical function theory on their unital hearts.

Acknowledgment

The author would like to thank the anonymous referee for polishing this text, making many passages much more clear and for writing useful remarks concerning the results of this article.

References

[1] S. Agmon, A. Douglis, L. Nirenberg, *Estimates near the boundary for solutions of elliptic partial differential equations satisfying general boundary conditions*, I, II, *Comm. Pure Applied Math.* **12**, 1959, 623–727, **17**, 1964, 35–92.

[2] V.I. Arnold, *Geometrical Methods In The Theory Of Ordinary Differential Equations*, Springer-Verlag (1988).

[3] M.F. Atiyah, R. Bott, *A Lefschetz fixed point formula for elliptic complexes: II. Applications*, Ann. of Math. (2) **88** (1968), 451–491.

[4] Z. Balanov and Y. Krasnov, *Complex structures in real algebras I. Two-dimensional commutative case*, Comm. Algebra **31** (2003), no. 9, 4571–4609.

[5] Z. Balanov, Y. Krasnov, W. Krawcewicz, *Complex structures in algebras and bounded solutions to quadratic ODEs*, Funct. Differ. Equ. **10** (2003), no. 1-2, 65–81.

[6] F. Brackx, R. Delanghe, F. Sommen, *Clifford Analysis,* Pitman Research Notes in Math; **76**, 1982, 308 pp.

[7] T. Date, *Classification and analysis of two-dimensional real homogeneous quadratic differential equation systems*, J. Differential Equations **3**, no. 2 (1979), 311–334.

[8] S. Eidelman, *Parabolic systems, North-Holland Publishing Company*, 1969, 469 pp.

[9] S.D. Eidelman, Y. Krasnov, *Operator method for solution of PDEs based on their symmetries.* Oper. Theory Adv. Appl., 157, Birkhäuser, Basel, 2005, pp. 107–137.

[10] S.D. Furta *On non-integrability of general systems of differential equations.* Z. Angew. Math. Phys. 47 (1996), no. 1, 112–131.

[11] R.P. Gilbert, J.L. Buchanan, *First Order Elliptic Systems, Mathematics in Science and Engineering;* **163**, *Academic Press*, 1983, 281 pp.

[12] I.J. Good, *A simple generalization of analytic function theory, Exposition. Math,* **6**, no. 4, 1988, 289–311.

[13] A. Goriely *A brief history of Kovalevskaya exponents and modern developments.* Regul. Chaotic Dyn. 5 (2000), no. 1, 3–15.

[14] P. Hartman, *Ordinary Differential Equations*, Classics in Applied Mathematics **38**, Society for Industrial and Applied Mathematics (SIAM), Philadelphia, PA (2002).

[15] M. Hirsch and S. Smale, *Differential Equations, Dynamical Systems and Bifurcations of Vector Fields*, Academic Press, 1974.

[16] P.W. Ketchum, *Analytic functions of hypercomplex variables*, *Trans. Amer. Mat. Soc.*, **30**, # 4, pp 641–667, 1928.

[17] M.K. Kinyon, A.A. Sagle, *Quadratic dynamical systems and algebras*, J. Differential Equations **117** no. 1 (1995), 67–126.

[18] Y. Krasnov, *Commutative algebras in Clifford analysis*, Progress in analysis, Vol. I, II (Berlin, 2001), 349–359, World Sci. Publishing, River Edge, NJ, (2003).

[19] Y. Krasnov, *The structure of monogenic functions,* Clifford Algebras and their Applications in Mathematical Physics, vol 2, Progr. Phys., 19, Birkhäuser, Boston pp. 247–272, 2000.

[20] J. Llibre, X. Zhang, *Polynomial first integrals of quadratic systems*, Rocky Mountain J. Math. **31** (2001), no. 4, 1317–1371.

[21] L. Markus, *Quadratic differential equations and non-associative algebras*, Contributions to the theory of nonlinear oscillations , Vol. V (L. Cesari, J.P. LaSalle, and S. Lefschetz, eds.), Princeton Univ. Press, Princeton (1960), 185–213. J. Math. Pures Appl. **8** (1960), 187–216.

[22] M. Mencinger, *On stability of the origin in quadratic systems of ODEs via Markus approach*, Nonlinearity **16** (2003), no. 1, 201–218.

[23] P.J. Olver, *Equivalence, Invariants, and Symmetry*, Cambridge University Press, Cambridge (1995).

[24] V.P. Palamodov, *Linear differential operators with constant coefficients*, Springer-Verlag, New York-Berlin 1970, 444 pp.

[25] P.S. Pedersen, *Cauchy's integral theorem on a finitely generated, real, commutative, and associative algebra.* Adv. Math. 131 (1997), no. 2, 344–356.

[26] I.G. Petrovsky, *Partial Differential Equations, CRC Press, Boca Raton*, 1996.

[27] R.D. Schafer, *The Peirce decomposition*, Section 3.2 in An Introduction to Nonassociative Algebras. New York: Dover (1996), 32–37.

[28] I.R. Shafarevich, *Basic Algebraic Geometry, IV*, Springer, Berlin-Heidelberg-New York (1977).

[29] G.E. Shilov, *Integral curves of a homogeneous equation of the first order*, (Russian) Uspehi Matem. Nauk (N.S.) **5** (1950), no. 5(39), 193–203.

[30] K.S. Sibirsky, *Introduction to Topological Dynamics.* Translated from the Russian by L.F. Boron. Noordhoff International Publishing, Leiden (1975).

[31] R.A. Smith, *Orbital stability for differential equations*, J. Differential Equations **69** (1987), 265–287.

[32] F. Sommen, N. Van Acker, *Monogenic differential operators, Results in Math.* Vol. **22**, 1992, pp. 781–798.

[33] A. Tsygvintsev, *On the existence of polynomial first integrals of quadratic homogeneous systems of ordinary differential equations*, J. Phys. A **34** (2001), no. 11, 2185–2193.

[34] I. Vekua, *Generalized analytic functions. London. Pergamon*, 1962.

[35] S. Walcher, *Algebras and Differential Equations*, Hadronic Press, Palm Harbor, FL, (1991).

[36] B. Zalar, M. Mencinger, *On stability of critical points of Riccati differential equations in nonassociative algebras*, Glas. Mat. Ser. III **38**(58) (2003), no. 1, 19–27.

Yakov Krasnov
Department of Mathematics, Bar-Ilan University
52900 Ramat-Gan, Israel
e-mail: `krasnov@math.biu.ac.il`

Quaternionic and Clifford Analysis
Trends in Mathematics, 207–220

Necessary and Sufficient Conditions for Associated Pairs in Quaternionic Analysis

Le Hung Son and Nguyen Thanh Van

Abstract. This paper deals with the initial value problem of the type

$$\frac{\partial w}{\partial t} = L\left(t, x, w, \frac{\partial w}{\partial x_i}\right) \tag{1}$$

$$w(0, x) = \varphi(x) \tag{2}$$

where t is the time, L is a linear differential operator of first order in Quaternionic Analysis and φ is a regular function taking values in the Quaternionic Algebra. The necessary and sufficient conditions on the coefficients of the operator L under which L is associated to the generalized Cauchy-Riemann operator of the Quaternionic Analysis are proved.
This criterion makes it possible to construct the operator L for which the initial problem (1),(2) is solvable for an arbitrary initial regular function φ and the solution is also regular for each t.

1. Preliminaries and notations

Let $\mathcal{H}$ be the Quaternionic Algebra with the basis formed by e_0, e_1, e_2, e_3 where $e_0 = 1, e_3 = e_1e_2 = e_{12}$.

Suppose that Ω is a bounded domain of $\mathbb{R}^3$. A function f defined in Ω and taking values in the Quaternionic Algebra $\mathcal{H}$ can be presented as

$$f = \sum_{j=0}^{3} f_j e_j,$$

where $f_j(x)$ are the real-valued functions. If all $f_j(x) \in C^k(\Omega)$, we note that $f \in C^k(\Omega, \mathcal{H})$.

We introduce the generalized Cauchy-Riemann operator

$$D = \sum_{k=0}^{2} e_k \frac{\partial}{\partial x_k}.$$

Definition 1. *A function $f \in C^1(\Omega, \mathcal{H})$ is said to be regular in Ω if f satisfies*

$$Df = 0.$$

Remark 1. If $f \in C^2(\Omega, \mathcal{H})$ is a regular function, then f is harmonic in Ω.

2. The necessary and sufficient conditions for associated pairs

Suppose that $f = \sum_{j=0}^{3} f_j e_j$ is a twice continuously differentiable function with respect to the space-like variables x_0, x_1, x_2. Now assume that f is regular. This means that $Df = 0$. It is easy to verify that the condition $Df = 0$ is equivalent to

$$\sum_{i=0}^{2} A_i \frac{\partial f}{\partial x_i} = 0,$$

where

$$A_0 = \begin{bmatrix} 1 & 0 & 0 & 0 \\ 0 & 1 & 0 & 0 \\ 0 & 0 & 1 & 0 \\ 0 & 0 & 0 & 1 \end{bmatrix}, \quad A_1 = \begin{bmatrix} 0 & -1 & 0 & 0 \\ 1 & 0 & 0 & 0 \\ 0 & 0 & 0 & -1 \\ 0 & 0 & 1 & 0 \end{bmatrix}, \quad A_2 = \begin{bmatrix} 0 & 0 & -1 & 0 \\ 0 & 0 & 0 & 1 \\ 1 & 0 & 0 & 0 \\ 0 & -1 & 0 & 0 \end{bmatrix}$$

$$\frac{\partial f}{\partial x_i} = \begin{pmatrix} \frac{\partial f_0}{\partial x_i} \\ \frac{\partial f_1}{\partial x_i} \\ \frac{\partial f_2}{\partial x_i} \\ \frac{\partial f_3}{\partial x_i} \end{pmatrix}.$$

We define an operator ℓ as follows

$$\ell f = \sum_{i=0}^{2} A_i \frac{\partial f}{\partial x_i}. \tag{3}$$

It is clear that $Df = 0$ if and only if $\ell f = 0$. Next, we identify the function f with $f := \begin{pmatrix} f_0 \\ f_1 \\ f_2 \\ f_3 \end{pmatrix}$ and introduce a differential operator L as follows

$$Lf = \sum_{j=0}^{2} B_j \frac{\partial f}{\partial x_j} + Cf + K, \tag{4}$$

where $B_j = [b_{\alpha\beta}^{(j)}]$, $C = [c_{\alpha\beta}]$, $K = \begin{pmatrix} d_0 \\ d_1 \\ d_2 \\ d_3 \end{pmatrix}$, $b_{\alpha\beta}^{(j)}$, $c_{\alpha\beta}$, d_α, $(\alpha, \beta = 0, 1, 2, 3)$ are the real-valued functions which are supposed to depend at least continuously on the time t and the space-like variables x_0, x_1, x_2. A pair of operators ℓ, L is

said to be *associated* (see [9]) if $\ell f = 0$ implies $\ell(Lf) = 0$ (for each t in case the coefficient of L depend on t). Now we formulate the necessary and sufficient conditions on the coefficients of operator L under which L is associated to the operator ℓ (in other words, L is associated to the generalized Cauchy-Riemann operator of Quaternionic Analysis). Assume that the functions $b^{(j)}_{\alpha\beta}$, $c_{\alpha\beta}$, d_α ($j = 0, 1, 2, \alpha, \beta = 0, 1, 2, 3$) are continuously differentiable with respect to the space-like variables x_0, x_1, x_2 and differentiable in t.
Let us put

$$P_j = [p^{(j)}_{\alpha\beta}] = A_j B_j, \quad j = 0, 1, 2 \tag{5}$$

$$Q_{ij} = [q^{(ij)}_{\alpha\beta}] = A_i B_j + A_j B_i, \quad 0 \le i < j \le 2 \tag{6}$$

$$R_j = [r^{(j)}_{\alpha\beta}] = \sum_{i=0}^{2} A_i \frac{\partial B_j}{\partial x_i} + A_j C, \quad j = 0, 1, 2, \quad \alpha, \beta = 0, 1, 2, 3. \tag{7}$$

Then we get the following Theorem

Theorem 1. *The operator L is associated to the operator ℓ if and only if the following conditions are satisfied*

i) *The functions* $h^{(\alpha)} = \sum_{i=0}^{3} c_{i\alpha} e_i$, $\alpha = 0, 1, 2, 3$, *and* $g = \sum_{i=0}^{3} d_i e_i$ *are regular.*

ii) $$\begin{cases} r^{(1)}_{i0} = r^{(0)}_{i1}, & r^{(2)}_{i0} = r^{(0)}_{i2} \\ r^{(1)}_{i1} = -r^{(0)}_{i0}, & r^{(2)}_{i1} = -r^{(0)}_{i3} \\ r^{(1)}_{i2)} = r^{(0)}_{i3}, & r^{(2)}_{i2} = -r^{(0)}_{i0} \\ r^{(1)}_{i3} = -r^{(0)}_{i2}, & r^{(2)}_{i3} = r^{(0)}_{i1} \\ i = 0, 1, 2, 3. & \end{cases}$$

iii) $$\begin{cases} q^{(01)}_{i0} = p^{(0)}_{i1} - p^{(1)}_{i1}, & q^{(02)}_{i0} = p^{(0)}_{i2} - p^{(2)}_{i2}, & q^{(12)}_{i0} = -p^{(1)}_{i3} + p^{(2)}_{i3} \\ q^{(01)}_{i1} = -p^{(0)}_{i0} + p^{(1)}_{i0}, & q^{(02)}_{i1} = -p^{(0)}_{i3} + p^{(2)}_{i3}, & q^{(12)}_{i1} = -p^{(1)}_{i2} + p^{(2)}_{i2} \\ q^{(01)}_{i2} = p^{(0)}_{i3} - p^{(1)}_{i3}, & q^{(02)}_{i2} = -p^{(0)}_{i0} + p^{(2)}_{i0}, & q^{(12)}_{i2} = p^{(1)}_{i1} - p^{(2)}_{i1} \\ q^{(01)}_{i3} = -p^{(0)}_{i2} + p^{(1)}_{i2}, & q^{(02)}_{i3} = p^{(0)}_{i1} - p^{(2)}_{i1}, & q^{(12)}_{i3} = p^{(1)}_{i0} - p^{(2)}_{i0}. \end{cases}$$

Proof. We get

$$\begin{aligned} \ell(Lf) &= \sum_{i=0}^{2} A_i \frac{\partial(Lf)}{\partial x_i} \\ &= \sum_{i=0}^{2} A_i \frac{\partial}{\partial x_i} \left(\sum_{j=0}^{2} B_j \frac{\partial f}{\partial x_j} + Cf + K \right) \\ &= \sum_{i=0}^{2} A_i \frac{\partial}{\partial x_i} \left(\sum_{j=0}^{2} B_j \frac{\partial f}{\partial x_j} \right) + \sum_{i=0}^{2} A_i \frac{\partial(Cf)}{\partial x_i} + \sum_{i=0}^{2} A_i \frac{\partial K}{\partial x_i} \end{aligned}$$

$$
\begin{aligned}
&= \sum_{i=0}^{2}\sum_{j=0}^{2} A_i B_j \frac{\partial^2 f}{\partial x_i \partial x_j} + \sum_{i=0}^{2}\sum_{j=0}^{2} A_i \frac{\partial B_j}{\partial x_i}\frac{\partial f}{\partial x_j} \\
&\quad + \left(\sum_{i=0}^{2} A_i \frac{\partial C}{\partial x_i}\right) f + \sum_{i=0}^{2} A_i C \frac{\partial f}{\partial x_i} + \sum_{i=0}^{2} A_i \frac{\partial K}{\partial x_i} \\
&= \sum_{i=0}^{2} A_i B_i \frac{\partial^2 f}{\partial x_i^2} + \sum_{0\le i<j\le 2} (A_i B_j + A_j B_i)\frac{\partial^2 f}{\partial x_i \partial x_j} \\
&\quad + \sum_{j=0}^{2}\left(\sum_{i=0}^{2} A_i \frac{\partial B_j}{\partial x_i} + A_j C\right)\frac{\partial f}{\partial x_j} \\
&\quad + \left(\sum_{i=0}^{2} A_i \frac{\partial C}{\partial x_i}\right) f + \sum_{i=0}^{2} A_i \frac{\partial K}{\partial x_i}. \qquad (8)
\end{aligned}
$$

By (5), (6) and (7), we have that (8) can be rewritten as follows

$$
\begin{aligned}
l(Lf) &= \sum_{i=0}^{2} P_i \frac{\partial^2 f}{\partial x_i^2} + \sum_{0\le i<j\le 2} Q_{ij}\frac{\partial^2 f}{\partial x_i \partial x_j} + \sum_{j=0}^{2} R_j \frac{\partial f}{\partial x_j} \\
&\quad + \left(\sum_{i=0}^{2} A_i \frac{\partial C}{\partial x_i}\right) f + \sum_{i=0}^{2} A_i \frac{\partial K}{\partial x_i}. \qquad (9)
\end{aligned}
$$

Denote

$$
\begin{aligned}
M &= \sum_{i=0}^{2} P_i \frac{\partial^2 f}{\partial x_i^2} + \sum_{0\le i<j\le 2} Q_{ij}\frac{\partial^2 f}{\partial x_i \partial x_j} = \begin{pmatrix} m_0 \\ m_1 \\ m_2 \\ m_3 \end{pmatrix}, \\
N &= \sum_{j=0}^{2} R_j \frac{\partial f}{\partial x_j} = \begin{pmatrix} n_0 \\ n_1 \\ n_2 \\ n_3 \end{pmatrix}, \\
S &= \left(\sum_{i=0}^{2} A_i \frac{\partial C}{\partial x_i}\right) f, \\
T &= \sum_{i=0}^{2} A_i \frac{\partial K}{\partial x_i}.
\end{aligned}
$$

Then we obtain

$$
l(Lf) = M + N + S + T. \qquad (10)
$$

We get

$$\begin{aligned}
m_i = {} & p_{i0}^{(0)} \frac{\partial^2 f_0}{\partial x_0^2} + p_{i1}^{(0)} \frac{\partial^2 f_1}{\partial x_0^2} + p_{i2}^{(0)} \frac{\partial^2 f_2}{\partial x_0^2} + p_{i3}^{(0)} \frac{\partial^2 f_3}{\partial x_0^2} \\
& + p_{i0}^{(1)} \frac{\partial^2 f_0}{\partial x_1^2} + p_{i1}^{(1)} \frac{\partial^2 f_1}{\partial x_1^2} + p_{i2}^{(1)} \frac{\partial^2 f_2}{\partial x_1^2} + p_{i3}^{(1)} \frac{\partial^2 f_3}{\partial x_1^2} \\
& + p_{i0}^{(2)} \frac{\partial^2 f_0}{\partial x_2^2} + p_{i1}^{(2)} \frac{\partial^2 f_1}{\partial x_2^2} + p_{i2}^{(2)} \frac{\partial^2 f_2}{\partial x_2^2} + p_{i3}^{(2)} \frac{\partial^2 f_3}{\partial x_2^2} \\
& + q_{i0}^{(01)} \frac{\partial^2 f_0}{\partial x_0 \partial x_1} + q_{i1}^{(01)} \frac{\partial^2 f_1}{\partial x_0 \partial x_1} + q_{i2}^{(01)} \frac{\partial^2 f_2}{\partial x_0 \partial x_1} + q_{i3}^{(01)} \frac{\partial^2 f_3}{\partial x_0 \partial x_1} \\
& + q_{i0}^{(02)} \frac{\partial^2 f_0}{\partial x_0 \partial x_2} + q_{i1}^{(02)} \frac{\partial^2 f_1}{\partial x_0 \partial x_2} + q_{i2}^{(02)} \frac{\partial^2 f_2}{\partial x_0 \partial x_2} + q_{i3}^{(02)} \frac{\partial^2 f_3}{\partial x_0 \partial x_2} \\
& + q_{i0}^{(12)} \frac{\partial^2 f_0}{\partial x_1 \partial x_2} + q_{i1}^{(12)} \frac{\partial^2 f_1}{\partial x_1 \partial x_2} + q_{i2}^{(12)} \frac{\partial^2 f_2}{\partial x_1 \partial x_2} + q_{i3}^{(12)} \frac{\partial^2 f_3}{\partial x_1 \partial x_2}. \qquad (11)
\end{aligned}$$

Similarly,

$$\begin{aligned}
n_i = {} & r_{i0}^{(0)} \frac{\partial f_0}{\partial x_0} + r_{i1}^{(0)} \frac{\partial f_1}{\partial x_0} + r_{i2}^{(0)} \frac{\partial f_2}{\partial x_0} + r_{i3}^{(0)} \frac{\partial f_3}{\partial x_0} \\
& + r_{i0}^{(1)} \frac{\partial f_0}{\partial x_1} + r_{i1}^{(1)} \frac{\partial f_1}{\partial x_1} + r_{i2}^{(1)} \frac{\partial f_2}{\partial x_1} + r_{i3}^{(1)} \frac{\partial f_3}{\partial x_1} \\
& + r_{i0}^{(2)} \frac{\partial f_0}{\partial x_2} + r_{i1}^{(2)} \frac{\partial f_1}{\partial x_2} + r_{i2}^{(2)} \frac{\partial f_2}{\partial x_2} + +r_{i3}^{(2)} \frac{\partial f_3}{\partial x_2}. \qquad (12)
\end{aligned}$$

Suppose that f is regular , then

$$\begin{cases}
\frac{\partial f_0}{\partial x_0} - \frac{\partial f_1}{\partial x_1} - \frac{\partial f_2}{\partial x_2} = 0 \\
\frac{\partial f_0}{\partial x_1} + \frac{\partial f_1}{\partial x_0} + \frac{\partial f_3}{\partial x_2} = 0 \\
\frac{\partial f_0}{\partial x_2} + \frac{\partial f_2}{\partial x_0} - \frac{\partial f_3}{\partial x_1} = 0 \\
\frac{\partial f_1}{\partial x_2} - \frac{\partial f_2}{\partial x_1} - \frac{\partial f_3}{\partial x_0} = 0.
\end{cases} \qquad (13)$$

It follows from (13) that

$$\begin{cases}
\frac{\partial f_0}{\partial x_0} = \frac{\partial f_1}{\partial x_1} + \frac{\partial f_2}{\partial x_2} \\
\frac{\partial f_1}{\partial x_0} = -\frac{\partial f_0}{\partial x_1} - \frac{\partial f_3}{\partial x_2} \\
\frac{\partial f_2}{\partial x_0} = -\frac{\partial f_0}{\partial x_2} + \frac{\partial f_3}{\partial x_1} \\
\frac{\partial f_3}{\partial x_0} = \frac{\partial f_1}{\partial x_2} - \frac{\partial f_2}{\partial x_1}
\end{cases} \qquad (14)$$

and

$$\begin{cases}
\frac{\partial^2 f_0}{\partial x_0^2} = \frac{\partial^2 f_1}{\partial x_0 \partial x_1} + \frac{\partial^2 f_2}{\partial x_0 \partial x_2} \\
\frac{\partial^2 f_0}{\partial x_1^2} = -\frac{\partial^2 f_1}{\partial x_0 \partial x_1} - \frac{\partial^2 f_3}{\partial x_1 \partial x_2} \\
\frac{\partial^2 f_0}{\partial x_2^2} = -\frac{\partial^2 f_2}{\partial x_0 \partial x_2} + \frac{\partial^2 f_3}{\partial x_1 \partial x_2},
\end{cases} \qquad (15)$$

and the similar expressions for the other $\frac{\partial^2 f_i}{\partial x_j^2}, i = 1, 2, 3; \quad j = 0, 1, 2.$

Hence we get 3 remaining systems having the form of (15). Thus, one has a total of 12 equations. Substituting the above 12 equations into (11), and after a calculation, we obtain

$$\begin{aligned} m_i = &\left(-p_{i1}^{(0)} + p_{i1}^{(1)} + q_{i0}^{(01)}\right)\frac{\partial^2 f_0}{\partial x_0 \partial x_1} + \left(-p_{i2}^{(0)} + p_{i2}^{(2)} + q_{i0}^{(02)}\right)\frac{\partial^2 f_0}{\partial x_0 \partial x_2} \\ &+ \left(p_{i3}^{(1)} - p_{i3}^{(2)} + q_{i0}^{(12)}\right)\frac{\partial^2 f_0}{\partial x_1 \partial x_2} + \left(p_{i0}^{(0)} - p_{i0}^{(1)} + q_{i1}^{(01)}\right)\frac{\partial^2 f_1}{\partial x_0 \partial x_1} \\ &+ \left(p_{i3}^{(0)} - p_{i3}^{(2)} + q_{i1}^{(02)}\right)\frac{\partial^2 f_1}{\partial x_0 \partial x_2} + \left(p_{i2}^{(1)} - p_{i2}^{(2)} + q_{i1}^{(12)}\right)\frac{\partial^2 f_1}{\partial x_1 \partial x_2} \\ &+ \left(-p_{i3}^{(0)} + p_{i3}^{(1)} + q_{i2}^{(01)}\right)\frac{\partial^2 f_2}{\partial x_0 \partial x_1} + \left(p_{i0}^{(0)} - p_{i0}^{(2)} + q_{i2}^{(02)}\right)\frac{\partial^2 f_2}{\partial x_0 \partial x_2} \\ &+ \left(-p_{i1}^{(1)} + p_{i1}^{(2)} + q_{i2}^{(12)}\right)\frac{\partial^2 f_2}{\partial x_1 \partial x_2} + \left(p_{i2}^{(0)} - p_{i2}^{(1)} + q_{i3}^{(01)}\right)\frac{\partial^2 f_3}{\partial x_0 \partial x_1} \\ &+ \left(-p_{i1}^{(0)} + p_{i1}^{(2)} + q_{i3}^{(02)}\right)\frac{\partial^2 f_3}{\partial x_0 \partial x_2} + \left(-p_{i0}^{(1)} + p_{i0}^{(2)} + q_{i3}^{(12)}\right)\frac{\partial^2 f_3}{\partial x_1 \partial x_2}. \end{aligned} \tag{16}$$

Analogously, substituting the relation (14) into (12), one gets

$$\begin{aligned} n_i = &(-r_{i1}^{(0)} + r_{i0}^{(1)})\frac{\partial f_0}{\partial x_1} + (-r_{i2}^{(0)} + r_{i0}^{(2)})\frac{\partial f_0}{\partial x_2} + (r_{i0}^{(0)} + r_{i1}^{(1)})\frac{\partial f_1}{\partial x_1} \\ &+ (r_{i3}^{(0)} + r_{i1}^{(2)})\frac{\partial f_1}{\partial x_2} + (-r_{i3}^{(0)} + r_{i2}^{(1)})\frac{\partial f_2}{\partial x_1} + (r_{i0}^{(0)} + r_{i2}^{(2)})\frac{\partial f_2}{\partial x_2} \\ &+ (r_{i2}^{(0)} + r_{i3}^{(1)})\frac{\partial f_3}{\partial x_1} + (-r_{i1}^{(0)} + r_{i3}^{(2)})\frac{\partial f_3}{\partial x_2}. \end{aligned} \tag{17}$$

(*) The sufficient condition

Suppose that the conditions (i), (ii), and (iii) of the Theorem are satisfied. From the relation (i), it follows that $S = T = 0$. Because of (ii) it leads to $n_i = 0, i = 0,1,2,3$. Using the condition (iii) it implies $m_i = 0, i = 0,1,2,3$. This means that $M = N = 0$.

Hence $l(Lf) = M + N + S + T = 0$ for all regular functions f. The sufficient condition is proved.

(*) The necessary condition

Assume that (l, L) is an associated pair, i.e., if $lf = 0$, then $l(Lf) = 0$. We will choose 22 regular functions as follows.

First, choose $f^{(1)} = 0$, then (10) passes into T. Because $l(Lf) = 0$, then $T = 0$. This means that $g = \sum_{i=0}^{3} d_i e_i$ is a regular function. Thus the term T can be omitted in (10). Next, we choose $f^{(2)}$ as an arbitrary Quaternionic constant, $f^{(2)} \neq 0$. For this choice (10) implies $S = 0$. Since $f^{(2)}$ is arbitrary, then $\sum_{i=0}^{2} A_i \frac{\partial C}{\partial x_i} = 0$. In other words $h^{(\alpha)} = \sum_{i=0}^{3} c_{i\alpha} e_i$, $\alpha = 0,1,2,3$ are regular functions. Hence S vanishes in (10). Now, choose $f^{(3)} = x_0 + x_1 e_1$, then (10) leads to $N = 0$, so $n_i = 0, i = 0, 1, 2, 3$. But in fact $n_i = r_{i0}^{(0)} + r_{i1}^{(1)}$. Therefore, we get $r_{i1}^{(1)} = -r_{i0}^{(0)}$. Note that the equality is the same as the condition 3rd of the relation (i).

By a similar method, choose

$$\begin{aligned} f^{(4)} &= x_1 - x_0 e_1, \quad f^{(5)} = x_0 e_2 + x_1 e_3, \quad f^{(6)} = x_1 e_2 - x_0 e_3, \\ f^{(7)} &= x_0 + x_2 e_2, \quad f^{(8)} = x_0 e_1 - x_2 e_3, \quad f^{(9)} = x_2 - x_0 e_2, \\ f^{(10)} &= x_2 e_1 + x_0 e_3 \end{aligned}$$

and substitute these functions into (10) we obtain $N = 0$ for all $f^{(i)}$, $i = 4, \ldots, 10$. From this, we have the remaining equalities which are contained in the condition (ii). Hence N can be omitted in (10).

Now we choose $f^{(11)} = (x_0^2 - x_1^2) + 2x_0 x_1 e_1$ and replace f in (10) by $f^{(11)}$. It follows that $M = 0$. This means

$$m_i = -p_{i1}^{(0)} + p_{i1}^{(1)} + q_{i0}^{(01)} = 0, \quad i = 0, 1, 2, 3.$$

The equality leads to

$$q_{i0}^{(01)} = p_{i1}^{(0)} - p_{i1}^{(1)}. \tag{18}$$

Note that (18) is the same as the first condition of (iii). Similarly, choose

$$\begin{aligned} f^{(12)} &= (x_0^2 - x_2^2) + 2x_0 x_2 e_2, & f^{(13)} &= (x_1^2 - x_2^2) - 2x_1 x_2 e_3 \\ f^{(14)} &= -2x_0 x_1 + (x_0^2 - x_1^2) e_1 & f^{(15)} &= (x_0^2 - x_1^2) e_1 - 2x_0 x_2 e_3 \\ f^{(16)} &= (x_1^2 - x_2^2) e_1 - 2x_1 x_2 e_2 & f^{(17)} &= (x_0^2 - x_1^2) e_2 + 2x_0 x_1 e_3 \\ f^{(18)} &= -2x_0 x_2 + (x_0^2 - x_2^2) e_2 & f^{(19)} &= 2x_1 x_2 e_1 + (x_1^2 - x_2^2) e_2 \\ f^{(20)} &= -2x_0 x_1 + (x_0^2 - x_1^2) e_3 & f^{(21)} &= 2x_0 x_2 e_1 + (x_0^2 - x_2^2) e_3 \\ f^{(22)} &= 2x_1 x_2 + (x_1^2 - x_2^2) e_3, \end{aligned}$$

and substitute $f = f^{(j)}, j = 12, \ldots, 22$ into (10) one obtains $M = 0$. By a similar argument we get all the remaining equalities of the condition (iii). This completes the proof of the necessary condition. □

Remark 2. We can see that the conditions (ii) and (iii) of Theorem 1 can be written as follows

$$\begin{aligned} R_i &= R_0 A_i, \quad j = 1, 2 \\ Q_{ij} &= (P_i - P_j) A_j A_i, \quad 0 \le i < j \le 2. \end{aligned}$$

So we get

$$l(Lf) = \sum_{i=0}^{2} P_i \frac{\partial^2 f}{\partial x_i^2} + \sum_{0 \le i < j \le 2} (P_i - P_j) A_j A_i \frac{\partial^2 f}{\partial x_i \partial x_j} + \sum_{i=0}^{2} R_0 A_i \frac{\partial f}{\partial x_i}.$$

Note that $A_0 = E$, $A_i^2 = -E$ and $A_iA_j + A_jA_i = 0$, $i \neq j$, $i, j \in \{1, 2\}$. Then we easily obtain

$$l(Lf) = (P_0\frac{\partial}{\partial x_0} - P_1A_1\frac{\partial}{\partial x_1} - P_2A_2\frac{\partial}{\partial x_2} + R_0)(\sum_{i=0}^{2} A_i\frac{\partial f}{\partial x_i})$$
$$= V(lf),$$
$$\text{with } V = P_0\frac{\partial}{\partial x_0} - P_1A_1\frac{\partial}{\partial x_1} - P_2A_2\frac{\partial}{\partial x_2} + R_0,$$

and P_0, P_1, P_2, R_0 are given in (5) and (7)).

Therefore one gets the following Theorem

Theorem 2. *The operator L is associated to the operator ℓ if and only if*

$$lL = Vl,$$
$$\text{where } V = P_0\frac{\partial}{\partial x_0} - P_1A_1\frac{\partial}{\partial x_1} - P_2A_2\frac{\partial}{\partial x_2} + R_0.$$

Remark 3. If we replace the generalized Cauchy-Riemann operator by the Cauchy-Fueter operator

$$\mu = \sum_{k=0}^{3} e_k\frac{\partial}{\partial x_k}.$$

and consider the operators l, L which are given by

$$\ell f = \sum_{i=0}^{3} A_i\frac{\partial f}{\partial x_i}$$

where

$$A_0 = \begin{bmatrix} 1 & 0 & 0 & 0 \\ 0 & 1 & 0 & 0 \\ 0 & 0 & 1 & 0 \\ 0 & 0 & 0 & 1 \end{bmatrix}, \qquad A_1 = \begin{bmatrix} 0 & -1 & 0 & 0 \\ 1 & 0 & 0 & 0 \\ 0 & 0 & 0 & -1 \\ 0 & 0 & 1 & 0 \end{bmatrix},$$
$$A_2 = \begin{bmatrix} 0 & 0 & -1 & 0 \\ 0 & 0 & 0 & 1 \\ 1 & 0 & 0 & 0 \\ 0 & -1 & 0 & 0 \end{bmatrix}, \qquad A_3 = \begin{bmatrix} 0 & 0 & 0 & -1 \\ 0 & 0 & -1 & 0 \\ 0 & 1 & 0 & 0 \\ 1 & 0 & 0 & 0 \end{bmatrix}$$

and

$$Lf = \sum_{j=0}^{3} B_j\frac{\partial f}{\partial x_j} + Cf + K.$$

Putting

$$P_j = [p^{(j)}_{\alpha\beta}] = A_j B_j, \quad j = 0,1,2,3$$
$$Q_{ij} = [q^{(ij)}_{\alpha\beta}] = A_i B_j + A_j B_i, \quad 0 \le i < j \le 3.$$
$$R_j = [r^{(j)}_{\alpha\beta}] = \sum_{i=0}^{3} A_i \frac{\partial B_j}{\partial x_i} + A_j C, \quad j = 0,1,2,3, \quad \alpha,\beta = 0,1,2,3.$$

Then by an analogous method which is used in Section 2, we obtain the following Theorem

Theorem 3. *The operator L is associated to the operator ℓ if and only if the following conditions are satisfied*

i) *The functions $h^{(\alpha)} = \sum_{i=0}^{3} c_{i\alpha} e_i$, $\alpha = 0,1,2,3$, and $g = \sum_{i=0}^{3} d_i e_i$ are regular.*

ii)
$$\begin{cases} r^{(1)}_{i0} = r^{(0)}_{i1}, & r^{(2)}_{i0} = r^{(0)}_{i2}, \quad r^{(3)}_{i0} = r^{(0)}_{i3} \\ r^{(1)}_{i1} = -r^{(0)}_{i0}, & r^{(2)}_{i1} = -r^{(0)}_{i3}, \quad r^{(3)}_{i1} = r^{(0)}_{i2} \\ r^{(1)}_{i2)} = r^{(0)}_{i3}, & r^{(2)}_{i2} = -r^{(0)}_{i0}, \quad r^{(3)}_{i2} = -r^{(0)}_{i1} \\ r^{(1)}_{i3} = -r^{(0)}_{i2}, & r^{(2)}_{i3} = r^{(0)}_{i1}, \quad r^{(3)}_{i3} = -r^{(0)}_{i0} \\ i = 0,1,2,3. \end{cases}$$

iii)
$$\begin{cases} q^{(01)}_{i0} = p^{(0)}_{i1} - p^{(1)}_{i1}, & q^{(02)}_{i0} = p^{(0)}_{i2} - p^{(2)}_{i2}, & q^{(03)}_{i0} = p^{(0)}_{i3} - p^{(3)}_{i3} \\ q^{(12)}_{i0} = -p^{(1)}_{i3} + p^{(2)}_{i3}, & q^{(13)}_{i0} = p^{(1)}_{i2} - p^{(3)}_{i2}, & q^{(23)}_{i0} = -p^{(2)}_{i1} + p^{(3)}_{i1} \\ q^{(01)}_{i1} = -p^{(0)}_{i0} + p^{(1)}_{i0}, & q^{(02)}_{i1} = -p^{(0)}_{i3} + p^{(2)}_{i3}, & q^{(03)}_{i1} = p^{(0)}_{i2} - p^{(3)}_{i2} \\ q^{(12)}_{i1} = -p^{(1)}_{i2} + p^{(2)}_{i2}, & q^{(13)}_{i1} = -p^{(1)}_{i3} + p^{(3)}_{i3}, & q^{(23)}_{i1} = p^{(2)}_{i0} - p^{(3)}_{i0} \\ q^{(01)}_{i2} = p^{(0)}_{i3} - p^{(1)}_{i3}, & q^{(02)}_{i2} = -p^{(0)}_{i0} + p^{(2)}_{i0}, & q^{(03)}_{i2} = -p^{(0)}_{i1} + p^{(3)}_{i1} \\ q^{(12)}_{i2} = p^{(1)}_{i1} - p^{(2)}_{i1}, & q^{(13)}_{i2} = -p^{(1)}_{i0} + p^{(3)}_{i0}, & q^{(23)}_{i2} = -p^{(2)}_{i3} + p^{(3)}_{i3} \\ q^{(01)}_{i3} = -p^{(0)}_{i2} + p^{(1)}_{i2}, & q^{(02)}_{i3} = p^{(0)}_{i1} - p^{(2)}_{i1}, & q^{(03)}_{i3} = -p^{(0)}_{i0} + p^{(3)}_{i0} \\ q^{(12)}_{i3} = p^{(1)}_{i0} - p^{(2)}_{i0}, & q^{(13)}_{i3} = p^{(1)}_{i1} - p^{(3)}_{i1}, & q^{(23)}_{i3} = p^{(2)}_{i2} - p^{(3)}_{i2}. \end{cases}$$

Remark 4. In this case we also have the relations

$$R_i = R_0 A_i, \quad j = 1,2,3$$
$$Q_{ij} = (P_i - P_j) A_j A_i, \quad 0 \le i < j \le 3,$$

$A_0 = E$, $A_i^2 = -E$, $i = 1,2,3$ and $A_i A_j + A_j A_i = 0$, $i \ne j$, $i,j \in \{1,2,3\}$. Then the following Theorem is proved

Theorem 4. *The operator L is associated to the operator ℓ if and only if*

$$lL = Vl,$$

where $V = P_0 \dfrac{\partial}{\partial x_0} - P_1 A_1 \dfrac{\partial}{\partial x_1} - P_2 A_2 \dfrac{\partial}{\partial x_2} - P_3 A_3 \dfrac{\partial}{\partial x_3} + R_0.$

Remark 5. If we consider the Clifford algebra over 3 units, then one gets 8 basis elements $e_0, e_1, \ldots, e_7$, where $e_4 = e_{12}, e_5 = e_{13}, e_6 = e_{23}, e_7 = e_{123}$. Considering the generalized Cauchy-Riemann operator

$$D = \sum_{k=0}^{3} e_k \frac{\partial}{\partial x_k},$$

the operators l, L are given by

$$\ell f = \sum_{i=0}^{3} A_i \frac{\partial f}{\partial x_i}$$

where

$$A_0 = \begin{bmatrix} 1 & 0 & 0 & 0 & 0 & 0 & 0 & 0 \\ 0 & 1 & 0 & 0 & 0 & 0 & 0 & 0 \\ 0 & 0 & 1 & 0 & 0 & 0 & 0 & 0 \\ 0 & 0 & 0 & 1 & 0 & 0 & 0 & 0 \\ 0 & 0 & 0 & 0 & 1 & 0 & 0 & 0 \\ 0 & 0 & 0 & 0 & 0 & 1 & 0 & 0 \\ 0 & 0 & 0 & 0 & 0 & 0 & 1 & 0 \\ 0 & 0 & 0 & 0 & 0 & 0 & 0 & 1 \end{bmatrix}, \qquad A_1 = \begin{bmatrix} 0 & -1 & 0 & 0 & 0 & 0 & 0 & 0 \\ 1 & 0 & 0 & 0 & 0 & 0 & 0 & 0 \\ 0 & 0 & 0 & 0 & -1 & 0 & 0 & 0 \\ 0 & 0 & 0 & 0 & 0 & -1 & 0 & 0 \\ 0 & 0 & 1 & 0 & 0 & 0 & 0 & 0 \\ 0 & 0 & 0 & 1 & 0 & 0 & 0 & 0 \\ 0 & 0 & 0 & 0 & 0 & 0 & 0 & -1 \\ 0 & 0 & 0 & 0 & 0 & 0 & 1 & 0 \end{bmatrix},$$

$$A_2 = \begin{bmatrix} 0 & 0 & -1 & 0 & 0 & 0 & 0 & 0 \\ 0 & 0 & 0 & 0 & 1 & 0 & 0 & 0 \\ 1 & 0 & 0 & 0 & 0 & 0 & 0 & 0 \\ 0 & 0 & 0 & 0 & 0 & 0 & -1 & 0 \\ 0 & -1 & 0 & 0 & 0 & 0 & 0 & 0 \\ 0 & 0 & 0 & 0 & 0 & 0 & 0 & 1 \\ 0 & 0 & 0 & 1 & 0 & 0 & 0 & 0 \\ 0 & 0 & 0 & 0 & 0 & -1 & 0 & 0 \end{bmatrix}, \qquad A_3 = \begin{bmatrix} 0 & 0 & 0 & -1 & 0 & 0 & 0 & 0 \\ 0 & 0 & 0 & 0 & 0 & 1 & 0 & 0 \\ 0 & 0 & 0 & 0 & 0 & 0 & 1 & 0 \\ 1 & 0 & 0 & 0 & 0 & 0 & 0 & 0 \\ 0 & 0 & 0 & 0 & 0 & 0 & 0 & -1 \\ 0 & -1 & 0 & 0 & 0 & 0 & 0 & 0 \\ 0 & 0 & -1 & 0 & 0 & 0 & 0 & 0 \\ 0 & 0 & 0 & 0 & 1 & 0 & 0 & 0 \end{bmatrix},$$

and

$$Lf = \sum_{j=0}^{3} B_j \frac{\partial f}{\partial x_j} + Cf + K.$$

Then we obtain the following result

Theorem 5. *The operator L is associated to the operator ℓ if and only if*

$$lL = Vl,$$

where $V = P_0 \dfrac{\partial}{\partial x_0} - P_1 A_1 \dfrac{\partial}{\partial x_1} - P_2 A_2 \dfrac{\partial}{\partial x_2} - P_3 A_3 \dfrac{\partial}{\partial x_3} + R_0,$

P_0, P_1, P_2, P_3 and R_0 are defined as in Remark 3.

3. Example

3.1. The operator L is associated to the generalized Cauchy-Riemann operator

First, we choose $c_{\alpha\beta}$ as the arbitrary real-constants, $g = \sum_{i=0}^{3} d_i e_i$ is an arbitrary regular function and we choose the elements $b_{\alpha\beta}^{(0)}, \alpha, \beta = 0, 1, 2, 3$ of the matrix B_0 as follows

$$\begin{aligned}
b_{00}^{(0)} &= -(\gamma - c_{00})x_0 - c_{10}x_1 - c_{20}x_2 + \delta_{00}^{(0)}\\
b_{01}^{(0)} &= c_{01}x_0 + (\gamma - c_{11})x_1 - c_{21}x_2 + \delta_{01}^{(0)}\\
b_{02}^{(0)} &= c_{02}x_0 - c_{12}x_1 + (\gamma - c_{22})x_2 + \delta_{02}^{(0)}\\
b_{03}^{(0)} &= c_{03}x_0 - c_{13}x_1 - c_{23}x_2 + \delta_{03}^{(0)}\\
b_{10}^{(0)} &= c_{10}x_0 - (\gamma - c_{00})x_1 + c_{30}x_2 + \delta_{10}^{(0)}\\
b_{11}^{(0)} &= -(\gamma - c_{11})x_0 + c_{01}x_1 + c_{31}x_2 + \delta_{11}^{(0)}\\
b_{12}^{(0)} &= c_{12}x_0 + c_{02}x_1 + c_{32}x_2 + \delta_{12}^{(0)}\\
b_{13}^{(0)} &= c_{13}x_0 + c_{03}x_1 - (\gamma - c_{33})x_2 + \delta_{13}^{(0)}\\
b_{20}^{(0)} &= c_{20}x_0 - c_{30}x_1 - (\gamma - c_{00})x_2 + \delta_{20}^{(0)}\\
b_{21}^{(0)} &= c_{21}x_0 - c_{31}x_1 + c_{01}x_2 + \delta_{21}^{(0)}\\
b_{22}^{(0)} &= -(\gamma - c_{22})x_0 - c_{32}x_1 + c_{02}x_2 + \delta_{22}^{(0)}\\
b_{23}^{(0)} &= c_{23}x_0 + (\gamma - c_{33})x_1 + c_{03}x_2 + \delta_{23}^{(0)}\\
b_{30}^{(0)} &= c_{30}x_0 + c_{20}x_1 - c_{10}x_2 + \delta_{30}^{(0)}\\
b_{31}^{(0)} &= c_{31}x_0 + c_{21}x_1 + (\gamma - c_{11})x_2 + \delta_{31}^{(0)}\\
b_{32}^{(0)} &= c_{32}x_0 - (\gamma - c_{22})x_1 - c_{12}x_2 + \delta_{32}^{(0)}\\
b_{33}^{(0)} &= -(\gamma - c_{33})x_0 + c_{23}x_1 - c_{13}x_2 + \delta_{33}^{(0)},
\end{aligned}$$

where $\gamma, \delta_{\alpha\beta}^{(0)}, \alpha, \beta = 0, 1, 2, 3$ are the arbitrary real-constants.

Second, choose $B_1 = -A_1B_0$ and $B_2 = -A_2B_0$. Then it is easy to verify that all the conditions of the Theorem 1 are satisfied. In this way one obtains a class of the differential operators L which are associated to the generalized Cauchy-Riemann operator on the Quaternionic Algebra.

3.2. The operator L is associated to the Cauchy-Fueter operator

Take the arbitrary real-constants $c_{\alpha\beta}$ and the arbitrary regular function

$$g = \sum_{i=0}^{3} d_i e_i.$$

Further the elements $b^{(0)}_{\alpha\beta}$, $\alpha, \beta = 0, 1, 2, 3$ of the matrix B_0 are given by

$$\begin{aligned}
b^{(0)}_{00} &= \frac{1}{2}[-(\gamma - c_{00})x_0 - c_{10}x_1 - c_{20}x_2 - c_{30}x_3] + \delta^{(0)}_{00} \\
b^{(0)}_{01} &= \frac{1}{2}[c_{01}x_0 + (\gamma - c_{11})x_1 - c_{21}x_2 - c_{31}x_3] + \delta^{(0)}_{01} \\
b^{(0)}_{02} &= \frac{1}{2}[c_{02}x_0 - c_{12}x_1 + (\gamma - c_{22})x_2 - c_{32}x_3] + \delta^{(0)}_{02} \\
b^{(0)}_{03} &= \frac{1}{2}[c_{03}x_0 - c_{13}x_1 - c_{23}x_2 + (\gamma - c_{33})x_3] + \delta^{(0)}_{03} \\
b^{(0)}_{10} &= \frac{1}{2}[c_{10}x_0 - (\gamma - c_{00})x_1 + c_{30}x_2 - c_{20}x_3] + \delta^{(0)}_{10} \\
b^{(0)}_{11} &= \frac{1}{2}[-(\gamma - c_{11})x_0 + c_{01}x_1 + c_{31}x_2 - c_{21}x_3] + \delta^{(0)}_{11} \\
b^{(0)}_{12} &= \frac{1}{2}[c_{12}x_0 + c_{02}x_1 + c_{32}x_2 + (\gamma - c_{22})x_3] + \delta^{(0)}_{12} \\
b^{(0)}_{13} &= \frac{1}{2}[c_{13}x_0 + c_{03}x_1 - (\gamma - c_{33})x_2 - c_{23}x_3] + \delta^{(0)}_{13} \\
b^{(0)}_{20} &= \frac{1}{2}[c_{20}x_0 - c_{30}x_1 - (\gamma - c_{00})x_2 + c_{10}x_3] + \delta^{(0)}_{20} \\
b^{(0)}_{21} &= \frac{1}{2}[c_{21}x_0 - c_{31}x_1 + c_{01}x_2 - (\gamma - c_{11})x_3] + \delta^{(0)}_{21} \\
b^{(0)}_{22} &= \frac{1}{2}[-(\gamma - c_{22})x_0 - c_{32}x_1 + c_{02}x_2 + c_{12}x_3] + \delta^{(0)}_{22} \\
b^{(0)}_{23} &= \frac{1}{2}[c_{23}x_0 + (\gamma - c_{33})x_1 + c_{03}x_2 + c_{13}x_3] + \delta^{(0)}_{23} \\
b^{(0)}_{30} &= \frac{1}{2}[c_{30}x_0 + c_{20}x_1 - c_{10}x_2 - (\gamma - c_{00})x_3] + \delta^{(0)}_{30} \\
b^{(0)}_{31} &= \frac{1}{2}[c_{31}x_0 + c_{21}x_1 + (\gamma - c_{11})x_2 + c_{01}x_3] + \delta^{(0)}_{31} \\
b^{(0)}_{32} &= \frac{1}{2}[c_{32}x_0 - (\gamma - c_{22})x_1 - c_{12}x_2 + c_{02}x_3] + \delta^{(0)}_{32} \\
b^{(0)}_{33} &= \frac{1}{2}[-(\gamma - c_{33})x_0 + c_{23}x_1 - c_{13}x_2 + c_{03}x_3] + \delta^{(0)}_{33},
\end{aligned}$$

where γ, $\delta^{(0)}_{\alpha\beta}$, $\alpha, \beta = 0, 1, 2, 3$ are the arbitrary real-constants.
Next, choose

$$\begin{cases} B_1 = -A_1 B_0 \\ B_2 = -A_2 B_0 \\ B_3 = -A_3 B_0. \end{cases}$$

Then we can see that all the conditions of the Theorem 3 hold. So one gets a class of the differential operators L which are associated to the Cauchy-Fueter operator of Quaternionic Algebra.

4. The initial value problems with regular initial functions

The classical Cauchy-Kovalevskaya Theorem (in Complex Analysis) shows that the initial value problem is solvable provided L has holomorphic coefficients and the initial function is holomorphic, but in view of the H.Lewy example (see [4]), there exist linear first-order differential equations with infinitely differentiable coefficients not having any solutions. On the other hand, by the criterion which is given in Theorem 1 (and Theorem 3, respectively), we can construct operator L such that the initial value problem (1), (2) is solvable for each regular initial function φ. Because the components of regular functions are harmonic so the necessary interior estimate (see [10]) follows from the Poisson Integral Formula.

Finally, we get the following theorem

Theorem 6. *Suppose that the operator L is associated to the generalized Cauchy-Riemann operator (the Cauchy-Fueter, respectively) of Quaternionic Algebra. Then the initial value problem* (1), (2) *is solvable for any arbitrary initial regular function φ and the solution $u(t, x)$ is regular for each t.*

References

[1] F. Brackx, R. Delanghe and F. Sommen, *Clifford Algebra*, Research Notes in Math., Pitman, London, 1982.

[2] Nguyen Minh Chuong, L. Nirenberg and W. Tutschke, *Abstract and Applied Analysis*, World Scientific, Singapore, 2004.

[3] H. Florian, N. Ortner, F.J. Schnitzer and W. Tutschke, *Functional-analytic and Comlplex Methods, their Interactions and Applications to Partial Differential Equations*, World Scientific, Singapore, 2001.

[4] H. Lewy, *An Example of a Smooth Linear Partial Differential Equations Without Solution*, Ann.Math., **66**, 155–158 (1957).

[5] E. Obolashvili, *Partial Differential Equations in Clifford Analysis*, Addison Wesley Longman, Harlow, 1988.

[6] Le Hung Son and W. Tutschke, *First Order Differential Operator Associated to the Cauchy-Riemann Operator in the Plane*, Complex Variables, **48**, 797–801 (2003).

[7] Le Hung Son and Nguyen Thanh Van, *Differential Associated Operators in a Clifford Analysis and Their Applications*, Complex Analysis and its Applications, Procceedings of 15th ICFIDCAA, Osaka Municipal Universities Press, Vol. **2(2)**, 325–332 (2007).

[8] A. Sudbery, *Quaternionic Analysis*, Math. Proc. Camb. Phil. Soc., **85**, 199-225 (1979).

[9] W. Tutschke, *Solution of Initial Value Problems in Classes of Generalized Analytic Functions*, Teubner Leipzig and Springer Verlag, 1989.

[10] W. Tutschke, *Associated Partial Differential Operators – Applications to well- and ill-posed Problems*, Contained in [2], 373–383.

[11] Nguyen Thanh Van, *First Order Differential Operators Transforming Regular Function of Quarternionic Analysis into Themselves*, Method of Complex and Clifford Analysis, SAS International Publications Delhi, 363–367 (2006).

[12] Nguyen Thanh Van, *First Order Differential Operators Associated to the Cauchy-Riemann Operator in a Clifford Algebra*, Accepted for Publication in the Proceedings of the 14th ICFIDCAA, Hue, Vietnam (2006).

[13] Nguyen Thanh Van, *First Order Differential Operators Associated to the Cauchy-Riemann Operator in a Quaternionic Algebra*, Preprint of ICTP, IC/135/2006.

[14] W. Walter, *An Elementary Proof of the Cauchy-Kowalevsky Theorem*, Amer. Math. Monthly, 92, 115–125 (1985).

Le Hung Son
Faculty of Applied Mathematics
Hanoi University of Technology
Hanoi, Vietnam
e-mail: `sonlehung2003@yahoo.com`

Nguyen Thanh Van
Faculty of Mathematics, Mechanics and Informatics
Hanoi University of Science
Hanoi, Vietnam
e-mail: `thanhvanao@yahoo.com`

Quaternionic and Clifford Analysis
Trends in Mathematics, 221–234

Hyperderivatives in Clifford Analysis and Some Applications to the Clifordian Cauchy-type Integrals

M.E. Luna-Elizarrarás*, M.A. Macías-Cedeño and M. Shapiro*

Abstract. We introduce the notion of the derivative as the limit of a quotient where the numerator and the denominator represent a kind of the "increments" of a function and of the independent variable respectively. The directional derivative is introduced where a direction means a hyperplane in $\mathbb{R}^{m+1}$ for a Clifford algebra $C\ell_{0,m}$. The latter applies for obtaining a formula showing how to exchange the integral sign and the hyperderivative of the Cliffordian Cauchy-type integral as a hyperholomorphic function.

Mathematics Subject Classification (2000). Primary 30G35; Secondary 32A10.

Keywords. Hyperderivative, m-dimensional directional hyperderivative, Cliffordian Cauchy-type integrals.

1. Introduction and rudiments of Clifford analysis

1.1. In one-dimensional complex analysis there are several equivalent approaches to the introducing of the main object of interest, the class of holomorphic (called also analytic, regular, etc) functions. Among them are: the derivative of a complex function as the limit of a certain quotient, the Cauchy-Riemann conditions, the complex differentiability as a special case of real differentiability in $\mathbb{R}^2$; but many others as well. Clifford analysis pretends to be a proper generalization of it onto higher dimensions but the corresponding class of functions (called in this work hyperholomorphic functions) is defined, almost allways, in terms of the Cauchy-Riemann conditions.

In the present work we introduce the notion of the derivative in terms of the limit of a quotient where the numerator and the denominator have a good

*The research was partially supported by CONACYT projects as well as by Instituto Politécnico Nacional in the framework of COFAA and SIP programs.

reason to be seen as the increments of a function and of the independent variable respectively. We introduce also a directional derivative where the direction means a hyperplane in $\mathbb{R}^{m+1}$. It is remarkable that there is a deep analogy among the relations of both notions in Clifford analysis and in its one-dimensional complex analysis counterpart. The directional derivative becomes crucial when one wants to realize in which sense the density of the Cauchy-type integral should be derivatived if one tries to exchange the integral sign and the hyperderivative of the Cauchy-type integral as a hyperholomorphic function.

1.1.1. We are aware about a handful of directly preceding papers only. The topic began in the famous Sudbery paper [12] in the framework of quaternionic analysis which was followed by [9] again dealing with quaternionic analysis, and by [3], [8] now in the Clifford analysis context.

1.2. Given $m \in \mathbb{N}$, let $C\ell_{0,m}$ denote the corresponding real Clifford algebra with signature $(0, m)$. An exhaustive information about Clifford algebras can be found in many sources, some of them are [6], [10], [1], [2]. In order to fix the notation we give here a necessary minimum of it. We denote by $\mathbf{e}_0 = 1$ and $\mathbf{e}_1, \mathbf{e}_2, \dots, \mathbf{e}_m$ respectively the real unit and the "imaginary units " of $C\ell_{0,m}$ meaning that

$$\mathbf{e}_k^2 = -\mathbf{e}_0 \qquad (k = 1, 2, \dots, m),$$
$$\mathbf{e}_k\mathbf{e}_\ell + \mathbf{e}_\ell\mathbf{e}_k = -2\delta_{k\ell}\mathbf{e}_0, \quad (k, \ell = 1, 2, \dots, m),$$

where $\delta_{k\ell}$ is the Kronecker symbol. As a real linear space, $C\ell_{0,m}$ is 2^m-dimensional with a basis $\{\mathbf{e}_A : A \subseteq \{1, 2, \dots, m\}\}$ where $\mathbf{e}_A := \mathbf{e}_{h_1}\mathbf{e}_{h_2} \cdots \mathbf{e}_{h_r}, 1 \le h_1, < \cdots < h_r \le m, \mathbf{e}_\emptyset = \mathbf{e}_0 = 1$. Hence the Clifford numbers (i.e., the elements of the Clifford algebra) are of the form

$$a = \sum_A a_A \mathbf{e}_A \,,$$

with $a_A \in \mathbb{R}$. The conjugate of a is defined by

$$Z(a) := \overline{a} = \sum_A a_A \, \overline{\mathbf{e}}_A \,,$$

where

$$\overline{\mathbf{e}}_A := \overline{\mathbf{e}}_{h_r}\overline{\mathbf{e}}_{h_{r-1}} \cdots \overline{\mathbf{e}}_{h_1}; \quad \overline{\mathbf{e}}_k := -\mathbf{e}_k \, (k = 1, \dots, m) \,, \quad \overline{\mathbf{e}}_0 = \mathbf{e}_0 = 1 \,;$$

and for $a, b \in C\ell_{0,m}$ it holds:

$$\overline{a\,b} = \overline{b}\,\overline{a} \,. \tag{1.2.1}$$

1.3. Let $\Omega \subset \mathbb{R}^{m+1}$ be a domain. We'll work with $C\ell_{0,m}$-valued functions $f : \Omega \to C\ell_{0,m}$, hence they are of the form

$$f(x) = \sum_A f_A(x)\, \mathbf{e}_A \,,$$

where $f_A(x) \in \mathbb{R}$ and $x \in \Omega$.

If all the components of f belongs to some function class (C^k, L_p, etc) then we say that f itself belongs to the class. We do not touch a much more delicate question about additional structures of those classes, namely, whether they form Clifford modules, or is it possible to endow them with a norm, etc.

1.4. The (generalized) Cauchy-Riemann operator is defined on $C^1(\Omega, C\ell_{0,m})$ by

$$D := \sum_{\ell=0}^{m} \mathbf{e}_\ell \frac{\partial}{\partial x_\ell}, \tag{1.4.1}$$

and its conjugate by

$$\overline{D} := \sum_{\ell=0}^{m} \overline{\mathbf{e}}_\ell \frac{\partial}{\partial x_\ell}. \tag{1.4.2}$$

They possess a crucial property of factorizing the Laplace operator in $\mathbb{R}^{m+1}$, i.e.,

$$D\,\overline{D} = \overline{D}\,D = \Delta_{\mathbb{R}^{m+1}}. \tag{1.4.3}$$

Let $a \in C\ell_{0,m}$ then *the right multiplication operator* $M^a : C\ell_{0,m} \to C\ell_{0,m}$ is defined by

$$M^a[b] := b\,a,$$

for $b \in C\ell_{0,m}$.

Extending the last definition onto the functions $f \in C^1(\Omega, C\ell_{0,m})$ the *right Cauchy-Riemann operator* is defined by

$$D_r := \sum_{\ell=0}^{m} M^{\mathbf{e}_\ell} \frac{\partial}{\partial x_\ell}, \tag{1.4.4}$$

and its conjugate by

$$\overline{D}_r := \sum_{\ell=0}^{m} M^{\overline{\mathbf{e}}_\ell} \frac{\partial}{\partial x_\ell}. \tag{1.4.5}$$

From (1.2.1) it holds:

$$D_r = Z\overline{D}Z, \quad \overline{D}_r = ZDZ.$$

Hence operators (1.4.4) and (1.4.5) preserve the property of factorizing the Laplace operator:

$$D_r\,\overline{D}_r = \overline{D}_r\,D_r = \Delta_{\mathbb{R}^{m+1}}.$$

1.5. Definition. *A function $f \in C^1(\Omega, C\ell_{0,m})$ is called left-hyperholomorphic if*

$$D\,[f]\,(x) = \frac{\partial f}{\partial x_0}(x) + \mathbf{e}_1 \frac{\partial f}{\partial x_1}(x) + \cdots + \mathbf{e}_m \frac{\partial f}{\partial x_m}(x) = 0 \tag{1.5.1}$$

in Ω. We set $\mathfrak{M}(\Omega) :=$ ker D. Similarly, f is right-hyperholomorphic if $D_r[f] = 0$, and we set $\mathfrak{M}_r(\Omega) :=$ ker D_r.

1.5.1. One of the most important examples of a hyperholomorphic function is the function

$$E(x) = \frac{\overline{x}}{A_{m+1}|x|^{m+1}}, \tag{1.5.2}$$

where A_{m+1} is the surface area of the unitary sphere $\mathbb{S}^m$ in $\mathbb{R}^{m+1}$. It is known as the *Cauchy kernel* and it is the fundamental solution of the Cauchy-Riemann operators (1.4.1) and (1.4.4).

1.6. We shall need several differential forms in what follows. As usual, the volume element in $\mathbb{R}^{m+1}$ is the real-valued $(m+1)$-form given by

$$dV := dx_0 \wedge dx_1 \wedge \cdots \wedge dx_m \,. \tag{1.6.1}$$

Then one sets:

$$\sigma_x := d\hat{x}_0 - \mathbf{e}_1 d\hat{x}_1 + \cdots + (-1)^m \mathbf{e}_m d\hat{x}_m \,, \tag{1.6.2}$$

where $d\hat{x}_k$, is the differential m-form which is obtained from (1.6.1) omitting the factor dx_k, for $k = 0, 1, \ldots, m$. One may call it a Cliffordian representation of the (m-dimensional) surface element since if Γ is a smooth surface in $\mathbb{R}^{m+1}$, then

$$\sigma_x = n_x dS_x \,,$$

where $n_x := \sum_{\ell=0}^{m} \mathbf{e}_\ell \, n_{\ell,x}$ and $\vec{n} := (n_0, n_1, \ldots, n_m)$ is the outward-pointing unit normal at the point $x \in \Gamma$, dS_x is the elementary surface element in $\mathbb{R}^{m+1}$.

Finally, we set, following [8], τ_x to be an $(m-1)$-differential form given by

$$\tau_x := -\mathbf{e}_1 d\widehat{x}_{0,1} + \mathbf{e}_2 d\widehat{x}_{0,2} + \cdots + (-1)^{m-1} \mathbf{e}_m d\widehat{x}_{0,m} \,. \tag{1.6.3}$$

Here $d\widehat{x}_{0,\ell}$ denotes the $(m-1)$-differential form $d\hat{x}_0$ with the factor dx_ℓ, $\ell = 1, \ldots, m$, omitted.

1.7. In order to facilitate their usage, we present below some basic integral formulas.

1.7.1. Cauchy-integral formula. *Let Ω be a bounded domain in $\mathbb{R}^{m+1}$ and let Γ be its boundary which we assume to have a sufficient smoothness. Then for $f \in \mathfrak{M}(\Omega) \cap C(\overline{\Omega}, C\ell_{0,m})$ there holds:*

$$\frac{1}{A_{m+1}} \int_\Gamma \frac{\overline{y-x}}{|y-x|^{m+1}} \, \sigma_y f(y) = \begin{cases} f(x), & \text{if } x \in \Omega \,, \\ 0, & \text{if } x \in \mathbb{R}^{m+1} \setminus \overline{\Omega} \,. \end{cases} \tag{1.7.1}$$

1.7.2. Sokhotski-Plemelj formulas. *Set $\Omega^+ := \Omega$, $\Omega^- := \mathbb{R}^{m+1} \setminus \overline{\Omega^+}$ and let Γ be additionally a Lyapunov surface, then for $f \in C^{0,\mu}(\Gamma, C\ell_{0,m})$, the set of Hölder functions with $0 < \mu \leq 1$, and for $x^0 \in \Gamma$ there holds:*

$$\lim_{\Omega^\pm \ni x \to x^0} \frac{1}{A_{m+1}} \int_\Gamma \frac{\overline{y-x}}{|y-x|^{m+1}} \, \sigma_y f(y) = \pm \frac{1}{2} f(x^0) + \frac{1}{A_{m+1}} \int_\Gamma \frac{\overline{y-x^0}}{|y-x^0|^{m+1}} \, \sigma_y f(y). \tag{1.7.2}$$

Note that both formulas hold under much more general conditions but we shall not use this in the sequel.

2. Hyperholomorphy, hyperdifferentiability and hyperderivability in Clifford analysis

In the previous section we dealt with the global concept of hyperholomorphy for a function $f \in C^1(\Omega, C\ell_{0,m})$, the one related to the whole domain of it. Now we concentrate on a local version of hyperholomorphy.

2.1. Definition. *A function $f : \Omega \to C\ell_{0,m}$ is called left-hyperholomorphic at $x^0 \in \Omega$ if f is left-hyperholomorphic in some open neighborhood $V(x^0) \subset \Omega$ of x^0.*

2.1.1. For $m = 1$ there holds:

$$C\ell_{0,1} = \mathbb{C}; \quad D = \sum_{\ell=0}^{1} \mathbf{e}_\ell \frac{\partial}{\partial x_\ell} = \frac{\partial}{\partial x_0} + \mathbf{e}_1 \frac{\partial}{\partial x_1} =: 2\frac{\partial}{\partial \overline{z}}$$

with $z = x_0 + \mathbf{e}_1 x_1$, the usual complex variable where $\mathbf{e}_1$ stands for the imaginary unit i. Hence the hyperholomorphy, global or local, in this case becomes a usual holomorphy of complex functions expressed in terms of the Cauchy-Riemann conditions.

2.2. Definition. *The function $f \in C^1(\Omega, C\ell_{0,m})$ is called left-hyperdifferentiable in Ω if for any $x \in \Omega$ there is a Clifford number denoted by $f'(x)$, such that*

$$d(\tau_x\, f(x)) = \sigma_x\, f'(x)\,. \tag{2.2.1}$$

The Clifford number $f'(x)$ is named the left-hyperderivative of f at x.

2.2.1. Again taking $m = 1$ we get from (2.2.1) that $df = f'(z)dz$, with $f'(z)$ being the usual derivative of a complex function, which explains the terminology introduced; of course here the complex derivative at a point is understood as a proportionality coefficient.

2.3. In complex analysis case, the holomorphy is equivalent to the complex differentiability. Let us show that this extends onto the hyperholomorphy and the hyperdifferentiability for any Clifford algebra $C\ell_{0,m}$.

In [8] it is proved that for $x \in \Omega$ and for the left side of (2.2.1) one has:

$$d(\tau_x f(x)) = \frac{1}{2}\,\sigma_x \overline{D}[f](x) - \frac{1}{2}\,\overline{\sigma}_x D[f](x)\,. \tag{2.3.1}$$

In particular, for $m = 1$ the formula (2.3.1) gives:

$$df(z^0) = \frac{\partial f}{\partial z}(z^0)dz + \frac{\partial f}{\partial \overline{z}}(z^0)d\overline{z}\,,$$

hence (2.3.1) is its extension onto the general situation of Clifford algebras $C\ell_{0,m}$. In particular (2.3.1), see [8], implies immediately

2.3.1. Theorem. *Let $f \in C^1(\Omega, C\ell_{0,m})$. Then f is left-hyperholomorphic at $x^0 \in \Omega$ if and only if f is left-hyperdifferentiable at x^0 and for such functions*

$$f'(x^0) = \frac{1}{2}\,\overline{D}[f](x^0)\,. \tag{2.3.2}$$

In particular, for $m = 1$ the formula (2.3.2) gives: $f'(z) = \partial f/\partial\overline{z}$; i.e., for a holomorphic function f its "formal derivative" $\partial f/\partial\overline{z}$ becomes an "authentic derivative" $f'(z)$; thus we may reasonably interpret (2.3.2) in the same way for the general situation of Clifford algebras $C\ell_{0,m}$.

2.3.2. Note that (2.3.1) has its right-hand-sided analogue

$$d(f(x)\tau_x) = \frac{1}{2}\overline{D}_r[f](x)\sigma_x - \frac{1}{2}D_r[f](x)\overline{\sigma}_x\,, \tag{2.3.3}$$

and both are combined in the following formula:

$$\begin{aligned} d(f(x)\tau_x g(x)) = \frac{1}{2}\{&\overline{D}_r[f](x)\sigma_x g(x) - D_r[f](x)\overline{\sigma}_x g(x) \\ &+ (-1)^{m-1}f(x)\sigma_x\overline{D}[g](x) + (-1)^m f(x)\overline{\sigma}_x D[g](x)\}\,. \end{aligned} \tag{2.3.4}$$

We wonder if (2.3.4) has an interpretation in the above terms which does not reduce to a simple combination of the left and the right cases.

2.4. There exists one more approach, in one complex variable, that in terms of the limit of the quotient of the increments both of the function and of the variable. Our aim now is to obtain an analogue for Clifford analysis case.

Let us define a non-degenerate m-dimensional parallelepiped with vertex $x^0 \in \mathbb{R}^{m+1}$ and edge vectors $\{v_1, v_2, \ldots, v_m\} \subset \mathbb{R}^{m+1}$ by

$$\Pi := \left\{x^0 + \sum_{\ell=1}^{m} t_\ell\, v_\ell \in \mathbb{R}^{m+1} \mid (t_1, \ldots, t_m) \in [0,1]^m\right\}$$

and its boundary by

$$\partial\Pi := \left\{x^0 + \sum_{\ell=1}^{m} t_\ell\, v_\ell \in \mathbb{R}^{m+1} \mid (t_1, \ldots, t_m) \in \partial[0,1]^m\right\}.$$

2.5. Theorem. *Let $f \in \Omega \to C\ell_{0,m}$ be left-hyperholomorphic at x^0 and let $f'(x^0)$ be the left-hyperderivative. Then for every sequence $\{\Pi_k\}_{k=1}^{\infty}$ of non-degenerate oriented m-parallelepiped with vertex x^0 the equality*

$$\lim_{k\to\infty}\left[\left(\int_{\Pi_k}\sigma_x\right)^{-1}\left(\int_{\partial\Pi_k}\tau_x\cdot f(x)\right)\right] = f'(x^0)\,, \tag{2.5.1}$$

is true if $\lim\limits_{k\to\infty} \operatorname{diam}\Pi_k = 0$.

Proof. From Stokes theorem and (2.3.1)

$$\int_{\partial\Pi_k}\tau_x\cdot f(x) = \int_{\Pi_k} d(\tau_x\cdot f(x)) = \int_{\Pi_k}\left(\frac{1}{2}\,\sigma_x\overline{D}[f](x) - \frac{1}{2}\,\overline{\sigma}_x D[f](x)\right).$$

Since f is hyperholomorphic

$$\int_{\partial\Pi_k}\tau_x\cdot f(x) = \int_{\Pi_k}\left(\frac{1}{2}\,\sigma_x\overline{D}[f](x)\right). \tag{2.5.2}$$

From (2.5.2) and since $f \in C^1$ it holds:

$$\begin{aligned}
&\lim_{k\to\infty}\left[\left(\int_{\Pi_k}\sigma_x\right)^{-1}\left(\int_{\partial\Pi_k}\tau_x\cdot f(x)\right)\right] \\
&= \lim_{k\to\infty}\left[\left(\int_{\Pi_k}\sigma_x\right)^{-1}\cdot\left(\int_{\Pi_k}\left(\frac{1}{2}\sigma_x\overline{D}[f](x)\right)\right)\right] \qquad (2.5.3)\\
&= \lim_{k\to\infty}\left[\left(\int_{\Pi_k}\sigma_x\right)^{-1}\cdot\left(\int_{\Pi_k}\sigma_x\left(\frac{1}{2}\overline{D}[f](x)\right)\right)\right] = f'(x^0)\,. \qquad \square
\end{aligned}$$

2.5.1. Above, the hyperderivative was defined as a proportionality coefficient of the two differential forms which is a rather formal approach if one recalls the usual definition of the complex (or even real) derivative as the limit of a certain quotient. The preceding theorem says that, at least for hyperholomorphic functions, the hyperderivative is the limit of a quotient; what is more the numerator of the quotient is a kind of the increment of the function meanwhile the denominator represents the increment of the independent variable. The one-dimensional case supports this reasoning.

2.5.2. In case $m = 1$ the 1-dimensional parallelepiped

$$\Pi := \{z \in \mathbb{C} \,|\, z = z_0 + tv_1\,, t \in [0,1]\}\,, \qquad (2.5.4)$$

is a segment connecting the points z_0 and v_1, and its boundary is the discrete set $\partial\Pi := \{z_0, z\}$. The differential forms (1.6.2) and (1.6.3) are:

$$\sigma_x = dx_1 - \mathbf{e}_1 dx_0 \qquad \text{and} \qquad \tau_x = -\mathbf{e}_1\,. \qquad (2.5.5)$$

Since $\sigma_x = -\mathbf{e}_1 dz$ then

$$\begin{aligned}
&\left[\left(\int_\Pi\sigma_x\right)^{-1}\left(\int_{\partial\Pi}\tau_x\cdot f(x)\right)\right] = \left[\left(\int_\Pi -\mathbf{e}_1\,dz\right)^{-1}\left(\int_{\partial\Pi} -\mathbf{e}_1\cdot f(x)\right)\right] \\
&= \left[\left(\int_\Pi dz\right)^{-1}\left(\int_{\partial\Pi} f(x)\right)\right] = \frac{\displaystyle\int_{\{z_0,z\}} f(\zeta)}{\displaystyle\int_{z_0}^{z} d\zeta} =: \frac{f(z)-f(z_0)}{z-z_0}\,, \qquad (2.5.6)
\end{aligned}$$

where ζ is the integration variable. If the function f is complex differentiable at z_0 then

$$f'(z_0) = \lim_{z\to z_0}\frac{\displaystyle\int_{\{z_0,z\}} f(\zeta)}{\displaystyle\int_{z_0}^{z} d\zeta}. \qquad (2.5.7)$$

for any 1-dimensional parallelepiped in every direction shrinking to z_0, i.e., the general formula (2.5.1) reduced to a well-known fact.

3. An m-dimensional directional hyperderivative in Clifford analysis

3.1. The "directions" that we are going to consider are given by hyperplanes $L \subset \mathbb{R}^{m+1}$ with equation

$$\gamma(x) = \sum_{\ell=0}^{m} n_\ell x_\ell + d = 0\,, \tag{3.1.1}$$

where $d \in \mathbb{R}$ and $\vec{n} = (n_0, n_1, \dots, n_m)$ is the unit normal vector to L.

Since for every L, there exists the hyperplane L_0 with unit normal vector $\overrightarrow{n^0}$; such that L_0 passes through the origin of $\mathbb{R}^{m+1}$ and L is parallel to L_0, then $\mathbf{n}^0 = \sum_{\ell=0}^{m} n_\ell^0\, \mathbf{e}_\ell$ indicates the direction of L.

3.2. Definition. *Let $x_0 \in L \cap \Omega$. The function $f : \Omega \to C\ell_{0,m}$ is said to be left-hyperderivable at x_0 along L, if for any sequence $\{\Pi_k\}_{k=1}^{\infty}$, $\Pi_k \subset L$, such that $\lim_{k\to\infty} diam \Pi_k = 0$ of non-degenerate m-parallelepipeds with vertex at x_0, the limit*

$$\lim_{k\to\infty} \left[\left(\int_{\Pi_k} \sigma_x \right)^{-1} \left(\int_{\partial\Pi_k} \tau_x \cdot f(x) \right) \right], \tag{3.2.1}$$

exists and does not depend on a choice of the sequence $\{\Pi_k\}_{k=1}^{\infty}$.

If exists, the limit is called the left m-dimensional directional hyperderivative along a hyperplane L and it will be denoted by $f'_L(x_0)$.

3.2.1. Notice that, of course, a fine point in this definition is that the family of parallelepipeds Π_k is fully contained in L, in contrast to the conditions in Theorem 2.5 where the parallelepipeds are free to move in $\mathbb{R}^{m+1}$.

3.3. Theorem. *Let $V(x^0)$ be an $(m+1)$-dimensional neighborhood of $x^0 \in \mathbb{R}^{m+1}$. Let $f \in C^1(V(x^0); C\ell_{0,m})$. Then f is left-hyperderivable at x^0 along any hyperplane $L \ni x^0$.*

Proof. First we are going to establish a relation between the surface forms σ_x and $\overline{\sigma}_x$ along L. Applying (2.3.1) to γ we have:

$$d(\tau_x \gamma(x)) = \frac{1}{2}\, \sigma_x \overline{D}[\gamma](x) - \frac{1}{2}\, \sigma_x D[\gamma](x) = 0\,;$$

therefore

$$\begin{aligned} \overline{\sigma}_x &= \sigma_x \overline{D}[\gamma](x) \left(D[\gamma](x)\right)^{-1} \\ &= \sigma_x \left(\overline{D}[\gamma](x)\right)^2 \\ &= \sigma_x (\overline{\mathbf{n}^0})^2\,. \end{aligned} \tag{3.3.1}$$

Now take $\{\Pi_k\}_{k=1}^{\infty}$, with $\Pi_k \subset L$ and $\lim_{k\to\infty} \operatorname{diam}\Pi_k = 0$. Then (2.3.1) and (3.3.1) imply:

$$\begin{aligned}
&\lim_{k\to\infty}\left[\left(\int_{\Pi_k}\sigma_x\right)^{-1}\left(\int_{\partial\Pi_k}\tau_x\cdot f(x)\right)\right] \\
&= \lim_{k\to\infty}\left[\left(\int_{\Pi_k}\sigma_x\right)^{-1}\left(\int_{\Pi_k} d(\tau_x\cdot f(x))\right)\right] \\
&= \frac{1}{2}\lim_{k\to\infty}\left[\left(\int_{\Pi_k}\sigma_x\right)^{-1}\cdot\int_{\Pi_k}\left(\sigma_x\overline{D}[f](x)-\overline{\sigma}_x D[f](x)\right)\right] \\
&= \frac{1}{2}\lim_{k\to\infty}\left[\left(\int_{\Pi_k}\sigma_x\right)^{-1}\cdot\left(\int_{\Pi_k}\sigma_x\left(\overline{D}[f](x)-(\overline{\mathbf{n}^0})^2 D[f](x)\right)\right)\right].
\end{aligned}$$

Appealing to the fact that $f\in C^1(V(x^0), C\ell_{0,m})$ we conclude that the limit in the left side exists, and thus the proof is completed. □

Of course it also gives immediately

3.3.1. Corollary. *Under the conditions of Theorem* 3.3 *there holds*:

$$f_L'(x^0) = \frac{1}{2}\left(\overline{D}[f](x^0) - (\overline{\mathbf{n}^0})^2 D[f](x^0)\right). \tag{3.3.2}$$

3.3.2. Corollary. *Let $f\in C^1(V(x^0); C\ell_{0,m})$. Then f is left-hyperholomorphic at x^0 iff $f_L'(x^0)$ does not depend on the hyperplane L.*

3.4. It is well known that for a function $f:\Omega\subset\mathbb{R}^{m+1}\to\mathbb{R}^p$, the directional derivative at $x^0\in\Omega$ along the direction given by a vector $\vec{u}\in\mathbb{R}^{m+1}$, where $|\,\vec{u}\,|=1$, is defined by the limit

$$\frac{\partial f}{\partial\vec{u}}(x^0) := \lim_{t\to 0}\frac{f(x^0+t\vec{u})-f(x^0)}{t}, \tag{3.4.1}$$

where t is real. In particular for $\vec{u}$ being a unit vector of the coordinate axis x_ℓ the limit (3.4.1) agrees with the partial derivative $\frac{\partial f}{\partial x_\ell}(x^0)$ for every ℓ. This keeps being true for $m=1$, i.e., in $\mathbb{R}^2$, but now the complex structure of $\mathbb{C}$ in $\mathbb{R}^2$ offers a "complex approach" to define the directional derivative. In effect, let ξ be a complex number of modulus 1; set

$$z_\xi := \{z\in\mathbb{C} | z = z_0 + t\,\xi, t\in\mathbb{R}\}.$$

Then if the limit

$$\lim_{z_\xi\ni z\to z_0}\frac{f(z)-f(z_0)}{z-z_0}, \tag{3.4.2}$$

exists, it is called the (complex) directional derivative of f at z_0 along the direction z_ξ and it is denoted by $\dfrac{\partial f}{\partial z_\xi}(z_0)$. Since

$$\lim_{z_\xi \ni z \to z_0} \frac{f(z)-f(z_0)}{z-z_0} = \lim_{t\to 0} \frac{f(z_0+t\xi)-f(z_0)}{t\xi}$$
$$= \bar{\xi} \lim_{t\to 0} \frac{f(z_0+t\vec{\xi})-f(z_0)}{t},$$

both directional derivatives, in $\mathbb{R}^2$, exist simultaneously only and whenever it happens they differ by a constant factor

$$\frac{\partial f}{\partial \vec{\xi}}(z_0) = \xi \cdot \frac{\partial f}{\partial z_\xi}(z_0). \tag{3.4.3}$$

Hence if $\xi = 1$ then

$$\frac{\partial f}{\partial z_\xi}(z_0) = \frac{\partial f}{\partial x_0}(z_0), \tag{3.4.4}$$

and if $\xi = i$ then

$$\frac{\partial f}{\partial z_\xi}(z_0) = -i\frac{\partial f}{\partial x_1}(z_0). \tag{3.4.5}$$

The latter equation exhibit the difference between the two definitions of the directional derivatives because in it, the complex structure is manifested. Note also that for the complex directional derivative in $\mathbb{C}$, the following formula is known for functions of class C^1:

$$\frac{\partial f}{\partial z_\xi}(z_0) = \frac{\partial f}{\partial z}(z_0) + e^{-2i\arg\xi}\frac{\partial f}{\partial \bar{z}}(z_0). \tag{3.4.6}$$

3.4.1. Let $\Gamma \subset \mathbb{C}$ be a smooth curve and $f : \Gamma \to \mathbb{C}$. Then the derivative of f along Γ, f'_Γ, is defined by the limit

$$f'_\Gamma(\zeta) := \lim_{\Gamma \ni z \to \zeta} \frac{f(z)-f(\zeta)}{z-\zeta}, \tag{3.4.7}$$

for $\zeta \in \Gamma$. In general, one cannot speak about the directional derivatives in this situation so assume additionally that f is of class C^1 in a small neighborhood of Γ. Let ξ define the tangent direction at $\zeta \in \Gamma$, hence there exists $\dfrac{\partial f}{\partial z_\xi}(\zeta)$. It is easy to prove now that $f'_\Gamma(\zeta)$ exists as well and $f'_\Gamma(\zeta) = \dfrac{\partial f}{\partial z_\xi}(\zeta)$.

What does the results obtained in Section 3.3 mean for $m = 1$?

When $m = 1$, the hyperplane L is a straight line in $\mathbb{R}^2$ with equation $\gamma = n_0x_0 + n_1x_1 + d$. Definition 3.2 becomes

$$\lim_{L\ni z\to z_0} \left[\left(\int_{z_0}^{z} -\mathbf{e}_1 dz \right)^{-1} \left(\int_{z_0,z} -\mathbf{e}_1 f(z) \right) \right] = \lim_{L\ni z\to z_0} \frac{f(z)-f(z_0)}{z-z_0}, \tag{3.4.8}$$

which is the definition of the complex directional derivative in the complex plane.

Let $\xi \in \mathbb{C}$ be such that $\mid \xi \mid = 1$ and that it determines the direction of L. Then $\mathbf{n}^0 = \mathbf{e}_1\xi$ (or $-\mathbf{e}_1\xi$) and $(\overline{\mathbf{n}^0})^2 = -\overline{\xi}^{\,2}$.

Therefore the formula (3.3.2) gives:

$$\begin{aligned}\frac{1}{2}\left(\overline{D}[f](x^0) - (\overline{\mathbf{n}^0})^2 D[f](x^0)\right) &= \frac{\partial f}{\partial z}(z_0) + \overline{\xi}^{\,2}\frac{\partial f}{\partial \overline{z}}(z_0) \\ &= \frac{\partial f}{\partial z}(z_0) + e^{-2\mathbf{e}_1 \arg\xi}\frac{\partial f}{\partial \overline{z}}(z_0)\,, \end{aligned} \tag{3.4.9}$$

which is the formula (3.4.6) for the complex directional derivative.

Thus the definition of the directional hyperderivative for $C\ell_{0,m}$-valued functions is a generalization of the complex approach to the notion of (one-dimensional) directional derivative of a complex function.

4. Hyperderivation of the Cliffordian Cauchy-type integral

4.1. As before let $\Omega \subset \mathbb{R}^{m+1}$ be a domain, which now is assumed to be simply connected and let $\Gamma := \left\{y \in \mathbb{R}^{m+1} | \varrho(y) = 0\right\}$ be its boundary, where $\varrho \in C^1(\mathbb{R}^{m+1}, \mathbb{R})$ and $\operatorname{grad} \varrho|\Gamma(y) \neq 0 \ \forall y \in \Gamma$.

Applying (2.3.4) to $E(y-x)$ and $f(y) \in C^1(\Gamma, C\ell_{0,n})$ one has:

$$\begin{aligned} d(E(y-x)\tau_y f(y)) = \frac{1}{2}\{&\overline{D}_{r,y}[E(y-x)]\sigma_y f(y) + (-1)^{m-1}E(y-x)\sigma_y\overline{D}[f](y) \\ &+ (-1)^m E(y-x)\overline{\sigma}_y D[f](y)\}\,, \end{aligned} \tag{4.1.1}$$

and after the integration over Γ there holds:

$$\begin{aligned} \int_\Gamma \overline{D}_{r,y}[E(y-x)]\sigma_y f(y) = (-1)^m &\int_\Gamma E(y-x)\sigma_y \overline{D}[f](y) \\ + (-1)^{m+1} &\int_\Gamma E(y-x)\overline{\sigma}_y D[f](y)\,. \end{aligned} \tag{4.1.2}$$

Because of

$$\overline{D}_{r,y}[E(y-x)] = -\,\overline{D}_{r,x}[E(y-x)] = -\overline{D}_x[E(y-x)]$$

(4.1.2) is

$$\begin{aligned} \overline{D}_x \int_\Gamma E(y-x)\sigma_y f(y) = (-1)^{m+1} &\int_\Gamma E(y-x)\sigma_y \overline{D}[f](y) \\ + (-1)^m &\int_\Gamma E(y-x)\overline{\sigma}_y D[f](y)\,. \end{aligned} \tag{4.1.3}$$

For $y \in \Gamma$ it holds that $d(\tau_y \varrho(y)) = 0$ and $\overline{\sigma}_y = \sigma_y \overline{D}[\varrho](y)\,(D[\varrho](y))^{-1}$. Therefore with $V|_\Gamma(y) := \overline{D}[\varrho](y)\,(D[\varrho](y))^{-1}$ we get that (4.1.3) can be written as

$$\overline{D}_x \int_\Gamma E(y-x)\sigma_y f(y) = \int_\Gamma E(y-x)\sigma_y \left((-1)^{m+1}\overline{D} + (-1)^m V|_\Gamma(y) D\right)[f](y)\,.$$

Thus we have arrived at the following conclusion.

4.2. Theorem. *Let $\Omega \subset \mathbb{R}^{m+1}$ be a simply connected domain with boundary $\Gamma := \{y \in \mathbb{R}^{m+1} | \varrho(y) = 0\}$, where $\varrho \in C^1(\mathbb{R}^{m+1}, \mathbb{R})$, grad $\varrho|\Gamma(y) \neq 0$ for all $y \in \Gamma$; and $f \in C^1(\Gamma, C\ell_{0,m})$. Then for all $x \notin \Gamma$*

$$\overline{D}_x \int_\Gamma E(y-x)\sigma_y f(y) = \int_\Gamma E(y-x)\sigma_y \left((-1)^{m+1}\overline{D} + (-1)^m V|_\Gamma(y) D\right)[f](y), \tag{4.2.1}$$

where $V|_\Gamma(y) := \overline{D}[\varrho](y)\,(D[\varrho](y))^{-1}$.

4.2.1. Let $T(y)$ be the tangent hyperplane to Γ at a point $y \in \Gamma$. Then (4.2.1) may be rewritten in more evident and palpable forms, first of all, as

$$\left(\int_\Gamma E(y-x)\sigma_y f(y)\right)' = \int_\Gamma E(y-x)\sigma_y\, f'_{T(y)}(y), \tag{4.2.2}$$

and finally as

$$(K[f])'(x) = K[f'_T](x), \tag{4.2.3}$$

where

$$K[f](x) := \int_\Gamma E(y-x)\sigma_y f(y)$$

is the Cliffordian Cauchy-type integral with density f. The formulas (4.2.1)–(4.2.3) say that under the conditions of Theorem 4.2, the hyperderivative of the Cliffordian Cauchy-type integral (the latter is a hyperholomorphic function, hence its hyperderivative is well defined) is again Cliffordian Cauchy-type integral but now with the density which is the directional hyperderivative along the tangent hyperplanes.

4.3. Obviously, for a hyperholomorphic function f there are well-defined hyperderivatives of any order $p \geq 1$: $f^{(p)} := (f^{(p-1)})'$, $f^{(0)} = f$; similarly for directional hyperderivatives. With this agreement, the following statements are simply obtained by induction.

4.3.1. Corollary. *Let $p \in \mathbb{N}$, $f \in C^p(\Gamma; C\ell_{0,m})$ and $\varrho \in C^p(\mathbb{R}^{m+1}; \mathbb{R})$. Then for every $y \in \Gamma$*

$$\overline{D}_x^p \left[\int_\Gamma E(y-x)\sigma_y f(y)\right] = \int_\Gamma E(y-x)\sigma_y \left((-1)^{m+1}V(y)D + (-1)^m\overline{D}\right)^p [f](y), \tag{4.3.1}$$

or equivalently

$$\left(\int_\Gamma E(y-x)\sigma_y f(y)\right)^{(p)} = \int_\Gamma E(y-x)\sigma_y\, f^{(p)}_{T(y)}(y), \tag{4.3.2}$$

$$(K[f])^{(p)}(x) = K[f^{(p)}_T](x). \tag{4.3.3}$$

4.3.2. Corollary. (Sohotski-Plemelj formulas for hyperderivatives of the Cliffordian Cauchy-type integral).

Let $f \in C^{p,\mu}(\Gamma; C\ell_{0,m})$, $\varrho \in C^{p,\mu}(\mathbb{R}^{m+1}; \mathbb{R})$. *Then for every* $x^0 \in \Gamma$ *the following both limits exist:*

$$\lim_{\Omega^{\pm} \ni y \to x^0} \left(\int_\Gamma E(y-x)\sigma_y f(y) \right)^{(p)}$$

and they are given by

$$\lim_{y^{\pm} \to x^0} \left(\int_\Gamma E(y-x)\sigma_y f(y) \right)^{(p)} = \pm \frac{1}{2} f^{(p)}_{T(x^0)}(x^0) + \int_\Gamma E(y-x^0)\sigma_y \, f^{(p)}_{T(y)}(y) \,. \tag{4.3.4}$$

4.4. The Cauchy kernel for $m = 1$ is given by

$$E(\zeta - z) = \frac{1}{2\pi} \cdot \frac{1}{\zeta - z} \,.$$

Then (4.2.1) has the form

$$\begin{aligned} &\frac{\partial}{\partial z} \left(\frac{1}{2\pi \mathbf{e}_1} \int_\Gamma \frac{1}{\zeta - z} f(\zeta) d\zeta \right) \\ &\quad = \frac{1}{2\pi \mathbf{e}_1} \int_\Gamma \frac{1}{\zeta - z} \left(\frac{\partial f}{\partial \zeta}(\zeta) - \frac{\partial \varrho}{\partial \zeta}(\zeta) \left(\frac{\partial \varrho}{\partial \overline{\zeta}}(\zeta) \right)^{-1} \frac{\partial f}{\partial \overline{\zeta}}(\zeta) \right) d\zeta \,. \end{aligned} \tag{4.4.1}$$

The Cauchy-Riemann operator can be seen as the "complexification" of the operator grad, thus

$$\frac{\partial \varrho}{\partial \zeta}(\zeta) \left(\frac{\partial \varrho}{\partial \overline{\zeta}}(\zeta) \right)^{-1} = e^{-2\mathbf{e}_1 \, arg \mathrm{grad} \varrho(\zeta)} \,. \tag{4.4.2}$$

Therefore (4.4.1) becomes

$$\left(\frac{1}{2\pi \mathbf{e}_1} \int_\Gamma \frac{1}{\zeta - z} f(\zeta) d\zeta \right)' = \frac{1}{2\pi \mathbf{e}_1} \int_\Gamma \frac{1}{\zeta - z} \, f'_{T(\zeta)} d\zeta \,, \tag{4.4.3}$$

which can be rewritten, in accordance with Subsection 3.4, as

$$\left(\frac{1}{2\pi \mathbf{e}_1} \int_\Gamma \frac{1}{\zeta - z} f(\zeta) d\zeta \right)' = \frac{1}{2\pi \mathbf{e}_1} \int_\Gamma \frac{1}{\zeta - z} \, f'_\Gamma(\zeta) d\zeta \,.$$

The latter formula is obtained by a simple integration by part for f living on Γ only and being derivable along Γ, without any additional assumptions on the smoothness of f. The general situation of Clifford analysis for an arbitrary m is much more complicated and it has required a more elaborated machinery which, hence, can be seen as a far-reaching generalization of the idea of integration by parts.

References

[1] R. Delanghe, F. Sommen, V. Souček. *Clifford algebra and spinor-valued functions.* Kluwer Academic Publishers. Mathematics and its Applications, 53, 1992.

[2] J.E. Gilbert, M.A.M. Murray. *Clifford algebras and Dirac operators in harmonic analysis.* Cambridge Studies in Adv. Math., 26, 1991.

[3] K. Gürlebeck, H.R. Malonek. *A Hypercomplex Derivative of Monogenic Functions in $\mathbb{R}^{n+1}$ and its Applications.* Complex Variables, 39, 1999, 199–228.

[4] K. Gürlebeck, W. Sprössig. *Quaternionic and Clifford Calculus for Physicists and Engineers.* John Wiley and Sons, 1997.

[5] V.V. Kravchenko, M.V. Shapiro. *Integral representations for spatial models of mathematical physics.* Addison-Wesley-Longman, Pitman Research Notes in Mathematics, 351, 1996.

[6] P. Lounesto. *Clifford algebras and Spinors.* Second edition, London Math. Soc. Lecture Note Series, 286, 2001.

[7] H. Malonek. *A New Hypercomplex Structure of the Euclidean Space $\mathbf{R}^{m+1}$ and the Concept of Hypercomplex Differentiability.* Complex Variables, 14, 1990, 25–33.

[8] H.R. Malonek. *Selected Topics in Hypercomplex Function Theory.* Clifford Algebras and Potential Theory, S.-L. Eriksson, ed., University of Joensun. Report Series 7, 2004, 111–150.

[9] I.M. Mitelman, M. Shapiro. *Differentiation of the Martinelli-Bochner Integrals and the Notion of Hyperderivability.* Math. Nachr., 172, 1995, 211–238.

[10] I.R. Porteous. *Clifford Algebras and the Classical Groups.* Cambridge University Press, Cambridge, 1995.

[11] M.V. Shapiro, N.L. Vasilevski. *Quaternionic ψ-Hyperholomorphic Functions, Singular Integral Operators and Boundary Value Problems I. ψ-Hyperholomorphic Function Theory.* Complex Variables, 27, 1995, 17–46.

[12] A. Sudbery. *Quaternionic Analysis.* Math. Proc. Camb. Phil. Soc. 85, 1979, 199–225.

M.E. Luna-Elizarrarás, M.A. Macías-Cedeño and M. Shapiro
Departamento Matemáticas
E.S.F.M. del I.P.N.
07338 México, D.F., México
e-mail: eluna@esfm.ipn.mx
marco.antonio.macias@hotmail.com
shapiro@esfm.ipn.mx

Quaternionic and Clifford Analysis
Trends in Mathematics, 235–258

Directional Quaternionic Hilbert Operators

Alessandro Perotti

Abstract. The paper discusses harmonic conjugate functions and Hilbert operators in the space of Fueter regular functions of one quaternionic variable. We consider left-regular functions in the kernel of the Cauchy–Riemann operator

$$\mathcal{D} = 2\left(\frac{\partial}{\partial \bar{z}_1} + j\frac{\partial}{\partial \bar{z}_2}\right) = \frac{\partial}{\partial x_0} + i\frac{\partial}{\partial x_1} + j\frac{\partial}{\partial x_2} - k\frac{\partial}{\partial x_3}.$$

Let J_1, J_2 be the complex structures on the tangent bundle of $\mathbb{H} \simeq \mathbb{C}^2$ defined by left multiplication by i and j. Let J_1^*, J_2^* be the dual structures on the cotangent bundle and set $J_3^* = J_1^* J_2^*$. For every complex structure $J_p = p_1 J_1 + p_2 J_2 + p_3 J_3$ ($p \in \mathbb{S}^2$ an imaginary unit), let $\overline{\partial}_p = \frac{1}{2}\left(d + pJ_p^* \circ d\right)$ be the Cauchy–Riemann operator w.r.t. the structure J_p. Let $\mathbb{C}_p = \langle 1, p\rangle \simeq \mathbb{C}$. If Ω satisfies a geometric condition, for every $\mathbb{C}_p$-valued function f_1 in a Sobolev space on the boundary $\partial\Omega$, we obtain a function $H_p(f_1) : \partial\Omega \to \mathbb{C}_p^{\perp}$, such that $f = f_1 + H_p(f_1)$ is the trace of a regular function on Ω. The function $H_p(f_1)$ is uniquely characterized by $L^2(\partial\Omega)$-orthogonality to the space of CR-functions w.r.t. the structure J_p. In this way we get, for every direction $p \in \mathbb{S}^2$, a bounded linear Hilbert operator H_p, with the property that $H_p^2 = id - S_p$, where S_p is the Szegö projection w.r.t. the structure J_p.

Mathematics Subject Classification (2000). Primary 30G35; Secondary 32A30.

Keywords. Quaternionic regular function, hyperholomorphic function, Hilbert operator, conjugate harmonic.

1. Introduction

The aim of this paper is to obtain some generalizations of the classical Hilbert transform used in complex analysis. We define a range of harmonic conjugate functions and Hilbert operators in the space of regular functions of one quaternionic variable.

Work partially supported by MIUR (PRIN Project "Proprietà geometriche delle varietà reali e complesse") and GNSAGA of INdAM.

Let Ω be a smooth bounded domain in $\mathbb{C}^2$. Let $\mathbb{H}$ be the space of real quaternions $q = x_0+ix_1+jx_2+kx_3$, where i, j, k denote the basic quaternions. We identify $\mathbb{H}$ with $\mathbb{C}^2$ by means of the mapping that associates the quaternion $q = z_1 + z_2 j$ with the pair $(z_1, z_2) = (x_0 + ix_1, x_2 + ix_3)$.

We consider the class $\mathcal{R}(\Omega)$ of *left-regular* (also called *hyperholomorphic*) functions $f : \Omega \to \mathbb{H}$ in the kernel of the Cauchy–Riemann operator

$$\mathcal{D} = 2\left(\frac{\partial}{\partial \bar{z}_1} + j\frac{\partial}{\partial \bar{z}_2}\right) = \frac{\partial}{\partial x_0} + i\frac{\partial}{\partial x_1} + j\frac{\partial}{\partial x_2} - k\frac{\partial}{\partial x_3}.$$

This differential operator is a variant of the original Cauchy–Riemann–Fueter operator (cf. for example [37] and [18, 19])

$$\frac{\partial}{\partial x_0} + i\frac{\partial}{\partial x_1} + j\frac{\partial}{\partial x_2} + k\frac{\partial}{\partial x_3}.$$

Hyperholomorphic functions have been studied by many authors (see for instance [1, 21, 24, 28, 34, 35]). Many of their properties can be easily deduced from known properties satisfied by Fueter-regular functions. However, regular functions in the space $\mathcal{R}(\Omega)$ have some characteristics that are more intimately related to the theory of holomorphic functions of two complex variables.

This space contains the identity mapping and any holomorphic map (f_1, f_2) on Ω defines a regular function $f = f_1 + f_2 j$. This is no longer true if we adopt the original definition of Fueter regularity. This definition of regularity is also equivalent to that of q-holomorphicity given by Joyce in [20], in the setting of hypercomplex manifolds.

The space $\mathcal{R}(\Omega)$ exhibits other interesting links with the theory of two complex variables. In particular, it contains the spaces of holomorphic maps with respect to any constant complex structure, not only the standard one.

Let J_1, J_2 be the complex structures on the tangent bundle $T\mathbb{H} \simeq \mathbb{H}$ defined by left multiplication by i and j. Let J_1^*, J_2^* be the dual structures on the cotangent bundle $T^*\mathbb{H} \simeq \mathbb{H}$ and set $J_3^* = J_1^* J_2^*$. For every complex structure $J_p = p_1 J_1 + p_2 J_2 + p_3 J_3$ (p a imaginary unit in the unit sphere $\mathbb{S}^2$), let d be the exterior derivative and

$$\overline{\partial}_p = \frac{1}{2}\left(d + pJ_p^* \circ d\right)$$

the Cauchy–Riemann operator with respect to the structure J_p. Let $\mathrm{Hol}_p(\Omega, \mathbf{H}) = \mathrm{Ker}\,\overline{\partial}_p$ be the space of holomorphic maps from (Ω, J_p) to $(\mathbf{H}, L_p)$, where L_p is the complex structure defined by left multiplication by p. Then every element of $\mathrm{Hol}_p(\Omega, \mathbf{H})$ is regular.

These subspaces do not fill the whole space of regular functions: it was proved in [27] that there exist regular functions that are not holomorphic for any p. This result is a consequence of an applicable criterion of J_p-holomorphicity, based on the energy-minimizing property of regular functions.

Other regular functions can be constructed by means of holomorphic maps with respect to non-constant almost complex structures on Ω (cf. [30]).

The classical Hilbert transform expresses one of the real components of the boundary values of a holomorphic function in terms of the other. We are interested in a quaternionic analogue of this relation, which links the boundary values of one of the complex components of a regular function $f = f_1 + f_2 j$ (f_1, f_2 complex functions) to those of the other.

In [21] and [32] some generalizations of the Hilbert transform to hyperholomorphic functions were proposed. In these papers the functions considered are defined on plane or spatial domains, while we are interested in domains of two complex variables. In the latter case, pseudoconvexity becomes relevant, since a domain in $\mathbb{C}^2$ is pseudoconvex if and only if every complex harmonic function on it is a complex component of a regular function (cf. [23] and [25]).

In the complex variable case, there is a close connection between harmonic conjugates and the Hilbert transform (see for example the monograph [6, §21]), given by harmonic extension and boundary restriction. Several generalizations of this relation to higher-dimensional spaces have been given (cf., e.g., [7, 8, 9, 12]), mainly in the framework of Clifford analysis, which can be considered as a generalization of quaternionic (and complex) analysis.

Our aim is to propose another variant of the quaternionic Hilbert operator, in which the complex structures J_p play a decisive role. Since these structures depend on a "direction" p in the unit sphere $\mathbb{S}^2$, we call it a *directional Hilbert operator* H_p.

The construction of H_p makes use of the rotational properties of regular functions (see §2.3), which were firstly studied in [37] in the context of Fueter-regularity. This allows to reduce the problem to the standard complex structure.

Let $\mathbb{C}_p = \langle 1, p \rangle$ be the copy of $\mathbb{C}$ in $\mathbb{H}$ generated by 1 and p and consider $\mathbb{C}_p$-valued function on the boundary $\partial\Omega$.

Assume that Ω satisfies a p-dependent geometric condition (see §3.1 for precise definitions), which is related to the pseudoconvexity property of Ω.

In Theorems 5 and 6 we show that for every $\mathbb{C}_p$-valued function f_1 in a Sobolev-type space $W^1_{\overline{\partial}_p}(\partial\Omega)$ and every fixed $q \in \mathbb{S}^2$ orthogonal to p, there exists a function $H_{p,q}(f_1) : \partial\Omega \to \mathbb{C}_p$ in the same space as f_1, such that $f = f_1 + H_{p,q}(f_1)q$ is the boundary value of a regular function on Ω. The function $H_{p,q}(f_1)$ is uniquely characterized by $L^2(\partial\Omega)$-orthogonality to the space of CR-functions with respect to the structure J_p. Moreover, $H_{p,q}$ is a bounded operator on the space $W^1_{\overline{\partial}_p}(\partial\Omega)$.

In Section 7 we prove our main result. We show how it is possible, for every fixed direction p, to choose a quaternionic regular harmonic conjugate of a $\mathbb{C}_p$-valued harmonic function in a way independent of the chosen orthogonal direction q. Taking restrictions to the boundary $\partial\Omega$, this construction permits to define the directional, p-dependent, Hilbert operator H_p.

In Theorem 10 we prove that even if the function $H_{p,q}(f_1)$ given by Theorem 6 depends on q, the product $H_{p,q}(f_1)q$ does not. Therefore we get a $\mathbb{C}_p$-antilinear, bounded operator

$$H_p : W^1_{\overline{\partial}_p}(\partial\Omega) \to W^1_{\overline{\partial}_p}(\partial\Omega, \mathbb{C}_p^{\perp}),$$

which exactly vanishes on the subspace $CR_p(\partial\Omega)$. Observe how the orthogonal decomposition of the codomain $\mathbb{H} = \mathbb{C}_p \oplus \mathbb{C}_p^\perp$ resembles the decomposition $\mathbb{C} = \mathbb{R} \oplus i\mathbb{R}$ which appears in the classical Hilbert transform.

The Hilbert operator H_p can be extended by right $\mathbb{H}$-linearity to the space $W^1_{\bar\partial_p}(\partial\Omega, \mathbb{H})$. The "*regular signal*" $R_p(f) := f + H_p(f)$ associated with f is always the trace of a regular function on Ω (Corollary 11). Moreover we show (Corollary 12) that $R_p(f)$ has a property similar to the one satisfied by analytic signals (cf. [31, Theorem 1.1]): f is the trace of a regular function on Ω if and only if $R_p(f) = 2f$ (modulo CR_p-functions).

The Hilbert operator H_p is also linked to the Szegö projection S_p with respect to J_p. In Theorem 13 we prove that $H_p^2 = id - S_p$ is the $L^2(\partial\Omega)$-orthogonal projection on the orthogonal complement of $CR_p(\partial\Omega)$.

When Ω is the unit ball B of $\mathbb{C}^2$, many of the stated results have a more precise formulation (see Theorem 7). The geometric condition is satisfied on the unit sphere $S = \partial B$ for every $p \in \mathbb{S}^2$. On S we are able to prove optimality of the boundary estimates satisfied by H_p.

In Section 6, we recall some applications of the harmonic conjugate construction to the characterization of the boundary values of pluriholomorphic functions. These functions are solutions of the PDE system

$$\frac{\partial^2 g}{\partial \bar z_i \partial \bar z_j} = 0 \quad \text{on } \Omega \quad (1 \le i, j \le 2)$$

(see for example [2, 3, 13, 14, 15] for properties of pluriholomorphic functions of two or more variables). The key point is that if $f = f_1 + f_2 j$ is regular, then f_1 is pluriholomorphic (and harmonic) if and only if f_2 is pluriharmonic, i.e., $\frac{\partial^2 f_2}{\partial z_i \partial \bar z_j} = 0$ on Ω $(1 \le i, j \le 2)$. Then known results about the boundary values of pluriharmonic functions (cf. [26]) can be applied to obtain a characterization of the traces of pluriholomorphic functions (Theorem 8).

2. Notations and definitions

2.1. Fueter regular functions

We identify the space $\mathbb{C}^2$ with the set $\mathbb{H}$ of quaternions by means of the mapping that associates the pair $(z_1, z_2) = (x_0 + ix_1, x_2 + ix_3)$ with the quaternion $q = z_1 + z_2 j = x_0 + ix_1 + jx_2 + kx_3 \in \mathbb{H}$. A quaternionic function $f = f_1 + f_2 j \in C^1(\Omega)$ is *(left) regular* (or *hyperholomorphic*) on Ω if

$$\mathcal{D}f = 2\left(\frac{\partial}{\partial \bar z_1} + j\frac{\partial}{\partial \bar z_2}\right) = \frac{\partial f}{\partial x_0} + i\frac{\partial f}{\partial x_1} + j\frac{\partial f}{\partial x_2} - k\frac{\partial f}{\partial x_3} = 0 \quad \text{on } \Omega.$$

We will denote by $\mathcal{R}(\Omega)$ the space of regular functions on Ω.

With respect to this definition of regularity, the space $\mathcal{R}(\Omega)$ contains the identity mapping and every holomorphic mapping (f_1, f_2) on Ω (with respect to the standard complex structure) defines a regular function $f = f_1 + f_2 j$. We recall some properties of regular functions, for which we refer to the papers of Sudbery[37], Shapiro and Vasilevski[34] and Nōno[24]:

1. The complex components are both holomorphic or both non-holomorphic.
2. Every regular function is harmonic.
3. If Ω is pseudoconvex, every complex harmonic function is the complex component of a regular function on Ω.
4. The space $\mathcal{R}(\Omega)$ of regular functions on Ω is a *right* $\mathbb{H}$-module with integral representation formulas.

A definition equivalent to regularity has been given by Joyce[20] in the setting of hypercomplex manifolds. Joyce introduced the module of *q-holomorphic* functions on a hypercomplex manifold.

A hypercomplex structure on the manifold $\mathbb{H}$ is given by the complex structures J_1, J_2 on $T\mathbb{H} \simeq \mathbb{H}$ defined by left multiplication by i and j. Let J_1^*, J_2^* be the dual structures on $T^*\mathbb{H} \simeq \mathbb{H}$. In complex co-ordinates

$$\begin{cases} J_1^* dz_1 = i\, dz_1, & J_1^* dz_2 = i\, dz_2 \\ J_2^* dz_1 = -d\bar{z}_2, & J_2^* dz_2 = d\bar{z}_1 \\ J_3^* dz_1 = i\, d\bar{z}_2, & J_3^* dz_2 = -i\, d\bar{z}_1 \end{cases}$$

where we make the choice $J_3^* = J_1^* J_2^*$, which is equivalent to $J_3 = -J_1 J_2$.

A function f is regular if and only if f is *q-holomorphic*, i.e.,

$$df + iJ_1^*(df) + jJ_2^*(df) + kJ_3^*(df) = 0.$$

In complex components $f = f_1 + f_2 j$, we can rewrite the equations of regularity as

$$\bar{\partial} f_1 = J_2^*(\partial \overline{f}_2).$$

The original definition of regularity given by Fueter (cf. [37] or [18]) differs from that adopted here by a real co-ordinate reflection. Let γ be the transformation of $\mathbb{C}^2$ defined by $\gamma(z_1, z_2) = (z_1, \bar{z}_2)$. Then a C^1 function f is regular on the domain Ω if and only if $f \circ \gamma$ is Fueter-regular on $\gamma^{-1}(\Omega)$, i.e., it satisfies the differential equation

$$\left(\frac{\partial}{\partial x_0} + i\frac{\partial}{\partial x_1} + j\frac{\partial}{\partial x_2} + k\frac{\partial}{\partial x_3}\right)(f \circ \gamma) = 0 \quad \text{on } \gamma^{-1}(\Omega).$$

2.2. Holomorphic functions with respect to a complex structure J_p

Let $J_p = p_1 J_1 + p_2 J_2 + p_3 J_3$ be the orthogonal complex structure on $\mathbb{H}$ defined by a unit imaginary quaternion $p = p_1 i + p_2 j + p_3 k$ in the sphere $\mathbb{S}^2 = \{p \in \mathbb{H} \mid p^2 = -1\}$. In particular, J_1 is the standard complex structure of $\mathbb{C}^2 \simeq \mathbb{H}$.

Let $\mathbb{C}_p = \langle 1, p \rangle$ be the complex plane spanned by 1 and p and let L_p be the complex structure defined on $T^*\mathbb{C}_p \simeq \mathbb{C}_p$ by left multiplication by p. If $f = f^0 + if^1 : \Omega \to \mathbb{C}$ is a J_p-*holomorphic* function, i.e., $df^0 = J_p^*(df^1)$ or, equivalently,

$df + iJ_p^*(df) = 0$, then f defines a regular function $\tilde{f} = f^0 + pf^1$ on Ω. We can identify $\tilde{f}$ with a holomorphic function

$$\tilde{f} : (\Omega, J_p) \to (\mathbb{C}_p, L_p).$$

We have $L_p = J_{\gamma(p)}$, where $\gamma(p) = p_1 i + p_2 j - p_3 k$. More generally, we can consider the space of holomorphic maps from (Ω, J_p) to $(\mathbb{H}, L_p)$

$$\mathrm{Hol}_p(\Omega, \mathbb{H}) = \{f : \Omega \to \mathbb{H} \text{ of class } C^1 \mid \overline{\partial}_p f = 0 \text{ on } \Omega\} = \mathrm{Ker}\, \overline{\partial}_p$$

where $\overline{\partial}_p$ is the Cauchy–Riemann operator with respect to the structure J_p

$$\overline{\partial}_p = \frac{1}{2}\left(d + pJ_p^* \circ d\right).$$

These functions will be called *J_p-holomorphic maps* on Ω.

For any positive orthonormal basis $\{1, p, q, pq\}$ of $\mathbb{H}$ $(p, q \in \mathbb{S}^2)$, let $f = f_1 + f_2 q$ be the decomposition of f with respect to the orthogonal sum

$$\mathbb{H} = \mathbb{C}_p \oplus (\mathbb{C}_p)q.$$

Let $f_1 = f^0 + pf^1$, $f_2 = f^2 + pf^3$, with f^0, f^1, f^2, f^3 the real components of f w.r.t. the basis $\{1, p, q, pq\}$. Then the equations of regularity can be rewritten in complex form as

$$\overline{\partial}_p f_1 = J_q^*(\partial_p \overline{f}_2),$$

where $\overline{f}_2 = f^2 - pf^3$ and $\partial_p = \frac{1}{2}\left(d - pJ_p^* \circ d\right)$. Therefore every $f \in \mathrm{Hol}_p(\Omega, \mathbb{H})$ is a regular function on Ω.

Remark 1.

1. The identity map belongs to the spaces $\mathrm{Hol}_i(\Omega, \mathbb{H}) \cap \mathrm{Hol}_j(\Omega, \mathbb{H})$, but not to $\mathrm{Hol}_k(\Omega, \mathbb{H})$.
2. For every $p \in \mathbb{S}^2$, $\mathrm{Hol}_{-p}(\Omega, \mathbb{H}) = \mathrm{Hol}_p(\Omega, \mathbb{H})$.
3. Every $\mathbb{C}_p$-valued regular function is a J_p-holomorphic function.

Proposition 1. *If $f \in \mathrm{Hol}_p(\Omega, \mathbb{H}) \cap \mathrm{Hol}_q(\Omega, \mathbb{H})$, with $p \neq \pm q$, then $f \in \mathrm{Hol}_r(\Omega, \mathbb{H})$ for every $r = \frac{\alpha p + \beta q}{\|\alpha p + \beta q\|}$ $(\alpha, \beta \in \mathbb{R})$ in the circle of $\mathbb{S}^2$ generated by p and q.*

Proof. Let $a = \|\alpha p + \beta q\|$. Then $a^2 = \alpha^2 + \beta^2 + 2\alpha\beta(p \cdot q)$, where $p \cdot q$ is the scalar product of the vectors p and q in $\mathbb{S}^2$. An easy computation shows that

$$pJ_q^* + qJ_p^* = -2(p \cdot q)Id.$$

From these identities we get that

$$\begin{aligned} rJ_r^*(df) &= a^{-2}(\alpha p + \beta q)(\alpha J_p^* + \beta J_q^*) \\ &= a^{-2}(\alpha^2 pJ_p^*(df) + \beta^2 qJ_q^*(df) + \alpha\beta(pJ_q^* + qJ_p^*)(df)) \\ &= a^{-2}(\alpha^2 pJ_p^*(df) + \beta^2 qJ_q^*(df) - 2\alpha\beta(p \cdot q)(df)) \\ &= a^{-2}(\alpha^2(-df) + \beta^2(-df) + 2\alpha\beta(p \cdot q)(-df)) = -df. \end{aligned}$$

Therefore $f \in \mathrm{Hol}_r(\Omega, \mathbb{H})$. □

In [27] it was proved that on every domain Ω there exist regular functions that are not J_p-holomorphic for any p. A similar result was obtained by Chen and Li[10] for the larger class of *q-maps* between hyperkähler manifolds.

This result is a consequence of a criterion of J_p-holomorphicity, which is obtained using the energy-minimizing property of regular functions.

2.3. Rotated regular functions

In [37] Proposition 5, Sudbery studied the action of rotations on Fueter-regular functions. Using that result and the reflection γ introduced in §2.1, we can obtain new regular functions by rotation.

Proposition 2. *Let $f \in \mathcal{R}(\Omega)$ and let $a \in \mathbb{H}$, $a \neq 0$. Let $r_a(z) = aza^{-1}$ be the three-dimensional rotation of $\mathbb{H}$ defined by a. Then the function*

$$f^a = r_{\gamma(a)} \circ f \circ r_a$$

is regular on $\Omega^a = r_a^{-1}(\Omega) = a^{-1}\Omega a$. Moreover, if $\gamma(r_a(i)) = p$, then $f \in \mathrm{Hol}_p(\Omega)$ if and only if $f^a \in \mathrm{Hol}_i(\Omega^a)$.

Proof. The first assertion is an immediate application of the cited result of Sudbery. Now let $p = \gamma(r_a(i))$, $p' = \gamma(p) = r_a(i)$ and $q = r_a(j)$ in $\mathbb{S}^2$. We first show that

$$r_a : (\mathbb{H}, J_1) \to (\mathbb{H}, L_{p'})$$

is holomorphic. Let $r_a(z) = aza^{-1} = x_0 + p'x_1 + qx_2 + p'qx_3 = (x_0 + p'x_1) + (x_2 + p'x_3)q = g_1 + g_2 q$, where g_1, g_2 are the $\mathbb{C}_{p'}$-valued holomorphic functions induced by z_1 and z_2. Then

$$p'J_1^*(dr_a) = p'J_1^*(dg_1) + p'J_1^*(dg_2)q = -dg_1 - dg_2\, q = -dr_a.$$

From this we get that also the map

$$r_{\gamma(a)}^{-1} = r_{\gamma(a)^{-1}} : (\mathbb{H}, J_1) \to (\mathbb{H}, L_p)$$

is holomorphic, since $r_{\gamma(a)^{-1}}(i) = \gamma(a)^{-1} i \gamma(a) = \gamma(aia^{-1}) = \gamma(r_a(i)) = p$. Now the commutative diagram

$$\begin{array}{ccc}
(\Omega, L_{\gamma(p)}) & \xrightarrow{\ f\ } & (\mathbb{H}, L_p) \\
{\scriptstyle r_a}\uparrow & & \downarrow{\scriptstyle r_{\gamma(a)}} \\
(\Omega^a, J_1) & \xrightarrow[\ f^a\ ]{} & (\mathbb{H}, J_1)
\end{array}$$

gives the stated equivalence, since $J_p = L_{\gamma(p)}$. □

Remark 2. The rotated function f^a has the following properties:

1. $(f^a)^{a^{-1}} = f$.
2. $f^{-a} = f^a$.
3. If $a \in \mathbb{S}^2$, then $(f^a)^a = f$.
4. If f is $\mathbb{C}_p$-valued on Ω, for $p = \gamma(r_a(i))$, then f^a is $\mathbb{C}$-valued on Ω^a.

2.4. Cauchy–Riemann operators

Let $\Omega = \{z \in \mathbb{C}^2 : \rho(z) < 0\}$ be a bounded domain with C^∞-smooth boundary in $\mathbb{C}^2$. We assume ρ of class C^∞ on $\mathbb{C}^2$ and $d\rho \neq 0$ on $\partial\Omega$. For every complex-valued function $g \in C^1(\overline{\Omega})$, we can define on a neighborhood of $\partial\Omega$ the normal components of ∂g and $\overline{\partial} g$

$$\partial_n g = \sum_k \frac{\partial g}{\partial z_k}\frac{\partial \rho}{\partial \bar{z}_k}\frac{1}{|\partial\rho|} \quad \text{and} \quad \overline{\partial}_n g = \sum_k \frac{\partial g}{\partial \bar{z}_k}\frac{\partial \rho}{\partial z_k}\frac{1}{|\partial\rho|},$$

where $|\partial\rho|^2 = \sum_{k=1}^2 \left|\frac{\partial\rho}{\partial z_k}\right|^2$. By means of the Hodge $*$-operator and the Lebesgue surface measure $d\sigma$, we can also write

$$\overline{\partial}_n g \, d\sigma = * \overline{\partial} g_{|\partial\Omega}.$$

In a neighborhood of $\partial\Omega$ we have the decomposition of $\overline{\partial} g$ in the tangential and the normal parts

$$\overline{\partial} g = \overline{\partial}_t g + \overline{\partial}_n g \frac{\overline{\partial}\rho}{|\overline{\partial}\rho|}.$$

Let $\mathcal{L}$ be the tangential Cauchy–Riemann operator

$$\mathcal{L} = \frac{1}{|\partial\rho|}\left(\frac{\partial\rho}{\partial\bar{z}_2}\frac{\partial}{\partial\bar{z}_1} - \frac{\partial\rho}{\partial\bar{z}_1}\frac{\partial}{\partial\bar{z}_2}\right).$$

The tangential part of $\overline{\partial} g$ is related to $\mathcal{L}g$ by the following formula

$$\overline{\partial}_t g \wedge d\zeta_{|\partial\Omega} = 2\mathcal{L}g \, d\sigma.$$

A complex function $g \in C^1(\partial\Omega)$ is a CR-function if and only if $\mathcal{L}g = 0$ on $\partial\Omega$. Notice that $\overline{\partial} g$ has coefficients of class $L^2(\partial\Omega)$ if and only if both $\overline{\partial}_n g$ and $\mathcal{L}g$ are of class $L^2(\partial\Omega)$.

If $g = g_1 + g_2 j$ is a regular function of class C^1 on Ω, then the equations $\overline{\partial}_n g_1 = -\overline{\mathcal{L}(g_2)}$, $\overline{\partial}_n g_2 = \overline{\mathcal{L}(g_1)}$ hold on $\partial\Omega$. Conversely, a harmonic function f of class $C^1(\Omega)$ is regular if it satisfies these equations on $\partial\Omega$ (cf. [28]). If Ω has connected boundary, it is sufficient that one of the equations is satisfied.

In place of the standard complex structure J_1, we can take on $\mathbb{C}^2$ a different complex structure J_p and consider the corresponding Cauchy–Riemann operators. We will denote by $\partial_{p,n}$ and $\overline{\partial}_{p,n}$ the normal components of ∂_p and $\overline{\partial}_p$ respectively, by $\overline{\partial}_{p,t}$ the tangential component of $\overline{\partial}_p$ and by $\mathcal{L}_p$ the tangential Cauchy–Riemann operator with respect to the structure J_p. Then we have the relations

$$\overline{\partial}_p g = \overline{\partial}_{p,t}\, g + \overline{\partial}_{p,n} g \frac{\overline{\partial}_p \rho}{|\overline{\partial}_p \rho|}, \qquad \overline{\partial}_{p,t}\, g \wedge d\zeta_{|\partial\Omega} = 2\mathcal{L}_p g \, d\sigma,$$

$$\overline{\partial}_{p,n} g \, d\sigma = * \overline{\partial}_p g_{|\partial\Omega}.$$

The space

$$CR_p(\partial\Omega) = \operatorname{Ker} \mathcal{L}_p = \{g : \partial\Omega \to \mathbb{C}_p \mid \mathcal{L}_p g = 0\}$$

has elements the CR-functions on $\partial\Omega$ with respect to the operator $\overline{\partial}_p$.

Remark 3. The operators $\overline{\partial}_p$, $\partial_{p,n}$, $\overline{\partial}_{p,n}$ and $\mathcal{L}_p$ are $\mathbb{C}_p$-linear and they map $\mathbb{C}_p$-valued functions of class C^1 to continuous $\mathbb{C}_p$-valued functions.

The relation between the Cauchy–Riemann operators $\overline{\partial}$ and $\overline{\partial}_p$ can be expressed by means of the rotations introduced in Proposition 2.

Proposition 3. *Let $a \in \mathbb{H}$, $a \neq 0$. If $p = \gamma(r_a(i))$ and $g : \overline{\Omega} \to \mathbb{C}_p$ is of class $C^1(\overline{\Omega})$, then $\overline{\partial} g^a = (\overline{\partial}_p g)^a$. Moreover $\overline{\partial}_n g^a = (\overline{\partial}_{p,n} g)^a$ and $\mathcal{L} g^a = (\mathcal{L}_p g)^a$ on $\partial\Omega^a$. In particular, $g \in CR_p(\partial\Omega)$ if and only if $g^a \in CR(\partial\Omega^a)$.*

Proof. Let $p' = \gamma(p)$, $a' = \gamma(a)$. We have

$$2(\overline{\partial}_p g)^a = dr_{a'} \circ (dg + pJ_p^*(dg)) \circ dr_a = dg^a + dr_{a'} \circ L_p \circ J_p^*(dg) \circ dr_a,$$

while

$$2\overline{\partial} g^a = dg^a + L_i \circ J_1^*(dg^a) = dg^a + L_i \circ dg^a \circ J_1.$$

The last term is

$$L_i \circ dg^a \circ J_1 = J_1^*(dr_{a'}) \circ dg \circ (dr_a \circ J_1) = (dr_{a'} \circ L_p) \circ dg \circ (L_{p'} \circ dr_a),$$

since $r_a : (\mathbb{H}, J_1) \to (\mathbb{H}, L_{p'})$ and $r_{a'} : (\mathbb{H}, L_p) \to (\mathbb{H}, J_1)$ are holomorphic, as seen in the proof of Proposition 2. Therefore it suffices to notice that $J_p^*(dg) = dg \circ L_{p'}$ and this is true because $J_p = L_{p'}$. For the second statement, we have

$$* \,\overline{\partial} g^a{}_{|\partial\Omega^a} = \overline{\partial}_n g^a d\sigma^a$$

where $d\sigma^a$ is the Lebesgue measure on $\partial\Omega^a$. On the other hand,

$$* \,(\overline{\partial}_p g)^a{}_{|\partial\Omega} = (* \,\overline{\partial}_p g_{|\partial\Omega})^a = (\overline{\partial}_{p,n} g \, d\sigma)^a = (\overline{\partial}_{p,n} g)^a d\sigma^a.$$

From the first part it follows that $\overline{\partial}_n g^a = (\overline{\partial}_{p,n} g)^a$. Then also the tangential parts are in the same relation and this implies that $\mathcal{L} g^a = (\mathcal{L}_p g)^a$ on $\partial\Omega^a$. □

3. Quaternionic harmonic conjugation

3.1. L^2 boundary estimates

Let $p \in \mathbb{S}^2$. Given a $\mathbb{C}_p$-valued function $f = f^0 + pf^1$, with f^0, f^1 real functions of class $L^2(\partial\Omega)$, we define the $L^2(\partial\Omega)$-norm of f as

$$\|f\| = (\|f^0\|^2 + \|f^1\|^2)^{1/2},$$

and the $L^2(\partial\Omega)$-product of f and $g = g^0 + pg^1$ as

$$(f, g) = (f^0, g^0)_{L^2(\partial\Omega)} + (f^1, g^1)_{L^2(\partial\Omega)}.$$

We will denote by $L^2(\partial\Omega, \mathbb{C}_p)$ the space of functions $f = f^0 + pf^1$, $f^0, f^1 \in L^2(\partial\Omega)$ real-valued functions.

In the following we shall assume that Ω satisfies a $L^2(\partial\Omega)$-estimate for some $p \in \mathbb{S}^2$: there exists a positive constant C_p such that

$$|(f, \mathcal{L}_p g)| \leq C_p \|\partial_{p,n} f\| \, \|\overline{\partial}_{p,n} g\| \tag{1}$$

for every $\mathbb{C}_p$-valued harmonic functions f, g on Ω, of class C^1 on $\overline{\Omega}$.

From Proposition 3 and the invariance of the Laplacian w.r.t. rotations, it follows that Ω satisfies (1) if and only if the rotated domain $\Omega^a = r_a^{-1}(\Omega)$, with $p = \gamma(r_a(i))$, satisfies the estimate with $p = i$:

$$|(f, \mathcal{L}g)| \le C_p \|\partial_n f\| \, \|\overline{\partial}_n g\| \tag{2}$$

for all complex-valued harmonic functions f, g on Ω^a, of class C^1 on $\overline{\Omega^a}$.

Proposition 4. *On the unit ball B of $\mathbb{C}^2$, the estimate* (1) *is satisfied with constant $C_p = 1$ for every $p \in \mathbb{S}^2$.*

Proof. From rotational symmetry of B, it is sufficient to prove the estimate for the case $p = i$, the standard complex structure. In this case, the proof was given in [29]. For convenience of the reader, we repeat here the proof.

We denote $\mathcal{L}_i$ by $\mathcal{L}$, $\partial_{i,n}$ by ∂_n and $\overline{\partial}_{i,n}$ by $\overline{\partial}_n$. Let $S = \partial B$. The space $L^2(S)$ is the sum of the pairwise orthogonal spaces $\mathcal{H}_{s,t}$, whose elements are the harmonic homogeneous polynomials of degree s in z_1, z_2 and t in $\bar{z}_1, \bar{z}_2$ (cf. for example Rudin[33, §12.2]). The spaces $\mathcal{H}_{s,t}$ can be identified with the spaces of the restrictions of their elements to S (*spherical harmonics*).

It suffices to prove the estimate for a pair of polynomials $f \in \mathcal{H}_{s,t}$, $g \in \mathcal{H}_{l,m}$, since the orthogonal subspaces $\mathcal{H}_{s,t}$ are eigenspaces of the operators ∂_n and $\overline{\partial}_n$. We can restrict ourselves to the case $s = l+1 > 0$ and $m = t+1 > 0$, since otherwise the product $(f, \mathcal{L}g)$ is zero. We have

$$|(f, \mathcal{L}g)|^2 \le \|f\|^2 \|\mathcal{L}g\|^2 = \|f\|^2 (\mathcal{L}^* \mathcal{L}g, g) = \|f\|^2 (-\overline{\mathcal{L}}\mathcal{L}g, g) = \|f\|^2 (l+1)m\|g\|^2$$

since the $L^2(S)$-adjoint $\mathcal{L}^*$ is equal to $-\overline{\mathcal{L}}$ (cf. [33, §18.2.2]) and $\overline{\mathcal{L}}\mathcal{L} = -(l+1)m\, Id$ when $m > 0$. On the other hand,

$$\|\partial_n f\| \|\overline{\partial}_n g\| = (l+1)m\|f\|\|g\|.$$

and the estimate is proved. □

Remark 4. It was proved in [29] that the estimate (2) implies the pseudoconvexity of Ω with respect to the standard structure. It can be shown that the same holds for a complex structure J_p. We conjecture that in turn the estimate (1) is always valid on a (strongly) pseudoconvex domain in $\mathbb{C}^2$ (w.r.t. J_p).

A domain Ω biholomorphic to B in the standard structure (e.g., an ellipsoid with defining function $\rho = c_1^2|z_1|^2 + c_2^2|z_2|^2 - 1$) satisfies estimate (2) but it does not necessarily satisfies estimate (1) for $p \neq i$, since the domain Ω^a can be not pseudoconvex.

3.2. Harmonic conjugate

We now prove some results about the existence of quaternionic harmonic conjugates in the space of $\mathbb{C}_p$-valued functions of class $L^2(\partial\Omega)$. We consider the following Sobolev-type Hilbert subspace of $L^2(\partial\Omega, \mathbb{C}_p)$:

$$\begin{aligned} W^1_{\overline{\partial}_p}(\partial\Omega) &= \{f \in L^2(\partial\Omega, \mathbb{C}_p) \mid \overline{\partial}_p f \in L^2(\partial\Omega, \mathbb{C}_p)\} \\ &= \{f \in L^2(\partial\Omega, \mathbb{C}_p) \mid \overline{\partial}_{p,n} f \text{ and } \mathcal{L}_p f \in L^2(\partial\Omega, \mathbb{C}_p)\} \end{aligned}$$

with product

$$(f,g)_{W^1_{\overline{\partial}_p}} = (f,g) + (\overline{\partial}_{p,n} f, \overline{\partial}_{p,n} g) + (\mathcal{L}_p f, \mathcal{L}_p g).$$

Here and in the following we always identify $f \in L^2(\partial\Omega)$ with its harmonic extension on Ω. We will use also the space

$$\overline{W}^1_{p,n}(\partial\Omega) = \{f \in L^2(\partial\Omega, \mathbb{C}_p) \mid \overline{\partial}_{p,n} f \in L^2(\partial\Omega, \mathbb{C}_p)\} \supset W^1_{\overline{\partial}_p}(\partial\Omega)$$

with product

$$(f,g)_{\overline{W}^1_{p,n}} = (f,g) + (\overline{\partial}_{p,n} f, \overline{\partial}_{p,n} g),$$

and the conjugate space

$$W^1_{p,n}(\partial\Omega) = \{f \in L^2(\partial\Omega, \mathbb{C}_p) \mid \partial_{p,n} f \in L^2(\partial\Omega)\}$$

with product

$$(f,g)_{W^1_{p,n}} = (f,g) + (\partial_{p,n} f, \partial_{p,n} g).$$

These spaces are vector spaces over $\mathbb{R}$ and over $\mathbb{C}_p$.

For every $\alpha > 0$, the spaces $W^1_{\overline{\partial}_p}(\partial\Omega)$, $\overline{W}^1_{p,n}(\partial\Omega)$ and $W^1_{p,n}(\partial\Omega)$ contain, in particular, every $\mathbb{C}_p$-valued function f of class $C^{1+\alpha}(\partial\Omega)$. Indeed, under this regularity condition f has a harmonic extension of class (at least) C^1 on $\overline{\Omega}$.

Let S_p be the Szegö projection from $L^2(\partial\Omega, \mathbb{C}_p)$ onto the (closure of the) subspace of holomorphic functions with respect to the structure J_p, continuous up to the boundary. We have the following orthogonal decomposition

$$W^1_{\overline{\partial}_p}(\partial\Omega) = CR_p(\partial\Omega) \oplus CR_p(\partial\Omega)^\perp = \operatorname{Ker} S_p^\perp \oplus \operatorname{Ker} S_p,$$

where $S_p^\perp = Id - S_p$.

In the case of the standard complex structure ($p = i$), we will denote the space $W^1_{\overline{\partial}_i}(\partial\Omega)$ simply by $W^1_{\overline{\partial}}(\partial\Omega)$ and the same for the spaces $\overline{W}^1_{i,n}(\partial\Omega) = \overline{W}^1_n(\partial\Omega)$ and $W^1_{i,n}(\partial\Omega) = W^1_n(\partial\Omega)$.

Remark 5. From Proposition 3 it follows that if $p = \gamma(r_a(i))$, then

$$W^1_{\overline{\partial}_p}(\partial\Omega)^a := \{f^a \mid f \in W^1_{\overline{\partial}_p}(\partial\Omega)\} = W^1_{\overline{\partial}}(\partial\Omega^a).$$

Similar relations hold for the other function spaces and the correspondence $f \mapsto f^a$ is an isometry between these spaces. The Szegö projection S_p on Ω is related to the standard Szegö projection S on Ω^a by $S_p(f)^a = S(f^a)$.

Theorem 5. *Assume that the boundary $\partial\Omega$ is connected and that the domain Ω satisfies estimate* (1)*. Given $f_1 \in \overline{W}^1_{p,n}(\partial\Omega)$, for every $q \in \mathbb{S}^2$ orthogonal to p, there exists $f_2 \in L^2(\partial\Omega, \mathbb{C}_p)$, unique up to a CR_p-function, such that $f = f_1 + f_2 q$ is the trace of a regular function on Ω. Moreover, f_2 satisfies the estimate*

$$\inf_{f_0} \|f_2 + f_0\|_{L^2(\partial\Omega)} \le C_p \|f_1\|_{\overline{W}^1_{p,n}(\partial\Omega)},$$

where the infimum is taken among the CR_p-functions $f_0 \in L^2(\partial\Omega, \mathbb{C}_p)$. The constant C_p is the same occurring in the estimate (1).

Theorem 6. *Assume that $\partial\Omega$ is connected and that Ω satisfies estimate* (1). *Given $f_1 \in W^1_{\overline{\partial}_p}(\partial\Omega)$, for every $q \in \mathbb{S}^2$ orthogonal to p, there exists $H_{p,q}(f_1) \in W^1_{\overline{\partial}_p}(\partial\Omega)$ such that $f = f_1 + H_{p,q}(f_1)q$ is the trace of a regular function on Ω. Moreover, $H_{p,q}(f_1)$ satisfies the estimate*

$$\|H_{p,q}(f_1)\|_{W^1_{\overline{\partial}_p}} \leq \sqrt{C_p^2 + 1}\,\|f_1\|_{W^1_{\overline{\partial}_p}}$$

with the same constant C_p given in (1). *The operator $H_{p,q}$ is a $\mathbb{C}_p$-antilinear bounded operator of the space $W^1_{\overline{\partial}_p}(\partial\Omega)$, with kernel the subspace $CR_p(\partial\Omega)$.*

We will show in Section 5 that when $\Omega = B$, the unit ball, then a sharper estimate can be proved.

4. Proof of Theorems 5 and 6

4.1. An existence principle

We recall a powerful existence principle in Functional Analysis proved by Fichera in the 50's (cf. [16, 17] and [11, §12]).

Let M_1 and M_2 be linear homomorphisms from a vector space V over the real (or complex) numbers into the Banach spaces B_1 and B_2, respectively.

Let us consider the following problem: given a linear functional Ψ_1 defined on B_1, find a linear functional Ψ_2 defined on B_2 such that

$$\Psi_1(M_1(v)) = \Psi_2(M_2(v)) \quad \forall\, v \in V.$$

Fichera's existence principle is the following:

Theorem (Fichera). *A necessary and sufficient condition for the existence, for any $\Psi_1 \in B_1^*$, of a linear functional Ψ_2 defined on B_2 such that*

$$\Psi_1(M_1(v)) = \Psi_2(M_2(v)) \quad \forall\, v \in V$$

is that there exists a positive constant C such that, for all $v \in V$,

$$\|M_1(v)\| \leq C\|M_2(v)\|.$$

Moreover, we have the following dual estimate with the same constant C:

$$\inf_{\Psi_0 \in \mathcal{N}} \|\Psi_2 + \Psi_0\| \leq C\|\Psi_1\|,$$

where $\mathcal{N}$ is the subspace of B_2^* composed of the functionals Ψ_0 that are orthogonal to the range of M_2, i.e., $\mathcal{N} = \{\Psi_0 \in B_2^* \mid \Psi_0(M_2(v)) = 0\ \forall v \in V\}$.

The theorem can be applied only if the kernel of M_2 is contained in the kernel of M_1. If this condition is not satisfied, the vector Ψ_1 has to satisfy the compatibility conditions:

$$\Psi_1(M_1(v)) = 0 \quad \forall\, v \in \operatorname{Ker}(M_2).$$

As mentioned in [11], this result includes important existence theorems, like, e.g., the Hahn–Banach theorem and the Lax–Milgram lemma.

4.2. Proof of Theorem 5

Given two orthogonal imaginary units p, q, there exists a unique rotation $r_{a'}$ that fixes the reals and maps p to i and q to j. Let $a = \gamma(a')$. Then $p = \gamma(r_a(i))$ and the domain Ω^a satisfies the estimate (2) of §3.1. The rotated function f_1^a belongs to the space $\overline{W}^1_n(\partial\Omega^a)$.

Now we state and prove the theorem for the standard structure $p = i$ (cf. [26, Theorem 3]) and then we will show how this is sufficient to get the general result.

Theorem. *Suppose that the estimate (2) is satisfied. For every $f_1 \in \overline{W}^1_n(\partial\Omega)$, there exists $f_2 \in L^2(\partial\Omega)$, unique up to a CR-function, such that $f = f_1 + f_2 j$ is the trace of a regular function on Ω. Moreover, f_2 satisfies the estimate*

$$\inf_{f_0} \|f_2 + f_0\|_{L^2(\partial\Omega)} \le C\|f_1\|_{\overline{W}^1_n(\partial\Omega)},$$

where the infimum is taken among the CR-functions $f_0 \in L^2(\partial\Omega)$.

Proof. We apply the existence principle to the following setting. Let $V = \mathrm{Harm}^1(\Omega)$ be the space of complex-valued harmonic functions on Ω, of class C^1 on $\overline{\Omega}$.

By means of the identification of $L^2(\partial\Omega)$ with its dual, we get dense, continuous injections $W^1_n(\partial\Omega) \subset L^2(\partial\Omega) = L^2(\partial\Omega)^* \subset W^1_n(\partial\Omega)^*$.

Let $A = \overline{CR(\partial\Omega)}$ be the closed subspace of $L^2(\partial\Omega)$ whose elements are conjugate CR-functions. It was shown by Kytmanov in [22, §17.1] that the set of the harmonic extensions of elements of A is the kernel of ∂_n.

Let $B_1 = \left(W^1_n(\partial\Omega)/A\right)^*$ and $B_2 = L^2(\partial\Omega)$. Let $M_1 = \pi \circ \mathcal{L}$, $M_2 = \overline{\partial}_n$, where π is the quotient projection $\pi : L^2 \to L^2/A = \left(L^2/A\right)^* \subset B_1$.

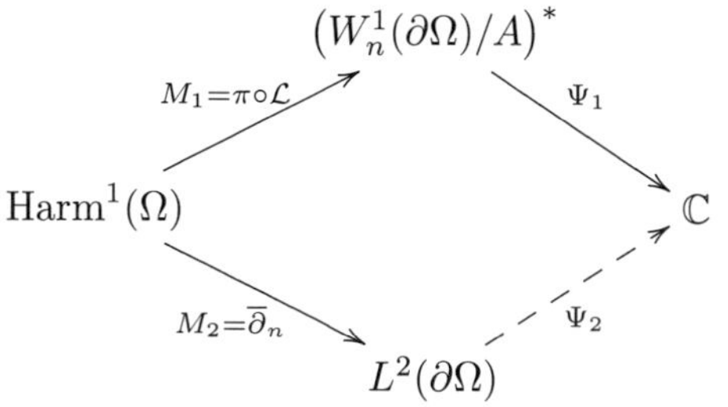

For every $g \in L^2(\partial\Omega)$, let $g^\perp$ denote the component of g in $A^\perp \subset L^2(\partial\Omega)$. A function $h_1 \in W^1_n(\partial\Omega)$ defines a linear functional $\Psi_1 \in B_1^* = W^1_n(\partial\Omega)/A$ such that

$$\Psi_1(\pi(g)) = (g^\perp, h_1)_{L^2} \quad \text{for every } g \in L^2(\partial\Omega).$$

If h is a CR-function on $\partial\Omega$,

$$(\mathcal{L}\phi, \bar{h}) = \frac{1}{2}\int_{\partial\Omega} h\overline{\partial}(\phi dz) = 0 \quad \text{and then} \quad (\mathcal{L}\phi)^\perp = \mathcal{L}\phi.$$

This implies that $\Psi_1(M_1(\phi)) = (\mathcal{L}\phi, h_1)$.

By the previous principle of Fichera, the existence of $h_2 \in L^2(\partial\Omega)$ such that

$$\Psi_1(M_1(\phi)) = (\mathcal{L}\phi, h_1)_{L^2} = \Psi_2(M_2(\phi)) = (\overline{\partial}_n\phi, h_2)_{L^2} \quad \forall\, \phi \in \mathrm{Harm}^1(\Omega)$$

is equivalent to the existence of $C > 0$ such that

$$\|\pi(\mathcal{L}\phi)\|_{(W^1_n(\partial\Omega)/A)^*} \le C\|\overline{\partial}_n\phi\|_{L^2(\partial\Omega)} \quad \forall\, \phi \in \mathrm{Harm}^1(\Omega). \tag{**}$$

The functional $\pi(\mathcal{L}\phi) \in L^2/A = \left(L^2/A\right)^* \subset B_1$ acts on $\pi(g) \in L^2/A$ in the following way:

$$\pi(\mathcal{L}\phi)(\pi(g)) = (g^\perp, \mathcal{L}\phi)_{L^2} = (g, \mathcal{L}\phi)_{L^2}$$

since $\mathcal{L}\phi \in A^\perp$. From the estimate (2) we imposed on Ω we get

$$\sup_{\|\pi(g)\|_{W^1_n(\partial\Omega)/A} \le 1} |(g, \mathcal{L}\phi)| \le C\|\overline{\partial}_n\phi\|_{L^2(\partial\Omega)} \quad \forall\, \phi \in \mathrm{Harm}^1(\Omega)$$

which is the same as estimate $(**)$. From the existence principle applied to $h_1 = \bar{f}_1 \in W^1_n(\partial\Omega)$, we get $f_2 = -h_2 \in L^2(\partial\Omega)$ such that

$$(\mathcal{L}\phi, \bar{f}_1)_{L^2} = -(\overline{\partial}_n\phi, f_2)_{L^2} \quad \forall\, \phi \in \mathrm{Harm}^1(\Omega).$$

Therefore

$$\frac{1}{2}\int_{\partial\Omega} f_1\overline{\partial}\phi \wedge d\zeta = -\int_{\partial\Omega} \bar{f}_2 * \overline{\partial}\phi \quad \forall\, \phi \in \mathrm{Harm}^1(\Omega)$$

and the result follows from the $L^2(\partial\Omega)$-version of Theorem 5 in [28], that can be proved as in [28] using the results given in [34, §3.7]. The estimate given by the existence principle is

$$\inf_{f_0 \in \mathcal{N}} \|f_2 + f_0\|_{L^2(\partial\Omega)} \le C\|\Psi_1\|_{W^1_n/A} \le C\|h_1\|_{W^1_n(\partial\Omega)} = C\|f_1\|_{\overline{W}^1_n(\partial\Omega)},$$

where $\mathcal{N} = \{f_0 \in L^2(\partial\Omega) \mid (\overline{\partial}_n\phi, f_0)_{L^2(\partial\Omega)} = 0 \ \forall\phi \in \mathrm{Harm}^1(\Omega)\}$ is the subspace of CR-functions in $L^2(\partial\Omega)$ (cf. [22, §17.1] and [11, §23]). □

We can now complete the proof of Theorem 5.

Proof of Theorem 5. Applying the preceding theorem to $f_1^a \in \overline{W}^1_n(\partial\Omega^a)$, we obtain a complex-valued function $g_2 \in L^2(\partial\Omega^a)$ such that $g = f_1^a + g_2 j$ is the trace of a regular function on Ω^a. We denote by the same symbols the extensions on the domains. Let $f_2 = (g_2)^{1/a}$ and $f = f_1 + f_2 q$. Then $f^a = r_{a'} \circ f \circ r_a = f_1^a + g_2 r_{a'}(q) = g$. Therefore $f \in \mathcal{R}(\Omega)$.

Given two functions $f_2, f_2' \in L^2(\Omega, \mathbb{C}_p)$ such that $f = f_1 + f_2 q$ and $f' = f_1 + f_2' q$ are regular on Ω, then $(f' - f)q = f_2 - f_2'$ is a $\mathbb{C}_p$-valued regular function and then it is J_p-holomorphic. Therefore f_2 is unique up to a CR_p-function.

The estimate for f on $\partial\Omega$ is a direct consequence of that satisfied by g on $\partial\Omega^a$. □

4.3. Proof of Theorem 6

Let $q \in \mathbb{S}^2$ be orthogonal to p and let f_2 be any function given by Theorem 5. Let $H_{p,q}(f_1)$ be the uniquely defined function $S_p^\perp(f_2) = f_2 - S_p(f_2)$. Notice that $f_1 \in CR_p(\partial\Omega)$ if and only if $f_2 \in CR_p(\partial\Omega)$ and therefore $H_{p,q}(f_1) = 0$ if and only if f_1 is a CR_p-function. Besides, for every f_1, we have

$$\|H_{p,q}(f_1)\|_{L^2(\partial\Omega)} = \|f_2 - S_p(f_2)\|_{L^2(\partial\Omega)} \leq \|f_2 + f_0\|_{L^2(\partial\Omega)}$$

for every CR_p-function f_0 on $\partial\Omega$. From Theorem 5 we get

$$\|H_{p,q}(f_1)\|_{L^2(\partial\Omega)} \leq C_p(\|f_1\|^2_{L^2(\partial\Omega)} + \|\overline{\partial}_{p,n} f_1\|^2_{L^2(\partial\Omega)})^{1/2}.$$

If $g = g_1 + g_2 j$ is a regular function of class C^1 on Ω, then the equations $\overline{\partial}_n g_1 = -\overline{\mathcal{L}(g_2)}$, $\overline{\partial}_n g_2 = \overline{\mathcal{L}(g_1)}$ hold on $\partial\Omega$ (cf. [28]). Then

$$\|\mathcal{L}g_2\|_{L^2(\partial\Omega)} = \|\overline{\partial}_n g_1\|_{L^2(\partial\Omega)}, \ \|\overline{\partial}_n g_2\|_{L^2(\partial\Omega)} = \|\mathcal{L}g_1\|_{L^2(\partial\Omega)}.$$

If g is regular, with trace of class $L^2(\partial\Omega)$, but not necessarily smooth up to the boundary, by taking its restriction to the boundary of $\Omega_\epsilon \subset \Omega$ and passing to the limit as ϵ goes to zero, we get the same norm equalities. Using rotations as in the proof of Theorem 5, we get

$$\|\mathcal{L}_p f_2\|_{L^2(\partial\Omega)} = \|\overline{\partial}_{p,n} f_1\|_{L^2(\partial\Omega)}, \ \|\overline{\partial}_{p,n} f_2\|_{L^2(\partial\Omega)} = \|\mathcal{L}_p f_1\|_{L^2(\partial\Omega)},$$

and then also

$$\|\mathcal{L}_p H_{p,q}(f_1)\|_{L^2(\partial\Omega)} = \|\overline{\partial}_{p,n} f_1\|_{L^2(\partial\Omega)}, \ \|\overline{\partial}_{p,n} H_{p,q}(f_1)\|_{L^2(\partial\Omega)} = \|\mathcal{L}_p f_1\|_{L^2(\partial\Omega)}.$$

Putting all together, we obtain

$$\begin{aligned}\|H_{p,q}(f_1)\|_{\overline{W}^1_{p,n}} &\leq C_p(\|f_1\|^2_{L^2} + \|\overline{\partial}_{p,n} f_1\|^2_{L^2})^{1/2} + \|\mathcal{L}_p f_1\|_{L^2} \\ &\leq \max\{1, C_p\}\|f_1\|_{W^1_{\overline{\partial}_p}}\end{aligned}$$

and finally the desired estimate

$$\begin{aligned}\|H_{p,q}(f_1)\|^2_{W^1_{\overline{\partial}_p}} &\leq C_p^2(\|f_1\|^2_{L^2} + \|\overline{\partial}_{p,n} f_1\|^2_{L^2}) + \|\mathcal{L}_p f_1\|^2_{L^2} + \|\overline{\partial}_{p,n} f_1\|^2_{L^2} \\ &\leq (C_p^2 + 1)\|f_1\|^2_{W^1_{\overline{\partial}_p}}.\end{aligned}$$

5. The case of the unit ball

On the unit ball B, an estimate sharper than the one given in Theorem 5 can be proved.

Theorem 7. *Given $f_1 \in \overline{W}^1_{p,n}(S)$, for every $q \in \mathbb{S}^2$ orthogonal to p, there exists $f_2 \in L^2(S, \mathbb{C}_p)$, unique up to a CR_p-function, such that $f = f_1 + f_2 q$ is the trace of a regular function on B. It satisfies the estimate*

$$\inf_{f_0 \in CR_p(S)} \|f_2 + f_0\|_{L^2(S)} \leq \|\overline{\partial}_{p,n} f_1\|_{L^2(S)}.$$

If $f_1 \in W^1_{\overline{\partial}_p}(S)$, for every $q \in \mathbb{S}^2$ orthogonal to p, there exists $H_{p,q}(f_1) \in W^1_{\overline{\partial}_p}(S)$ such that $f = f_1 + H_{p,q}(f_1)q$ is the trace of a regular function on B. Moreover, $H_{p,q}(f_1)$ satisfies the estimate

$$\|H_{p,q}(f_1)\|_{W^1_{\overline{\partial}_p}} \leq \left(2\|\overline{\partial}_{p,n} f_1\|^2_{L^2(S)} + \|\mathcal{L}_p f_1\|^2_{L^2(S)}\right)^{1/2}.$$

Proof. As in the proof of Theorem 5, it is sufficient to prove the thesis in the case of the standard complex structure. We use the same notation of Section §4.2. The space $W^1_n(S)/A$ is a Hilbert space also w.r.t. the product

$$(\pi(f), \pi(g))_{W^1_n/A} = (\partial_n f, \partial_n g).$$

This follows from the estimate $\|g^\perp\|_{L^2(S)} \leq \|\partial_n g\|_{L^2(S)}$, which holds for every $g \in W^1_n(S)$: if $g = \sum_{p\geq 0, q\geq 0} g_{p,q}$ is the orthogonal decomposition of g in $L^2(S)$, then

$$\begin{aligned}\|\partial_n g\|^2 &= \sum_{p>0,q\geq 0} \|p g_{p,q}\|^2 \\ &\geq \sum_{p>0,q\geq 0} \|g_{p,q}\|^2 = \|g^\perp\|^2.\end{aligned}$$

Then

$$\begin{aligned}\|\pi(g)\|^2_{W^1_n/A} &= \|g^\perp\|^2_{L^2} + \|\partial_n g\|^2_{L^2} \\ &\leq 2\|\partial_n g\|^2_{L^2}\end{aligned}$$

and therefore $\|\pi(g)\|_{W^1_n/A}$ and $\|\partial_n g\|_{L^2}$ are equivalent norms on $W^1_n(S)/A$. Now we can repeat the arguments of the proof of Section §4.2 and get the first estimate. The second estimate can be obtained in the same way as in the proof of Theorem 6. □

Remark 6. The last estimate in the statement of the previous Theorem is optimal: for example, if $f_1 = \bar{z}_1$, then $\overline{\partial}_n f_1 = \bar{z}_1$, $\mathcal{L} f_1 = -z_1$, $H_{i,j}(f_1) = \bar{z}_2$ and

$$\|H_{i,j}(f_1)\|^2_{W^{1}_{\overline{\partial}}} = \frac{3}{2} = 2\|\overline{\partial}_n f_1\|^2_{L^2(S)} + \|\mathcal{L} f_1\|^2_{L^2(S)},$$

since in the normalized measure ($Vol(S) = 1$) we have $\|z_1\| = \|z_2\| = 2^{-1/2}$.

Remark 7. The requirement that $\overline{\partial}_{p,n} f_1 \in L^2(S)$ cannot be relaxed. On the unit ball B, the estimate which is obtained from estimate (**) in §4.2 by taking the $L^2(S)$-norm also in the left-hand side is no longer valid (take for example $\phi \in \mathcal{H}_{k-1,1}(S)$). The necessity part of the existence principle gives that there exists $f_1 \in L^2(S)$ for which does not exist any $L^2(S)$ function f_2 such that $f_1 + f_2 j$ is the trace of a regular function on B. This means that the operation of quaternionic regular conjugation is not bounded in the harmonic Hardy space $h^2(B)$.

As it was shown in [29], a function $f_1 \in L^2(S)$ with the required properties is $f_1 = z_2(1 - \bar{z}_1)^{-1}$.

This phenomenon is different from what happens for pluriharmonic conjugation (cf. [36]) and in particular from the one-variable situation, which can be obtained by intersecting the domains with the complex plane $\mathbb{C}_j$ spanned by 1 and j. In this case f_1 and f_2 are real valued and $f = f_1 + f_2 j$ is the trace of a holomorphic function on $\Omega \cap \mathbb{C}_j$ with respect to the variable $\zeta = x_0 + x_2 j$.

6. Application to pluriholomorphic functions

In [29], Theorem 5 was applied in the case of the standard complex structure, to obtain a characterization of the boundary values of *pluriholomorphic* functions. These functions are solutions of the PDE system

$$\frac{\partial^2 g}{\partial \bar{z}_i \partial \bar{z}_j} = 0 \quad \text{on } \Omega \quad (1 \leq i, j \leq 2).$$

We refer to the works of Detraz[13], Dzhuraev[14, 15] and Begehr[2, 3] for properties of pluriholomorphic functions of two or more variables. The key point is that if $f = f_1 + f_2 j$ is regular, then f_1 is pluriholomorphic (and harmonic) if and only if f_2 is pluriharmonic., i.e., $\partial\overline{\partial} f_2 = 0$ on Ω.

We recall a characterization of the boundary values of pluriharmonic functions, proposed by Fichera in the 1980's and proved in Refs. [11] and [26]. Let

$$\mathrm{Harm}_0^1(\Omega) = \{\phi \in C^1(\overline{\Omega}) \mid \phi \text{ is harmonic on } \Omega,\ \overline{\partial}_n \phi \text{ is real on } \partial\Omega\}.$$

This space can be characterized by means of the Bochner-Martinelli operator of the domain Ω. Cialdea[11] proved the following result for boundary values of class L^2 (and more generally of class L^p).
Let $g \in L^2(\partial\Omega)$ be complex valued. Then g is the trace of a pluriharmonic function on Ω if and only if the following orthogonality condition is satisfied:

$$\int_{\partial\Omega} g * \overline{\partial}\phi = 0 \quad \forall \phi \in \mathrm{Harm}_0^1(\Omega).$$

If $f = f_1 + f_2 j : \partial\Omega \to \mathbb{H}$ is a function of class $L^2(\partial\Omega)$ and it is the trace of a regular function on Ω, then it satisfies the integral condition

$$\int_{\partial\Omega} f_1 \, \overline{\partial}\phi \wedge d\zeta = -2 \int_{\partial\Omega} \overline{f_2} * \overline{\partial}\phi \quad \forall \phi \in \mathrm{Harm}^1(\Omega).$$

If $\partial\Omega$ is connected, it can be proved that also the converse is true (cf. §4.2).

We can use this relation and the preceding result on pluriharmonic traces to obtain the following characterization of the traces of pluriholomorphic functions (cf. [29]). It generalizes some results obtained by Detraz [13] and Dzhuraev [14] on the unit ball (cf. also Refs. [2, 3, 4, 5, 15]).

Theorem 8. *Assume that Ω has connected boundary and satisfies the $L^2(\partial\Omega)$-estimate* (2). *Let $h \in \overline{W}_n^1(\partial\Omega)$. Then h is the trace of a harmonic pluriholomorphic function on Ω if and only if the following orthogonality condition is satisfied:*

$$\int_{\partial\Omega} h \, \overline{\partial}\phi \wedge d\zeta = 0 \quad \forall \phi \in \mathrm{Harm}_0^1(\Omega).$$

7. Directional Hilbert operators

In the complex one-variable case, there is a close connection between harmonic conjugates and the Hilbert transform (see for example the monograph [6, §21]). There are several extensions of this relation to higher-dimensional spaces (cf., e.g., [7, 8, 9, 12, 21, 32]), mainly in the framework of Clifford analysis, which can be considered as a generalization of quaternionic (and complex) analysis. In this section we apply the results obtained in §3 in order to introduce quaternionic Hilbert operators which depend on the complex structure J_p.

Let $L^2(\partial\Omega,\mathbb{C}_p^\perp)$ be the space of functions fq, $f\in L^2(\partial\Omega,\mathbb{C}_p)$, where $q\in\mathbb{S}^2$ is any unit orthogonal to p and let

$$W^1_{\overline{\partial}_p}(\partial\Omega,\mathbb{C}_p^\perp)=\{f\in L^2(\partial\Omega,\mathbb{C}_p^\perp)\mid\overline{\partial}_pf\in L^2(\partial\Omega,\mathbb{C}_p^\perp)\}.$$

Then $W^1_{\overline{\partial}_p}(\partial\Omega,\mathbb{C}_p^\perp)=\{fq\mid f\in W^1_{\overline{\partial}_p}(\partial\Omega)\}$ for any $q\in\mathbb{S}^2$ orthogonal to p. On these spaces we consider the products w.r.t. which the right multiplication by q is an isometry:

$$(f,g)_{L^2(\partial\Omega,\mathbb{C}_p^\perp)}=(fq,gq)_{L^2(\partial\Omega,\mathbb{C}_p)},$$
$$(f,g)_{W^1_{\overline{\partial}_p}(\partial\Omega,\mathbb{C}_p^\perp)}=(fq,gq)_{W^1_{\overline{\partial}_p}(\partial\Omega)}.$$

Proposition 9. *The above products are independent of $q\perp p$.*

Proof. Let $q'=aq+bpq\in\mathbb{C}_p^\perp$ be another element of $\mathbb{S}^2$ orthogonal to p, with $a,b\in\mathbb{R}$, $a^2+b^2=1$. If $fq=f^0+f^1p$, then $fq'=(af^0+bf^1)+(af^1-bf^0)p$. Similarly, $gq'=(ag^0+bg^1)+(ag^1-bg^0)p$, from which we get

$$\begin{aligned}(fq',gq')_{L^2(\partial\Omega,\mathbb{C}_p)}&=(af^0+bf^1,ag^0+bg^1)_{L^2}+(af^1-bf^0,ag^1-bg^0)_{L^2}\\&=(a^2+b^2)(f^0,g^0)_{L^2}+(a^2+b^2)(f^1,g^1)_{L^2}=(fq,gq)_{L^2(\partial\Omega,\mathbb{C}_p)}.\end{aligned}$$

The independence of the second product follows from that of the first. □

We will consider also the space of $\mathbb{H}$-valued functions

$$W^1_{\overline{\partial}_p}(\partial\Omega,\mathbb{H})=\{f\in L^2(\partial\Omega,\mathbb{H})\mid\overline{\partial}_pf\in L^2(\partial\Omega,\mathbb{H})\}$$

with norm

$$\|f\|_{W^1_{\overline{\partial}_p}(\partial\Omega,\mathbb{H})}=\left(\|f_1\|^2_{W^1_{\overline{\partial}_p}(\partial\Omega,\mathbb{C}_p)}+\|f_2\|^2_{W^1_{\overline{\partial}_p}(\partial\Omega,\mathbb{C}_p)}\right)^{1/2},$$

where $f=f_1+f_2q\in W^1_{\overline{\partial}_p}(\partial\Omega,\mathbb{C}_p)\oplus W^1_{\overline{\partial}_p}(\partial\Omega,\mathbb{C}_p^\perp)$, $f_i\in W^1_{\overline{\partial}_p}(\partial\Omega,\mathbb{C}_p)$ and q is any imaginary unit orthogonal to p. It follows from Proposition 9 that this norm does not depends on q.

Now we come to our main result. We show how it is possible, for every fixed direction p, to choose a quaternionic regular harmonic conjugate of a $\mathbb{C}_p$-valued harmonic function in a way independent of the orthogonal direction q. Taking restrictions to the boundary $\partial\Omega$ this construction permits to define a directional, p-dependent, Hilbert operator for regular functions.

Theorem 10. *Assume that $\partial\Omega$ is connected and that Ω satisfies estimate* (1)*. For every $\mathbb{C}_p$-valued function $f_1 \in W^{\frac{1}{2}}_{\overline{\partial}_p}(\partial\Omega)$, there exists $H_p(f_1) \in W^{\frac{1}{2}}_{\overline{\partial}_p}(\partial\Omega, \mathbb{C}_p^{\perp})$ such that $f = f_1 + H_p(f_1)$ is the trace of a regular function on Ω. Moreover, $H_p(f_1)$ satisfies the estimate*

$$\|H_p(f_1)\|_{W^{\frac{1}{2}}_{\overline{\partial}_p}(\partial\Omega,\mathbb{C}_p^{\perp})} \leq \sqrt{C_p^2+1}\,\|f_1\|_{W^{\frac{1}{2}}_{\overline{\partial}_p}(\partial\Omega)}$$

where C_p is the same constant as in estimate (1)*. The operator $H_p : W^{\frac{1}{2}}_{\overline{\partial}_p}(\partial\Omega) \to W^{\frac{1}{2}}_{\overline{\partial}_p}(\partial\Omega, \mathbb{C}_p^{\perp})$ is a right $\mathbb{C}_p$-linear bounded operator, with kernel $CR_p(\partial\Omega)$.*

Proof. Let $q, q' \in \mathbb{S}^2$ be two vectors orthogonal to p. We prove that

$$H_{p,q}(f_1)q = H_{p,q'}(f_1)q'.$$

Let $g = H_{p,q}(f_1)q - H_{p,q'}(f_1)q' \in W^{\frac{1}{2}}_{\overline{\partial}_p}(\partial\Omega, \mathbb{C}_p^{\perp})$. Then $gq \in W^{\frac{1}{2}}_{\overline{\partial}_p}(\partial\Omega)$ is the restriction of a $\mathbb{C}_p$-valued, regular function on Ω. But this implies that gq is a CR_p-function on $\partial\Omega$. On the other hand, gq belongs also to the space $CR_p(\partial\Omega)^{\perp}$, since $H_{p,q}(f_1) = S_p^{\perp}(f_2)$ and $H_{p,q'}(f_1)q'q = S_p^{\perp}(f_2')q'q$, with $q'q \in \mathbb{C}_p$, where f_2 and f_2' are functions given by Theorem 5. This implies that $gq = 0$ and then also g vanishes. Therefore we can put

$$H_p(f_1) = H_{p,q}(f_1)q \quad \text{for any } q \perp p.$$

The estimate is a direct consequence of what stated in Theorem 6. □

From Theorem 7, we immediately get the optimal estimate on the unit sphere:

$$\|H_p(f_1)\|_{W^{\frac{1}{2}}_{\overline{\partial}_p}(S,\mathbb{C}_p^{\perp})} \leq \left(2\|\overline{\partial}_{p,n} f_1\|^2_{L^2(S)} + \|\mathcal{L}_p f_1\|^2_{L^2(S)}\right)^{1/2}.$$

The operator H_p can be extended by right $\mathbb{H}$-linearity to the space $W^{\frac{1}{2}}_{\overline{\partial}_p}(\partial\Omega, \mathbb{H})$. If $f \in W^{\frac{1}{2}}_{\overline{\partial}_p}(\partial\Omega, \mathbb{H})$ and q is any imaginary unit orthogonal to p, let $f = f_1 + f_2 q \in W^{\frac{1}{2}}_{\overline{\partial}_p}(\partial\Omega, \mathbb{C}_p) \oplus W^{\frac{1}{2}}_{\overline{\partial}_p}(\partial\Omega, \mathbb{C}_p^{\perp})$, $f_i \in W^{\frac{1}{2}}_{\overline{\partial}_p}(\partial\Omega, \mathbb{C}_p)$. We set

$$H_p(f) = H_p(f_1) + H_p(f_2)q.$$

This definition is independent of q, because if $f = f_1 + f_2'q'$, then $(f_2 q - f_2' q')q$ is a CR_p-function and therefore $0 = H_p(-f_2 - f_2'q'q) = -H_p(f_2) - H_p(f_2')q'q \Rightarrow H_p(f_2)q = H_p(f_2')q'$. The operator H_p will be called a *directional Hilbert operator* on $\partial\Omega$.

Corollary 11. *The Hilbert operator $H_p : W^{\frac{1}{2}}_{\overline{\partial}_p}(\partial\Omega, \mathbb{H}) \to W^{\frac{1}{2}}_{\overline{\partial}_p}(\partial\Omega, \mathbb{H})$ is right $\mathbb{C}_p$-linear and $\mathbb{H}$-linear, its kernel is the space of $\mathbb{H}$-valued CR_p-functions and satisfies the estimate*

$$\|H_p(f)\|_{W^{\frac{1}{2}}_{\overline{\partial}_p}(\partial\Omega,\mathbb{H})} \leq \sqrt{C_p^2+1}\,\|f\|_{W^{\frac{1}{2}}_{\overline{\partial}_p}(\partial\Omega,\mathbb{H})}.$$

For every $f \in W^{\frac{1}{2}}_{\overline{\partial}_p}(\partial\Omega, \mathbb{H})$, the function $R_p(f) := f + H_p(f)$ is the trace of a regular function on Ω.

The "*regular signal*" $R_p(f) := f + H_p(f)$ associated with f has a property similar to the one satisfied by analytic signals (cf. [31, Theorem 1.1]).

Corollary 12. *Let $f \in W^1_{\overline{\partial}_p}(\partial\Omega, \mathbb{H})$. Then f is the trace of a regular function on Ω if and only if $R_p(f) = 2f$ (modulo CR_p-functions). Moreover, f is a CR_p-function if and only if $R_p(f) = f$.*

Proof. Let q be any imaginary unit orthogonal to p and let $f = f_1 + f_2 q$, $f_1, f_2 \in W^1_{\overline{\partial}_p}(\partial\Omega, \mathbb{C}_p)$. Then

$$R_p(f) = 2f \ (\mathrm{mod}\ CR_p) \Leftrightarrow \begin{cases} f_1 + H_p(f_2)q = 2f_1 \\ f_2 q + H_p(f_1) = 2f_2 q \end{cases} (\mathrm{mod}\ CR_p)$$
$$\Leftrightarrow \begin{cases} f_1 = H_p(f_2)q \\ f_2 = -H_p(f_1)q \end{cases} (\mathrm{mod}\ CR_p).$$

Therefore $R_p(f) = 2f \ (\mathrm{mod}\ CR_p) \Leftrightarrow f = f_1 + f_2 q = f_1 + H_p(f_1) = (f_2 + H_p(f_2))q \ (\mathrm{mod}\ CR_p)$, i.e., f is (the trace of) a regular function.

If $f_1, f_2 \in CR_p$, then $H_p(f_1) = H_p(f_2) = 0$ and therefore $R_p(f) = f$. Conversely, if $R_p(f) = f$, then from the first part we get $f = 2f (\mathrm{mod}\ CR_p)$ and so f is CR_p. □

We now study the relation between the Hilbert operator and the Szegö projection. When $f_1 \in W^1_{\overline{\partial}_p}(\partial\Omega)$, then $H_p(f_1) \in W^1_{\overline{\partial}_p}(\partial\Omega, \mathbb{C}_p^\perp)$ and therefore $H_p(H_p(f_1))$ is again in $W^1_{\overline{\partial}_p}(\partial\Omega)$.

Theorem 13. *Let $S_p : W^1_{\overline{\partial}_p}(\partial\Omega) \to CR_p(\partial\Omega) \subset W^1_{\overline{\partial}_p}(\partial\Omega)$ be the Szegö projection. Then $H_p^2 = id - S_p$. The same relation holds on the space $W^1_{\overline{\partial}_p}(\partial\Omega, \mathbb{H})$ if S_p is extended to $W^1_{\overline{\partial}_p}(\partial\Omega, \mathbb{H})$ in the same way as H_p. As a consequence, $R_p^2(f) = 2R_p(f)$ (modulo CR_p-functions) for every $f \in W^1_{\overline{\partial}_p}(\partial\Omega, \mathbb{H})$.*

Proof. For every $f_1 \in W^1_{\overline{\partial}_p}(\partial\Omega)$, the harmonic extension of $f = f_1 + H_{p,q}(f_1)q$ is regular. Then also $f' = (H_{p,q}(f_1) + H^2_{p,q}(f_1)q)q$ has regular extension and therefore the $\mathbb{C}_p$-valued function $f - f' = f_1 + H^2_{p,q}(f_1)$ is a CR_p-function. We have the decomposition

$$f_1 = (f_1 + H^2_{p,q}(f_1)) - H^2_{p,q}(f_1) \in CR_p(\partial\Omega) \oplus CR_p(\partial\Omega)^\perp,$$

that gives $f_1 + H^2_{p,q}(f_1) = S_p(f_1)$. But $H_{p,q} = -H_p q$ and then

$$H^2_{p,q}(f_1) = -H_{p,q}(H_p(f_1)q) = H_p(H_p(f_1)q)q.$$

By definition, $H_p^2(f_1) = H_p(H_p(f_1)) = H_p(-H_p(f_1)q)q$. Then $f_1 - H_p^2(f_1) = f_1 + H^2_{p,q}(f_1) = S_p(f_1)$. □

Remark 8. The Hilbert operator H_p can be expressed in terms of H_i using the rotations introduced in §4.2. It can be shown that $H_{p,q}(f_1)^a = H_{i,j}(f_1^a)$, from which it follows that

$$H_p(f_1)^a = H_i(f_1^a).$$

Theorem 10 says that even if the rotation vector a depends on p and q, the function $H_p(f_1) = H_i(f_1^a)^{1/a}$ only depends on p.

7.1. Examples

Let $\Omega = B$, $f = |z_1|^2 - |z_2|^2$.

1. $p = i$. We get

$$H_i(f) = \bar{z}_1\bar{z}_2 j \quad \text{and} \quad R_i(f) = |z_1|^2 - |z_2|^2 + \bar{z}_1\bar{z}_2 j$$

is a regular polynomial. We can check that

$$R_i^2(f) = 2(|z_1|^2 - |z_2|^2 + \bar{z}_1\bar{z}_2 j) = 2R_i(f).$$

2. $p = j$. We have

$$H_j(f) = (\bar{z}_1\bar{z}_2 - z_1z_2)j \quad \text{and} \quad R_j(f) = |z_1|^2 - |z_2|^2 + (\bar{z}_1\bar{z}_2 - z_1z_2)j.$$

Now

$$\begin{aligned} R_j^2(f) &= \frac{3}{2}\left(|z_1|^2 - |z_2|^2\right) + \frac{1}{2}\left(3\bar{z}_1\bar{z}_2 - 5z_1z_2\right)j \\ &= 2R_j(f) + \frac{1}{2}\left(|z_2|^2 - |z_1|^2\right) - \frac{1}{2}\left(z_1z_2 + \bar{z}_1\bar{z}_2\right)j \\ &= 2R_j(f) + CR_j\text{-function} \end{aligned}$$

In fact, the Hilbert operator H_j vanishes on $\frac{1}{2}\left(|z_2|^2 - |z_1|^2\right) - \frac{1}{2}(z_1z_2 + \bar{z}_1\bar{z}_2)j$.

3. $p = k$. In this case

$$H_k(f) = (\bar{z}_1\bar{z}_2 + z_1z_2)j, \quad R_k(f) = |z_1|^2 - |z_2|^2 + (\bar{z}_1\bar{z}_2 + z_1z_2)j \quad \text{and}$$

$$\begin{aligned} R_k^2(f) &= \frac{3}{2}\left(|z_1|^2 - |z_2|^2\right) + \frac{1}{2}\left(3\bar{z}_1\bar{z}_2 + 5z_1z_2\right)j \\ &= 2R_k(f) + \frac{1}{2}\left(|z_2|^2 - |z_1|^2\right) + \frac{1}{2}\left(z_1z_2 - \bar{z}_1\bar{z}_2\right)j \\ &= 2R_k(f) + CR_k\text{-function} \end{aligned}$$

Another example: let $g = z_1^2 \in \mathrm{Hol}_i(\mathbb{H})$.

1. $p = i$. Since g is holomorphic, we get $H_i(g) = 0$, $R_i(g) = g$.
2. $p = j$. We have

$$\begin{aligned} H_j(g) &= H_j(x_0^2 - x_1^2) + H_j(2x_0x_1)i \\ &= \frac{1}{8}\left(3z_1^2 - z_2^2 - 3\bar{z}_1^2 + \bar{z}_2^2\right) + \frac{1}{4}(z_1\bar{z}_2 - \bar{z}_1z_2)j \\ &\quad + \frac{1}{8}\left(3z_1^2 + z_2^2 + 3\bar{z}_1^2 + \bar{z}_2^2\right) - \frac{1}{4}(z_1\bar{z}_2 + \bar{z}_1z_2)j \\ &= \frac{3}{4}z_1^2 + \frac{1}{4}\bar{z}_2^2 - \frac{1}{2}\bar{z}_1z_2j. \end{aligned}$$

References

[1] R. Abreu-Blaya, J. Bory-Reyes, M. Shapiro, On the notion of the Bochner-Martinelli integral for domains with rectifiable boundary. *Complex Anal. Oper. Theory* **1** (2007), no. 2, 143–168.

[2] H. Begehr, Complex analytic methods for partial differential equations, *ZAMM* **76** (1996), Suppl. 2, 21–24.

[3] H. Begehr, Boundary value problems in $\mathbb{C}$ and $\mathbb{C}^n$, *Acta Math. Vietnam.* **22** (1997), 407–425.

[4] H. Begehr and A. Dzhuraev, *An Introduction to Several Complex Variables and Partial Differential Equations*, Addison Wesley Longman, Harlow, 1997.

[5] H. Begehr and A. Dzhuraev, Overdetermined systems of second order elliptic equations in several complex variables. In: *Generalized analytic functions* (Graz, 1997), Int. Soc. Anal. Appl. Comput., 1, Kluwer Acad. Publ., Dordrecht, 1998, pp. 89–109.

[6] S. Bell, *The Cauchy transform, potential theory and conformal mapping.* Studies in Advanced Mathematics. CRC Press, Boca Raton, FL, 1992.

[7] F. Brackx, R. Delanghe, F. Sommen, On conjugate harmonic functions in Euclidean space. Clifford analysis in applications. *Math. Methods Appl. Sci.* **25** (2002), no. 16-18, 1553–1562.

[8] F. Brackx, B. De Knock, H. De Schepper and D. Eelbode, On the interplay between the Hilbert transform and conjugate harmonic functions. *Math. Methods Appl. Sci.* **29** (2006), no. 12, 1435–1450.

[9] F. Brackx, H. De Schepper and D. Eelbode, A new Hilbert transform on the unit sphere in $\mathbb{R}^m$. *Complex Var. Elliptic Equ.* **51** (2006), no. 5-6, 453–462.

[10] J. Chen and J. Li, Quaternionic maps between hyperkhler manifolds, *J. Differential Geom.* **55** (2000), 355–384.

[11] A. Cialdea, On the Dirichlet and Neumann problems for pluriharmonic functions. In: *Homage to Gaetano Fichera*, Quad. Mat., 7, Dept. Math., Seconda Univ. Napoli, Caserta, 2000, pp. 31–78.

[12] R. Delanghe, On some properties of the Hilbert transform in Euclidean space. *Bull. Belg. Math. Soc. Simon Stevin* **11** (2004), no. 2, 163–180.

[13] J. Detraz, Problème de Dirichlet pour le système $\partial^2 f/\partial\overline{z}_i\partial\overline{z}_j = 0$. (French), *Ark. Mat.* **26** (1988), no. 2, 173–184.

[14] A. Dzhuraev, On linear boundary value problems in the unit ball of $\mathbb{C}^n$, *J. Math. Sci. Univ. Tokyo* **3** (1996), 271–295.

[15] A. Dzhuraev, Some boundary value problems for second order overdetermined elliptic systems in the unit ball of $\mathbb{C}^n$. In: *Partial Differential and Integral Equations* (eds.: H. Begehr et al.), Int. Soc. Anal. Appl. Comput., 2, Kluwer Acad. Publ., Dordrecht, 1999, pp. 37–57.

[16] G. Fichera, Alcuni recenti sviluppi della teoria dei problemi al contorno per le equazioni alle derivate parziali lineari. (Italian) In: *Convegno Internazionale sulle Equazioni Lineari alle Derivate Parziali*, Trieste, 1954, Edizioni Cremonese, Roma, 1955, pp. 174–227.

[17] G. Fichera, *Linear elliptic differential systems and eigenvalue problems.* Lecture Notes in Mathematics, 8 Springer-Verlag, Berlin-New York, 1965.

[18] K. Gürlebeck, K. Habetha and W. Sprössig, *Holomorphic Functions in the Plane and n-dimensional Space.* Birkhäuser, Basel, 2008.

[19] K. Gürlebeck and W. Sprössig, *Quaternionic Analysis and Elliptic Boundary Value Problems.* Birkhäuser, Basel, 1990.

[20] D. Joyce, Hypercomplex algebraic geometry, *Quart. J. Math. Oxford* **49** (1998), 129–162.

[21] V.V. Kravchenko and M.V. Shapiro, *Integral representations for spatial models of mathematical physics*, Harlow: Longman, 1996.

[22] A.M. Kytmanov, *The Bochner-Martinelli integral and its applications*, Birkhäuser Verlag, Basel, 1995.

[23] M. Naser, Hyperholomorphe Funktionen, *Sib. Mat. Zh.* **12**, 1327–1340 (Russian). English transl. in *Sib. Math. J.* **12**, (1971) 959–968.

[24] K. Nōno, α-hyperholomorphic function theory, *Bull. Fukuoka Univ. Ed. III* **35** (1985), 11–17.

[25] K. Nōno, Characterization of domains of holomorphy by the existence of hyperharmonic functions, *Rev. Roumaine Math. Pures Appl.* **31** n. 2 (1986), 159–161.

[26] A. Perotti, Dirichlet Problem for pluriharmonic functions of several complex variables, *Communications in Partial Differential Equations*, **24**, nn. 3&4, (1999), 707–717.

[27] A. Perotti, Holomorphic functions and regular quaternionic functions on the hyperkähler space $\mathbb{H}$. In: More Progresses in Analysis: *Proceedings of the 5th international ISAAC Congress*, (Catania 2005), eds. H.G.W. Begehr, F. Nicolosi, World Scientific, Singapore, 2008.

[28] A. Perotti, Quaternionic regular functions and the $\overline{\partial}$-Neumann problem in $\mathbb{C}^2$, *Complex Variables and Elliptic Equations* **52** No. 5 (2007), 439–453.

[29] A. Perotti, Dirichlet problem for pluriholomorphic functions of two complex variables, *J. Math. Anal. Appl.* **337/1** (2008), 107–115.

[30] A. Perotti, Every biregular function is biholomorphic, *Advances in Applied Clifford Algebras*, in press.

[31] T. Qian, Analytic signals and harmonic measures. *J. Math. Anal. Appl.* **314** (2006), no. 2, 526–536.

[32] R. Rocha-Chavez, M.V. Shapiro, L.M. Tovar Sanchez, On the Hilbert operator for α-hyperholomorphic function theory in $\mathbb{R}^2$. *Complex Var. Theory Appl.* **43**, no. 1 (2000), 1–28.

[33] W. Rudin, *Function theory in the unit ball of* $\mathbb{C}^n$, Springer-Verlag, New York, Heidelberg, Berlin 1980.

[34] M.V. Shapiro and N.L. Vasilevski, Quaternionic ψ-hyperholomorphic functions, singular integral operators and boundary value problems. I. ψ-hyperholomorphic function theory, *Complex Variables Theory Appl.* **27** no.1 (1995), 17–46.

[35] M.V. Shapiro and N.L. Vasilevski, Quaternionic ψ-hyperholomorphic functions, singular integral operators and boundary value problems. II: Algebras of singular integral operators and Riemann type boundary value problems, *Complex Variables Theory Appl.* **27** no.1 (1995), 67–96.

[36] E.L. Stout, H^p-functions on strictly pseudoconvex domains, *Amer. J. Math.* **98** n.3 (1976), 821–852.

[37] A. Sudbery, Quaternionic analysis, *Mat. Proc. Camb. Phil. Soc.* **85** (1979), 199–225.

Alessandro Perotti
Department of Mathematics
University of Trento
Via Sommarive, 14
I–38050 Povo Trento ITALY
e-mail: perotti@science.unitn.it

Quaternionic and Clifford Analysis
Trends in Mathematics, 259–275

Hilbert Transforms on the Sphere and Lipschitz Surfaces

Tao Qian

Abstract. Through a double-layer potential argument we define harmonic conjugates of the Cauchy type and prove their existence and uniqueness in Lipschitz domains. We further define inner and outer Hilbert transformations on Lipschitz surfaces and prove their boundedness in L^p, where the range for the index p depends on the Lipschitz constant of the boundary surface. The inner and outer Poisson kernels, the Cauchy type conjugate inner and outer Poisson kernels, and the kernels of the inner and outer Hilbert transformations on the sphere are obtained. We also obtain Abel sum expansions of the kernels. The study serves as a justification of the methods in a series of papers of Brackx et al. based on their method for computation of a certain type of harmonic conjugates.

Mathematics Subject Classification (2000). Primary 62D05, 30D10; Secondary 42B35.

Keywords. Poisson kernel, Conjugate Poisson kernel, Schwarz kernel, Hilbert transformation, Cauchy integral, double-layer potential, Clifford algebra.

1. Introduction

In the literature the usage of the terminology Hilbert transformation is not uniform. A series papers of Brackx et al., including [8], [5], [6], [7] and [4] define the terminology based on the harmonic conjugates obtained through their methods. Our definition below is based on the Cauchy singular integral that is consistent with [3] and [1].

A Hilbert transformation is a mapping from a function space to a function space on a co-dimension-1 surface with respect to the entire space. On the line with respect to the complex plane the expressions for the singular Cauchy transformation and the Hilbert transformation coincide. Consider $\mathbf{R}^m$, $m > 1$, as a

This work was supported by the Research Grants of University of Macau RG071/06-07S/08R/QT/FST and RG059/05-06S/08T/QT/FST.

co-dimension-1 surface of the Euclidean space $\mathbf{R}_1^m$. In the Clifford algebra setting (see §1), the singular Cauchy transformation on $\mathbf{R}^m$ coincides with the Hilbert transformation on $\mathbf{R}^m$, the latter being the linear combination $\sum_{k=1}^m R_i\mathbf{e}_i$, where R_i's are the Riesz transformations, and $\mathbf{e}_i$'s are the basis elements of the Clifford algebra (see Example 2.1). Likewise, on the curves and surfaces with zero curvature the two objects, viz. the singular Cauchy transformation and the Hilbert transformations, all coincide. In any but fixed higher-dimensional Euclidean space the kernels for the Cauchy integrals on co-dimension-1 surfaces are all the same. In contrast, for Hilbert transformations, as principal-valued singular integrals, their singular kernels vary from surface to surface. The simplest example is the unit circle in the complex plane. The inner and outer Hilbert transformations (see Example 2.2 in §2) on the unit circle coincide with the so-called *spherical Hilbert transform* (see [12]), given by

$$\tilde{H}f(e^{i\theta}) = p.v.\frac{i}{2\pi}\int_0^{2\pi} \cot\left(\frac{\theta - t}{2}\right) f(e^{it})dt.$$

Note that the kernel of the operator is not the singular Cauchy kernel in the complex plane (see next section).

The difference between our formulation of Hilbert transformations with that of Brackx et al. is that the latter defines this concept based on their methods of finding harmonic conjugates. Since harmonic conjugates are not unique, the certainty would need to be addressed. We, on the other hand, stick on the Cauchy integral and so to avoid ambiguity that may arise.

In §1 we will give a short introduction to the basic notation and terminology of Clifford algebra. Readers who are familiar with Clifford analysis may skip over this section. In §2 we define the terminology inner and outer Hilbert transformations through a double-layer potential argument. We present some examples. Based on the results in relation to existence of the solutions of the Dirichlet problems, we prove the existence and the L^p-boundedness of the inner and outer Hilbert transformations for $1 < p < \infty$ and $2 - \epsilon < p < \infty$, respectively for smooth and Lipschitz domains. In the latter case ϵ depends on the Lipschitz constant of the domain. In §3 we introduce the concept the Cauchy type harmonic conjugates, and prove their existence and uniqueness. The associated inner and outer conjugate Poisson kernels for Lipschitz domains are discussed. In §4 we deduce the inner and outer Poisson kernels and their Cauchy type conjugates on the unit sphere. The latter induce the kernels of the Hilbert transformations in the context. In §5 we deduce the Abel sum expansions of the kernels.

2. Preliminary

Let $\mathbf{e}_1, \ldots, \mathbf{e}_m$ be *basic elements* satisfying $\mathbf{e}_i\mathbf{e}_j + \mathbf{e}_j\mathbf{e}_i = -2\delta_{ij}$, where $\delta_{ij} = 1$ if $i = j$ and $\delta_{ij} = 0$ otherwise, $i, j = 1, 2, \ldots, m$. Let

$$\mathbf{R}^m = \{\underline{x} : \underline{x} = x_1\mathbf{e}_1 + \cdots + x_m\mathbf{e}_m : x_j \in \mathbf{R}, j = 1, 2, \ldots, m\}$$

be identical with the usual m-dimensional Euclidean space. We similarly define

$$\mathbf{C}^m = \{\underline{x} : \underline{x} = x_1\mathbf{e}_1 + \cdots + x_m\mathbf{e}_m : x_j \in \mathbf{C}, j = 1, 2, \ldots, m\}.$$

An element in $\mathbf{R}^m$ (or in $\mathbf{C}^m$) is called a *vector*. The real (complex) Clifford algebra generated by $\mathbf{e}_1, \ldots, \mathbf{e}_m$, denoted by $\mathbf{R}^{(m)}$ (or $\mathbf{C}^{(m)}$), is the Clifford algebra generated by $\mathbf{e}_1, \ldots, \mathbf{e}_m$, over the real (or complex) field $\mathbf{R}$ (or $\mathbf{C}$). A general element in $\mathbf{R}^{(m)}$ (**or** $\mathbf{C}^{(m)}$), therefore, is of the form $x = \sum_T x_T\mathbf{e}_T$, where $x_T \in \mathbf{R}$ (or $\mathbf{C}$), and $\mathbf{e}_T = \mathbf{e}_{i_1}\mathbf{e}_{i_2}\cdots\mathbf{e}_{i_l}$, being called *reduced products*, where T runs over all the ordered subsets of $\{1, \ldots, m\}$, namely

$$T = \{1 \le i_1 \cdots < i_l \le m\}, \quad 1 \le l \le m.$$

When $T = \emptyset$, we set $\mathbf{e}_T = \mathbf{e}_0 = 1$. We denote by $|T|$ the number of the basis elements in T. A general Clifford number x may be decomposed into

$$x = \sum_{l=0}^{m} x^{(l)}, \quad x^{(l)} = \sum_{|T|=l} x_T.$$

A Clifford number of the form $x^{(l)}$ is called a Clifford number of l-form. A Clifford number of 2-form is also called a *bi-vector*.

Set

$$\mathbf{R}_1^m(\mathit{or}\ \mathbf{C}_1^m) = \{x = x_0 + \underline{x} : x_0 \in \mathbf{R}(\mathit{or}\ \mathbf{C}), \underline{x} \in \mathbf{R}^m(\text{or } \mathbf{C}^m)\}.$$

Elements in $\mathbf{R}_1^m$ or $\mathbf{C}_1^m$ are called *para-vectors*. We define the Clifford conjugation for the para-vectors $x = x_0 + \underline{x}$ in $\mathbf{R}_1^m$ or $\mathbf{C}_1^m$ by $\overline{x} = x_0 - \underline{x}$. So the Clifford conjugate of a vector $\underline{x}$ is $\overline{\underline{x}} = -\underline{x}$. Note that Clifford conjugation does not change complex numbers. The conjugation and reversion of $\mathbf{e}_T = \mathbf{e}_{i_1}\cdots\mathbf{e}_{i_l}$ are defined, respectively, by $\overline{\mathbf{e}}_T = \overline{\mathbf{e}}_{i_l}\cdots\overline{\mathbf{e}}_{i_1}, \overline{\mathbf{e}}_j = -\mathbf{e}_j$ and $\widetilde{\mathbf{e}}_T = \mathbf{e}_{i_l}\cdots\mathbf{e}_{i_1}$. The conjugation and inversion are extended to the Clifford algebra $\mathbf{R}^{(m)}$ and $\mathbf{C}^{(m)}$ by linearity in $\mathbf{R}$ and $\mathbf{C}$, respectively. A sum of a 0-form and a 2-form is called a *para-bivectors*. The conjugation rule for para-vectors also applies to para-bivectors: If c is a scalar and x a bi-vector, then $\overline{c+x} = c - x$. We adopt the convention that any Clifford number $x \in \mathbf{C}^{(m)}$ has the decomposition $x = \mathrm{Sc}[x] + \mathrm{NSc}[x]$, where $\mathrm{Sc}[x] = x_0 \in \mathbf{C}$, the scalar part of x, and $\mathrm{NSc}[x]$ the non-scalar part. If $\underline{x}, \underline{y}$ are vectors,then

$$\underline{x}\underline{y} = -\langle\underline{x}, \underline{y}\rangle + \underline{x} \wedge \underline{y}, \tag{2.1}$$

where

$$Sc[\underline{x}\underline{y}] = -\langle\underline{x}, \underline{y}\rangle, \quad NSc[\underline{x}\underline{y}] = \underline{x} \wedge \underline{y} = \sum_{i<j}(x_iy_j - x_jy_i)\mathbf{e_i}\mathbf{e_j}.$$

It is easy to verify that $0 \ne \underline{x} \in \mathbf{R}^m$ or $0 \ne \underline{x} \in \mathbf{C}^m$ implies

$$\underline{x}^{-1} = \frac{\overline{\underline{x}}}{|\underline{x}|^2}.$$

The *open ball with center* 0 *and radius* 1 in $\mathbf{R}^m$ is denoted by B^m and the unit sphere in $\mathbf{R}^m$ is denoted by S^{m-1} whose surface area, denoted by σ_{m-1}, is of the value $2\pi^{\frac{m}{2}}/\Gamma(\frac{m}{2})$.

The natural inner product between x and y in $\mathbf{C}^{(m)}$, denoted by $\langle x, y\rangle$, is the complex number $\sum_T x_T\overline{y_T}$, where $x = \sum_T x_T\mathbf{e}_T$ and $y = \sum_T y_T\mathbf{e}_T$, and $\overline{y_T}$ is the complex conjugation. The norm associated with this inner product is

$$|x| = \langle x,\ x\rangle^{\frac{1}{2}} = (\sum_T |x_T|^2)^{\frac{1}{2}}.$$

In below we will study functions defined in subsets of $\mathbf{R}^m$ taking values in $\mathbf{C}^{(m)}$. So, they are of the form $f(\underline{x}) = \sum_T f_T(\underline{x})\mathbf{e}_T$, where f_T are complex-valued functions. We will use the *homogeneous Dirac operator* $\underline{D}$, where $\underline{D} = \frac{\partial}{\partial x_1}\mathbf{e}_1 + \cdots + \frac{\partial}{\partial x_m}\mathbf{e}_m$. We define the "left" and "right" roles of the operators $\underline{D}$ by

$$\underline{D}f = \sum_{i=1}^{m}\sum_T \frac{\partial f_T}{\partial x_i}\mathbf{e}_i\mathbf{e}_T \qquad \text{and} \qquad f\underline{D} = \sum_{i=1}^{m}\sum_T \frac{\partial f_T}{\partial x_i}\mathbf{e}_T\mathbf{e}_i.$$

If f has all continuous first-order partial derivatives and $\underline{D}f = 0$ in a domain (open and connected) Ω, then we say that f is *left-monogenic* in Ω; and, if $f\underline{D} = 0$ in Ω, we say that f is *right-monogenic* in Ω. The function theories for left-monogenic functions and for right-monogenic functions are parallel. In the sequel we will briefly write "left-monogenic" as "monogenic".

We call

$$E(\underline{x}) = \frac{\overline{\underline{x}}}{|\underline{x}|^m}$$

the *Cauchy kernel* in $\mathbf{R}^m$. It is easy to verify that $E(\underline{x})$ is a monogenic function in $\mathbf{R}^m \setminus \{0\}$.

There is a slightly different function theory for the *inhomogeneous Dirac operator* $D = \frac{\partial}{\partial x_0} + \underline{D}$, or Cauchy-Riemann operator, in $\mathbf{R}_1^m$. The standard complex analysis for Cauchy-Riemann equations is of this setting (see Example 2.2 below). The two settings can often be converted to each other but not always.

As holomorphic functions in function theory for one complex variable, monogenic function theory is central in Clifford analysis. In the framework of the latter the Cauchy theorem, Cauchy integral formula, Taylor and Laurent expansions of monogenic functions, etc. are all well established. For details we refer to [5], [13] and [10].

3. Hilbert transformations on Lipschitz domains

Singular integral theory on Lipschitz curves and surfaces has been well established ([9], [13], [16], [19]) that provides the foundation of this study. In particular, the Cauchy singular integral operator on Lipschitz curves and surfaces were proved to be L^p-bounded for $1 < p < \infty$ ([9]). This theme is closely related to the Hardy space theory ([13]). Let Ω be a bounded and connected Lipschitz domain in $\mathbf{R}^m$ with Lipschitz constant less than or equal to M. By this it means that Ω is a bounded and connected open set whose boundary $\partial\Omega$, denoted also by Σ in the sequel, may be covered by a finite number of balls in each of which the piece of the boundary of the domain under a suitable rotation and translation can become locally a piece of Lipschitz graph with a Lipschitz constant less than or equal to M ([15]). We further assume that the complement set $(\overline{\Omega})^c$ is unbounded and connected. Alternatively, we may assume that Ω is the open and connected domain above a Lipschitz graph Σ. In both cases Σ divides the whole space $\mathbf{R}^m$ into two parts, $\Omega^+ = \Omega$ and $\Omega^- = \mathbf{R}^m \setminus (\Sigma \cup \Omega)$.

Define, for a scalar-valued (i.e., complex-valued) function f in $L^p(\Sigma), 1 < p < \infty$, the Cauchy integrals

$$C_\Sigma^\pm f(\underline{x}) = C^\pm f(\underline{x}) = \frac{1}{\sigma_{m-1}} \int_\Sigma E(\underline{y} - \underline{x}) n^\pm(\underline{y}) f(\underline{y}) d\sigma(\underline{y}), \quad \underline{x} \in \Omega^\pm, \tag{3.1}$$

where $d\sigma(\underline{y})$ is the surface area measure and $n^\pm(\underline{y})$ are the outward- and inward-pointing normals of the surface Σ with respect to $\Omega^\pm$ at the point $\underline{y} \in \Sigma$, σ_{m-1} is the surface area of the $m-1$-dimensional unit sphere. By using (2.1) the above reduces to

$$\begin{aligned} C^\pm f(\underline{x}) &= \frac{1}{\sigma_{m-1}} \int_\Sigma \langle E(\underline{x} - \underline{y}), n^\pm(\underline{y}) \rangle f(\underline{y}) d\sigma(\underline{y}) \\ &+ \frac{1}{\sigma_{m-1}} \int_\Sigma E(\underline{y} - \underline{x}) \wedge n^\pm(\underline{y}) f(\underline{y}) d\sigma(\underline{y}), \quad \underline{x} \in \Omega^\pm. \end{aligned} \tag{3.2}$$

The Plemelj formula (see [9] or [16]) ensures that the non-tangential boundary values of $C^\pm f(\underline{x})$, denoted by $\mathbb{C}^\pm f(\underline{x})$, respectively, exist and are equal to

$$\mathbb{C}^\pm f(\underline{x}) = \frac{1}{2}[f(\underline{x}) \pm \mathbb{C} f(\underline{x})], \quad \text{a.e. } \underline{x} \in \Sigma, \tag{3.3}$$

where the operator denoted by $\mathbb{C}$ is the principal value Cauchy singular integral operator given by

$$\mathbb{C} f(\underline{x}) = \frac{2}{\sigma_{m-1}} \lim_{\epsilon \to 0+} \int_{|\underline{y} - \underline{x}| > \epsilon, \underline{y} \in \Sigma} E(\underline{y} - \underline{x}) n^+(\underline{y}) f(\underline{y}) d\sigma(\underline{y}), \quad \text{a.e. } \underline{x} \in \Sigma. \tag{3.4}$$

Using the conventional notation "p.v." instead of $\lim_{\epsilon \to 0+}$ in the last integral, by separating the integral into the scalar part and the 2-form part, we have

$$\begin{aligned} \mathbb{C} f(\underline{x}) &= \frac{2}{\sigma_{m-1}} \text{p.v.} \int_\Sigma E(\underline{y} - \underline{x}) n^+(\underline{y}) f(\underline{y}) d\sigma(\underline{y}) \\ &= \frac{2}{\sigma_{m-1}} \text{p.v.} \int_\Sigma \langle E(\underline{x} - \underline{y}), n^+(\underline{y}) \rangle f(\underline{y}) d\sigma(\underline{y}) \\ &+ \frac{2}{\sigma_{m-1}} \text{p.v.} \int_\Sigma (E(\underline{y} - \underline{x}) \wedge n^+(\underline{y})) f(\underline{y}) d\sigma(\underline{y}), \quad \text{a.e. } \underline{x} \in \Sigma. \end{aligned} \tag{3.5}$$

Known as Coifman-McIntosh-Meyer's Theorem ([9]), the operator $\mathbb{C}$ is L^p-bounded for $1 < p < \infty$, thus the operators $\mathbb{C}^\pm$ are L^p-bounded, too. It is easy to show that the operators $\mathbb{C}^\pm$ are *projections* with the characteristic properties ${\mathbb{C}^\pm}^2 = \mathbb{C}^\pm$, and, as consequence, $\mathbb{C}$ itself is a *reflection* operator, i.e., $\mathbb{C}^2 = I$, where I is the identity operator. Note that since the boundary data f is assumed to be scalar-valued, the Cauchy integrals $C^\pm f$, as well as the boundary values $\mathbb{C}^\pm f$, are all para-bivector-valued. In the complex plane and the $\mathbf{R}_1^n$ spaces the above-mentioned operators are para-vector-valued. (In the complex plane case the boundary data are assumed to be real-valued.)

There is a second reflection operator, N, representing the Clifford conjugation, namely,

$$Nf(\underline{x}) = \overline{f(\underline{x})}.$$

The operator N will be applied to the boundary values of the Cauchy integrals that are para-bivector-valued. The associated projections are the mappings $N^+ : f \to \mathrm{Sc}[f]$ and $N^- : f \to \mathrm{NSc}[f]$.

With the four pairs of the combinations $(\mathbb{C}, \mathbb{C}^{\pm})$ and $(N, N^{\pm})$ of the reflections and projections we can formulate the corresponding transmission problems (see [1]). The operator theory for $\mathbb{C}^-$ is similar to that of $\mathbb{C}^+$ with minor modifications dealing with the infinity. Below, we will mainly deal with $\mathbb{C}^+$.

We write

$$\mathbb{C}^+ f = \frac{1}{2}(I + \mathbb{C})f = u + v,$$

where the para-bivector-valued $\mathbb{C}^+ f$ has u as its scalar part, $u = \mathrm{Sc}[\mathbb{C}^+ f]$ and v as its 2-form part, $v = \mathrm{NSc}[\mathbb{C}^+ f]$. The above relation gives

$$u = \frac{1}{2}\left(I + N^+\mathbb{C}\right) f \quad \text{and} \quad v = \frac{1}{2} N^- \mathbb{C} f.$$

Therefore, at least formally,

$$f = 2(I + N^+\mathbb{C})^{-1} u.$$

Defining the mapping from u to v to be the *inner Hilbert transformation*, denoted by H^+, we have

$$\begin{aligned} v &= H^+ u \\ &= \frac{1}{2} N^- \mathbb{C} f \\ &= N^- \mathbb{C}(I + N^+\mathbb{C})^{-1} u. \end{aligned}$$

In the above formulation it is crucial to require the topological isomorphism property from u to f in the L^p space of the boundary that, as consequence, induces the existence and the L^p-boundedness of the Hilbert transformation from u to v. In other words, the double layer potential part $N^+\mathbb{C}$ should be comparatively smaller than the identity operator I. This requirement is met for $1 < p < \infty$ if the curve or surface is smooth. For Lipschtz curves or surfaces Σ, in general, this is met, however, only for $p_0 < p < \infty$, where the index $p_0 \in [1, \infty)$ depends on the Lipschitz constant of Σ. Whereas, for $1 < p \le p_0$, the closure in $L^p(\Sigma)$ of the images u forms a proper subspace of $L^p(\Sigma)$ ([15] and its references).
Similarly, $v = \frac{1}{2} N^- \mathbb{C} f = N^- \mathbb{C}(I - N^+\mathbb{C})^{-1} u$ is defined to be the *outer Hilbert transform* $H^- u$. To give the proper meaning of the above definition, as well as the operator formulas for H^+ and H^-, we are now to declare the existence of the inverse operator $(I \pm N^+\mathbb{C})^{-1}$ under certain conditions. Before we deal with the general theory it would be interesting to check some examples.

Example 2.1 Consider $\Sigma = \mathbf{R}^m$, where Ω is the upper-half-space $\mathbf{R}^m_{1,+} = \{x = x_0 + \underline{x} : x_0 > 0, \underline{x} \in \mathbf{R}^m\}$. In the case in (3.1), $n^{\pm}(\underline{y}) = \mp \mathbf{e_0} = \mp \mathbf{1}$ and $E(\underline{y} - \underline{x}) = \frac{\overline{\underline{y}-\underline{x}}}{|\underline{y}-\underline{x}|^{m+1}}$ has zero scalar part, while f and $d\sigma(\underline{y})$ both are scalar-valued, thus $N^+\mathbb{C}f = 0$. As consequence, $f = 2u$ and $v = \mathbb{C}u$. The situation for the outer Hilbert transform is the same. Hence, both the inner and outer Hilbert transformations coincide with the singular Cauchy integral transformation.

Example 2.2 Let $\Omega = \mathbf{D}$, the unit disc in the complex plane $\mathbf{C}$. We have

$$\begin{aligned} N^+\mathbb{C}f(e^{i\xi}) &= \text{p.v. Re}\left[\frac{1}{\pi i}\int_0^{2\pi}\frac{f(e^{it})}{e^{it}-e^{i\xi}}d(e^{it})\right] \\ &= \text{p.v.}\frac{1}{\pi}\int_0^{2\pi}\text{Re}\left[\frac{e^{it}}{e^{it}-e^{i\xi}}\right]f(e^{it})dt. \end{aligned}$$

Through a direct computation we have

$$\text{Re}[\frac{e^{it}}{e^{it}-e^{i\xi}}] = \frac{1}{2}.$$

Therefore,

$$N^+\mathbb{C}f(e^{i\xi}) = f_0 = I_0 f,$$

where f_0 denotes the average of $f(e^{it})$ over $[0, 2\pi]$ and I_0 the operator that maps f to f_0. We note that the operator norm

$$\|I_0\| = 1.$$

Simple computation gives

$$(I + I_0)^{-1}u = -\frac{u_0}{2} + u,$$

where $u_0 = I_0 u$.

Set

$$\tilde{H}f(e^{i\theta}) = N^-\mathbb{C}f(e^{i\theta}).$$

We have

$$\tilde{H}f = \text{p.v.}\frac{i}{2\pi}\int_0^{2\pi}\cot\left(\frac{\theta - t}{2}\right)f(e^{it})dt$$

that annihilates constants. Therefore,

$$\begin{aligned} H^+u &= N^-\mathbb{C}(I + I_0)^{-1}u \\ &= \tilde{H}(-\frac{u_0}{2} + u) \\ &= \tilde{H}u. \end{aligned}$$

Similarly, $(I - I_0)^{-1}$ is defined on the closed subspace

$$\left\{u \in L^2 \ : \ \int_0^{2\pi} u(e^{it})dt = 0\right\}$$

in which case there holds $(I - I_0)^{-1}u = u$, and

$$\begin{aligned} H^- u &= N^-\mathbb{C}(I - I_0)^{-1}u \\ &= \tilde{H}u. \end{aligned}$$

Note that in this example the inverse operator $(I - N^+\mathbb{C})^{-1}$ does not exist in the L^p spaces, but it does exist in the proper and closed subspace L_0^p of the L^p, defined by

$$L_0^p(\partial\mathbf{D}) = \{f \in L^p(\partial\mathbf{D}) \mid \int_0^{2\pi} f(e^{it})dt = 0\}.$$

Example 2.3 Consider $\Omega = B^m$, $\Sigma = S^{m-1}$, $m > 2$. For the higher-dimensional spheres, the double-layer potential $N^+\mathbb{C}$ is replaced by a non-trivial operator and the inner and the outer Hilbert transformations are distinguished (see, e.g., [5], [6] and [4]). On the sphere direct computation shows that the double layer potential reduces to

$$\langle E(\underline{x} - \underline{y}), n^+(\underline{y})\rangle = \frac{1}{2}\frac{1}{|\underline{x} - \underline{y}|^{m-2}}. \tag{3.6}$$

Therefore, by (3.5),

$$N^+\mathbb{C}f(\underline{x}) = \frac{1}{\sigma_{m-1}}\int_{S^{m-1}} \frac{f(\underline{y})}{|\underline{x} - \underline{y}|^{m-2}} d\sigma(\underline{y}).$$

Proposition 3.1. *The double-layer potential operator $N^+\mathbb{C}$ on the sphere is L^p-bounded, $1 \le p \le \infty$, with the operator norm being equal to* 1.

For a proof we refer to [2].

It turns out, however, what is crucial is not the bounds of the operator but the order of the singularity of the double-layer potential. On the sphere the existence and boundedness of the inverse operators in the L^p spaces, in fact, are guaranteed by the Fredholm theory. More generally, it may be shown that if Σ is C^∞, then

$$|\langle E(\underline{x} - \underline{y}), n^+(\underline{y})\rangle| \le \frac{C}{|\underline{x} - \underline{y}|^{m-2}}.$$

This estimate is consistent with (3.6). If Σ is $C^{1,\alpha}, 0 < \alpha < 1$, then

$$|\langle E(\underline{x} - \underline{y}), n^+(\underline{y})\rangle| \le \frac{C}{|\underline{y} - \underline{x}|^{m-1-\alpha}}. \tag{3.7}$$

In both cases the operator $N^+\mathbb{C}$ is compact and Fredholm theory may be used to show that $(I \pm N^+\mathbb{C})^{-1}$ exists and is a L^p-bounded operator for $1 < p < \infty$ (see [21], [15]). There, however, exists essential difficulty to use the Fredholm theory to C^1 and Lipschitz domains. As a matter of fact, the expected estimate of the kernel of the double-layer potential in the cases is only

$$|\langle E(\underline{x} - \underline{y}), n^+(\underline{y})\rangle| \le \frac{C}{|\underline{x} - \underline{y}|^{m-1}},$$

that is of the same singularity as the Cauchy singular integral kernel on the surfaces. Fabes, Jodeit and Riviére ([11]) were able to show that $N^+\mathbb{C}$ was compact for C^1 domains. For Lipschitz domains the operator $N^+\mathbb{C}$ is not necessarily compact in $L^p(\Sigma)$. New methods to prove the invertibility of $I + N^+\mathbb{C}$ had to be invented, and, it was done for the $p = 2$ case by Verchota ([21]), and for the optimal range of p's by Dahlberg and Kenig ([15]). For the details of the results see [15] and [13]. As a consequence of the mentioned results on existence and boundedness of $(I \pm N^+\mathbb{C})^{-1}$ we cite (see [2]).

Theorem 3.2. *The inner and outer Hilbert transformations $H^\pm$ for the Lipschitz domain $\Omega \subset R^m$, $m > 2$, both exist and are bounded from the $L^p(\Sigma)$ to $L^p(\Sigma)$, where $2 - \epsilon < p < \infty$, $\epsilon \in (0, 1]$ depends on the Lipschitz constant of Ω (for C^1 domain one may take $\epsilon = 1$). Moreover, for $u \in L^p(\Sigma)$,*

$$H^\pm u(\underline{x}) = \frac{2}{\sigma_{m-1}} p.v. \int_\Sigma [E(\underline{y} - \underline{x}) \wedge n^\pm(\underline{y})](I \pm N^+\mathbb{C})^{-1} u(\underline{y}) d\sigma(\underline{y}). \tag{3.8}$$

4. Poisson and conjugate Poisson kernels of the Cauchy type on Lipschitz surfaces

It is in question whether there exist explicit formulas for the kernels of the Hilbert transformations. This is related to whether explicit formulas exist for Poisson kernels and the harmonic conjugates of the Poisson kernels in the context. The answer is that in the general cases we can not expect to have explicit formulas. The explicit formulas and related issues for the sphere case are studied in [5], [6], [4], and then in [19].

Let U be a scalar-valued harmonic function in Ω. A harmonic function V is called *a harmonic conjugate of* U, if there hold (i) $\mathrm{Sc}[V] = 0$; and (ii) $\underline{D}(U+V) = 0$. By this definition, if $m > 2$, then the harmonic conjugate of a given harmonic function is not unique even modulo Clifford constants. Indeed, there exist non-constant harmonic functions V satisfying (i) but $\underline{D}(V) = 0$. For the determination we introduce

Definition 4.1. If V is a harmonic conjugate of U and there exists a scalar-valued boundary data f, allowing distributions, such that

$$U + V = C^+_\Sigma f,$$

then V is called a *Cauchy type harmonic conjugate,* or a *canonical harmonic conjugate of U in Ω.*

It is easy to see that if taking f to be the Dirac delta function at $\underline{y}$ on the boundary, then we conclude that

$$\frac{1}{\sigma_{m-1}} E(\underline{y} - \underline{x}) \wedge n^\pm(\underline{y})$$

is the Cauchy type harmonic conjugate of

$$\frac{1}{\sigma_{m-1}}\langle E(\underline{x}-\underline{y}), n^{\pm}(\underline{y})\rangle.$$

Note that a scalar-valued harmonic function may have more than one harmonic conjugate that differ by non-constant functions. On the other hand, there is only one harmonic conjugate of the Cauchy type. We have the following (see [2])

Theorem 4.2. *With the same assumptions on the Lipschitz domain Ω and on the range of the indices p's as in Theorem 3.2, let U be a scalar-valued harmonic function whose non-tangential maximal function*

$$u^*(\underline{x}) = \sup_{\underline{y}\in\Gamma_\alpha(\underline{x})} |U(\underline{y})|$$

is in the $L^p(\Sigma)$, where $\Gamma_\alpha(\underline{x})$ is the truncated cone of opening α whose axis is perpendicular to the tangent plane of Σ at $\underline{x}\in\Sigma$, then the Cauchy type harmonic conjugate of U exists and is unique.

Now we include a discussion on the Schwarz kernel and the associated Poisson and conjugate Poisson kernel of the Cauchy type. We seek for the integral representation of the operators S^+ such that $S^+u = C^+f$, where $u = \mathrm{Sc}[\mathbb{C}^+f]$. That is to seek for the kernel $S^+(\underline{x},\underline{y})$ such that

$$\begin{aligned} C^+f(\underline{x}) &= \frac{1}{\sigma_{m-1}}\int_\Sigma E(\underline{y}-\underline{x})n^+(\underline{y})f(\underline{y})d\sigma(\underline{y}) \\ &= \int_\Sigma S^+(\underline{x},\underline{y})u(\underline{y})d\sigma(\underline{y}) \quad = \quad S^+u(\underline{x}), \qquad \underline{x}\in\Omega. \end{aligned} \tag{4.1}$$

If the kernel $S^+(\underline{x},\underline{y})$ exists, then it is called the *inner Schwarz kernel.* The functions $P^+(\underline{x},\underline{y})$ and $Q^+(\underline{x},\underline{y})$ are called the *inner Poisson kernel* and the *conjugate inner Poisson kernel,* respectively, where

$$P^+ = \mathrm{Sc}[S^+], \quad Q^+ = \mathrm{NSc}[S^+].$$

The roles of P^+ and Q^+ are to give

$$\begin{aligned} U^+(\underline{x}) &= \int_\Sigma P^+(\underline{x},\underline{y})u(\underline{y})d\sigma(\underline{y}) \\ &= \frac{-2}{\sigma_{m-1}}\int_\Sigma \langle E(\underline{y}-\underline{x}), n^+(\underline{y})\rangle(I+N^+\mathbb{C})^{-1}u(\underline{y})d\sigma(\underline{y}) \end{aligned} \tag{4.2}$$

and

$$\begin{aligned} V^+(\underline{x}) &= \int_\Sigma Q^+(\underline{x},\underline{y})u(\underline{y})d\sigma(\underline{y}) \\ &= \frac{2}{\sigma_{m-1}}\int_\Sigma E(\underline{y}-\underline{x})\wedge n^+(\underline{y})(I+N^+\mathbb{C})^{-1}u(\underline{y})d\sigma(\underline{y}), \underline{x}\in\Omega, \end{aligned} \tag{4.3}$$

as the unique solutions of the Dirichlet problem inside Ω with the boundary data u and H^+u, respectively. $V^+(\underline{x})$ is, in fact, the Cauchy type harmonic conjugate of $U^+(\underline{x})$. The functions P^+ and Q^+ are the unique harmonic representations, respectively, of the Dirac δ_Σ function on the surface and of its Hilbert transform $H^+\delta_\Sigma$ in the distribution sense with the representation as the principal value singular kernel of the inner Hilbert transformation H^+. For Lipschitz domains the existence of the Poisson kernel is based on the Green function theory. For Lipschitz curves and surfaces it is a consequence of Plemelj's formula that the non-tangential boundary limit of the harmonic function $V^+(\underline{x}')$ is $H^+f(\underline{x})$. That is, for a.e. $\underline{x} \in \Sigma$,

$$\begin{aligned} \lim_{\underline{x}' \to \underline{x}} V^+(\underline{x}') &= H^+u(\underline{x}) = \text{p.v.} \int_\Sigma Q^+(\underline{x},\underline{y})u(\underline{y})d\sigma(\underline{y}) \\ &= \frac{2}{\sigma_{m-1}} p.v. \int_\Sigma [E(\underline{y}-\underline{x}) \wedge n^+(\underline{y})](I + N^+\mathbb{C})^{-1}u(\underline{y})d\sigma(\underline{y}). \end{aligned} \tag{4.4}$$

Since

$$V^+(\underline{x}') = \int_\Sigma P^+(\underline{x}',\underline{y})H^+f(\underline{y})d\underline{y},$$

we have

$$Q^+(\underline{x}',\underline{y}) = H^+_{\Sigma'}P^+(\underline{x}',\underline{y}), \quad \underline{x} \in \Sigma' \subset \Omega,\ \underline{y} \in \Sigma,$$

where the Hilbert transformation is with respect to an admissible Lipschitz surface Σ'.

The outer Cauchy integral C^- will correspond to the outer Poisson and conjugate outer Poisson kernels, and therefore the outer Schwarz kernel. The theory for them are similar.

5. Poisson and conjugate Poisson kernels on the unit sphere

The explicit inner Poisson kernel on the sphere is well known. There exist several methods to deduce the kernel. The harmonic conjugate of the inner Poisson kernel was first studied by Brackx et al. who gave the explicit formula of the kernel in the integral form. They further obtained a finite form of the formulas inductively in the the space dimension ([7]). Their methods, as a matter, yield the Cauchy type harmonic conjugates. Below we offer a different approach based on a double layer potential argument. We first formulate the result.

Theorem 5.1. *On the unit sphere the inner Poisson kernel and its Cauchy type harmonic conjugate are, respectively,*

$$P(\underline{x},\underline{\omega}) = \frac{1}{\sigma_{m-1}} \frac{1-|\underline{x}|^2}{|\underline{x}-\underline{\omega}|^m}, \tag{5.1}$$

and

$$Q(\underline{x},\underline{\omega}) = \frac{1}{\sigma_{m-1}}\left(\frac{2}{|\underline{x}-\underline{\omega}|^m} - \frac{m-2}{r^{m-1}}\int_0^r \frac{\rho^{m-2}}{|\rho\underline{\xi}-\underline{\omega}|^m}d\rho\right)\underline{x}\wedge\underline{\omega}, \quad 0<r<1. \tag{5.2}$$

Outline of the proof. From (3.3) and (3.5) we have

$$\mathbb{C}^\pm f = \frac{1}{2}(f \pm Sc[\mathbb{C}f]) \pm \frac{1}{2}NSc[\mathbb{C}f]. \tag{5.3}$$

For the inner Poisson kernel case and its conjugate we are working with the case "+" in the above formula. Comparing the corresponding formula in the case with the formula (3.2) and in view of harmonic extensions of the scalar and non-scalar part of the boundary values to inside part of the unit ball, we have

$$\frac{1}{\sigma_{m-1}}\frac{\langle \underline{\omega}-\underline{x},\underline{\omega}\rangle}{|\underline{\omega}-\underline{x}|^m}=\frac{1}{2}P(\underline{x},\underline{\omega})+\frac{1}{2}S(\underline{x},\underline{\omega}) \tag{5.4}$$

and

$$\frac{1}{\sigma_{m-1}}\frac{(\underline{x}-\underline{\omega})\wedge\underline{\omega}}{|\underline{x}-\underline{\omega}|^m}=\frac{1}{2}Q(\underline{x},\underline{\omega})+\frac{1}{2}\tilde{S}(\underline{x},\underline{\omega}),\quad |\underline{x}|<1,\quad |\underline{\omega}|<1,$$

where $S(\underline{x},\underline{\omega})$ is the single layer potential, given by

$$S(\underline{x},\underline{\omega})=\frac{1}{\sigma_{m-1}}\frac{1}{|\underline{x}-\underline{\omega}|^{m-2}},$$

and $\tilde{S}(\underline{x},\underline{\omega})$ is the Cauchy type harmonic conjugate of $S(\underline{x},\underline{\omega})$.

We then immediately obtain the formula for $P(\underline{x},\underline{\omega})$. To obtain $Q(\underline{x},\underline{\omega})$ we need to compute $\tilde{S}(\underline{x},\underline{\omega})$, that is given by the following lemma for $r<1$.

Lemma 5.2. *For* $r=|\underline{x}|<1$,

$$\begin{aligned}\sum_{k=0}^{\infty} r^k\frac{m-2}{m+k-2}P^{(k)}(\underline{\omega}^{-1}\underline{\xi}) &= \frac{m-2}{r^{m-2}}\int_0^r \rho^{m-3}E(\underline{\omega}-\rho\underline{\xi})\underline{\omega}d\rho \\ &= \frac{1}{|\underline{x}-\underline{\omega}|^{m-2}}+\frac{m-2}{r^{m-1}}\left(\int_0^r\frac{\rho^{m-2}}{|\rho\underline{\xi}-\underline{\omega}|^m}d\rho\right)\underline{x}\wedge\underline{\omega};\end{aligned} \tag{5.5}$$

and, for $r=|\underline{x}|>1$,

$$\begin{aligned}\sum_{k=1}^{\infty}\frac{m-2}{k}\frac{1}{r^{m-2+k}}P^{(-k)}(\underline{\omega}^{-1}\underline{\xi}) &= \frac{m-2}{r^{m-2}}\int_r^{\infty}\rho^{m-3}E(\underline{\omega}-\rho\underline{\xi})\overline{\underline{\omega}}d\rho \\ &= \frac{1}{|\underline{x}-\underline{\omega}|^{m-2}}-\frac{1}{r^{m-2}}-\frac{m-2}{r^{m-1}}\left(\int_r^{\infty}\frac{\rho^{m-2}}{|\rho\underline{\xi}-\underline{\omega}|^m}d\rho\right)\underline{x}\wedge\underline{\omega}.\end{aligned} \tag{5.6}$$

The proof of the lemma for the case $r<1$ is contained the work of Brackx et al., that for the case $r>1$ is given in [20].

Similarly, we have (see [20])

Theorem 5.3. *On the unit sphere the outer Poisson kernel and its Cauchy type harmonic conjugate are, respectively,*

$$P^{-}(\underline{x},\underline{\omega})=\frac{1}{\sigma_{m-1}}\frac{|\underline{x}|^2-1}{|\underline{x}-\underline{\omega}|^m}, \tag{5.7}$$

and

$$Q^{-}(\underline{x},\underline{\omega})=\frac{1}{\sigma_{m-1}}\left(-\frac{2}{|\underline{x}-\underline{\omega}|^m}+\frac{m-2}{r^{m-1}}\int_0^r\frac{\rho^{m-2}}{|\rho\underline{\xi}-\underline{\omega}|^m}d\rho\right)\underline{x}\wedge\underline{\omega},\quad r>1,\ \underline{\xi}\neq\underline{\omega}. \tag{5.8}$$

6. Abel sum formulas of the kernels

For $f \in L^2(S^{m-1})$, $\underline{x} = r\underline{\xi}, 0 \leq r < 1, \underline{y} = \underline{\omega} \in S^{m-1}$, we have (see [5])

$$C^+ f(\underline{x}) = \sum_{k=0}^{\infty} \frac{|\underline{x}|^k}{\sigma_{m-1}} \int_{S^{m-1}} C^+_{m,k}(\underline{\xi}, \underline{\omega}) f(\underline{\omega}) d\sigma(\underline{\omega}), \tag{6.1}$$

where

$$C^+_{m,k}(\underline{\xi}, \underline{\omega}) = \frac{m+k-2}{m-2} C_k^{(m-2)/2}(< \underline{\xi}, \underline{\omega} >) + C_{k-1}^{m/2}(\langle \underline{\xi}, \underline{\omega} \rangle) \underline{\xi} \wedge \underline{\omega}, \tag{6.2}$$

with $C_{-1}^{m/2}(\langle \underline{\xi}, \underline{\omega} \rangle) = 0$. The right-hand side of (6.2), in fact, is a function of $\underline{\omega}^{-1}\underline{x}$ ([19]), and thus we may write

$$P^{(k)}(\underline{\omega}^{-1}\underline{x}) = r^k C^+_{m,k}(\underline{\xi}, \underline{\omega}), \quad k = 0, 1, 2, \ldots,$$

and, consequently,

$$C^+ f(\underline{x}) = \sum_{k=0}^{\infty} \frac{1}{\sigma_{m-1}} \int_{S^{m-1}} P^{(k)}(\underline{\omega}^{-1}\underline{x}) f(\underline{\omega}) d\sigma(\underline{\omega}). \tag{6.3}$$

Similarly to (6.4), we have

$$\begin{aligned} C^- f(\underline{x}) &= \sum_{k=-1}^{-\infty} \frac{|\underline{x}|^{-m+2-k}}{\sigma_{m-1}} \int_{S^{m-1}} C^-_{m,|k|-1}(\underline{\xi}, \underline{\omega}) f(\underline{\omega}) d\sigma(\underline{\omega}) \\ &= \sum_{k=-1}^{-\infty} \frac{1}{\sigma_{m-1}} \int_{S^{m-1}} P^{(k)}(\underline{\omega}^{-1}\underline{x}) f(\underline{\omega}) d\sigma(\underline{\omega}), \end{aligned} \tag{6.4}$$

where

$$C^-_{m,|k|-1}(\underline{\xi}, \underline{\omega}) = \frac{|k|}{m-2} C_{|k|}^{(m-2)/2}(\langle \underline{\xi}, \underline{\omega} \rangle) - C_{|k|-1}^{m/2}(\langle \underline{\xi}, \underline{\omega} \rangle) \underline{\xi} \wedge \underline{\omega}. \tag{6.5}$$

Set

$$P^{(k)}(\underline{\omega}^{-1}\underline{x}) = r^{-m+2-k} C^-_{m,|k|-1}(\underline{\xi}, \underline{\omega}), \quad k = -1, -2, \ldots.$$

For the Fourier-Laplace expansion in the L^2 sense there holds

$$f(\underline{\xi}) = \sum_{k=-\infty}^{\infty} \frac{1}{\sigma_{m-1}} \int_{S^{m-1}} P^{(k)}(\underline{\omega}^{-1}\underline{\xi}) f(\underline{\omega}) d\sigma(\underline{\omega}). \tag{6.6}$$

This suggests that the series

$$\mathrm{Sc}\left[\sum_{k=-\infty}^{\infty} \frac{1}{\sigma_{m-1}} P^{(k)}(\underline{\omega}^{-1}\underline{\xi})\right]$$

plays the role of the Dirac-δ function.

Theorem 6.1. *The Abel sum expansions of the inner Poisson kernel and its Cauchy type harmonic conjugate are, respectively,*

$$P^+(\underline{x},\underline{\omega}) = \frac{1}{\sigma_{m-1}} \sum_{-\infty}^{\infty} r^{|k|} P^{(k)}(\underline{\omega}^{-1}\underline{\xi}), \quad \underline{x} = r\underline{\xi},\ r < 1, \tag{6.7}$$

and

$$Q^+(r\underline{\xi},\omega) = \frac{1}{\sigma_{m-1}} \left[\sum_{k=1}^{\infty} \frac{k}{m+k-2} r^k P^{(k)}(\underline{\omega}^{-1}\underline{\xi}) - \sum_{k=-\infty}^{-1} r^{|k|} P^{(k)}(\underline{\omega}^{-1}\underline{\xi}) \right], \ r < 1. \tag{6.8}$$

Proof. We set

$$A^+(r) = \frac{1}{\sigma_{m-1}} \left[\sum_{k=0}^{\infty} r^k P^{(k)}(\underline{\omega}^{-1}\underline{\xi}) \right]. \tag{6.9}$$

From the analysis given above, we have

$$A^+(r) = \frac{1}{2}P^+(r\underline{\xi},\underline{\omega}) + \frac{1}{2}S^+(r\underline{\xi},\underline{\omega}) + \frac{1}{2}\tilde{S^+}(r\underline{\xi},\underline{\omega}) + \frac{1}{2}Q^+(r\underline{\xi},\underline{\omega}), \quad r < 1, \tag{6.10}$$

where the last three entries are the harmonic representation of a half of the Cauchy singular integral of f, viz. $(1/2)\mathbb{C}f$. Similarly,

$$\begin{aligned} A^-(r) &= \frac{1}{\sigma_{m-1}} \left[\sum_{-\infty}^{-1} P^{(k)}(\underline{\omega}^{-1}\underline{x}) \right] \qquad (6.11) \\ &= \frac{1}{2}P^-(r\underline{\xi},\underline{\omega}) - \frac{1}{2}S^-(r\underline{\xi},\underline{\omega}) - \frac{1}{2}\tilde{S^-}(r\underline{\xi},\underline{\omega}) - \frac{1}{2}Q^-(r\underline{\xi},\underline{\omega}), \quad r > 1. \end{aligned}$$

It is easy to observe that the Kelvin inversion of A^-, denoted by $\mathbb{K}(A^-)$, satisfies the relation

$$\mathbb{K}(A^-)(r) = \frac{1}{2}P^+(r\underline{\xi},\underline{\omega}) - \frac{1}{2}S^+(r\underline{\xi},\underline{\omega}) - \frac{1}{2}\tilde{S^+}(r\underline{\xi},\underline{\omega}) - \frac{1}{2}Q^+(r\underline{\xi},\underline{\omega}), \quad r < 1.$$

We thus arrive

$$P^+(\underline{x},\underline{\omega}) = A^+(r) + \mathbb{K}(A^-)(r).$$

Applying Kelvin inversion term by term to the series expansion of A^- in (6.11) (the first equality), and using (6.9), we obtain the Abel sum (6.7) expansion for the Poisson kernel.

Next we deduce the Abel sum formula of the conjugate Poisson kernel $Q^+(\underline{x},\omega)$. In fact, by (5.5) in Lemma 1, in (6.10) all the entries but except

$(1/2)Q^+(r\underline{\xi},\underline{\omega})$ are already of the Abel sum form, therefore,

$$\frac{1}{2}Q^+(r\underline{\xi},\omega) = A^+(r) - \frac{1}{2}P^+(r\underline{\xi},\underline{\omega}) - \frac{1}{2}\frac{1}{\sigma_{m-1}}\left[\sum_{k=0}^{\infty} r^k \frac{m-2}{m+k-2}P^{(k)}(\underline{\omega}^{-1}\underline{\xi})\right]$$

$$= \frac{1}{\sigma_{m-1}}\left[\sum_{k=0}^{\infty} r^k P^{(k)}(\underline{\omega}^{-1}\underline{\xi}) - \frac{1}{2}\sum_{-\infty}^{\infty} r^{|k|}P^{(k)}(\underline{\omega}^{-1}\underline{\xi}) - \frac{1}{2}\sum_{k=0}^{\infty} r^k \frac{m-2}{m+k-2}P^{(k)}(\underline{\omega}^{-1}\underline{\xi})\right]$$

$$= \frac{1}{\sigma_{m-1}}\left[\frac{1}{2}\sum_{k=1}^{\infty}\frac{k}{m+k-2}r^k P^{(k)}(\underline{\omega}^{-1}\underline{\xi}) - \frac{1}{2}\sum_{k=-\infty}^{-1} r^{|k|}P^{(k)}(\underline{\omega}^{-1}\underline{\xi})\right].$$

We thus arrive (6.8). The proof is complete.

Theorem 6.2. *The Abel sum expansions of the outer Poisson kernel and its canonical harmonic conjugate are, respectively,*

$$P^-(r\underline{\xi},\underline{\omega}) = \frac{1}{\sigma_{m-1}}\sum_{-\infty}^{\infty} r^{-|k|-m+2}P^{(k)}(\underline{\omega}^{-1}\underline{\xi}), \quad r > 1, \tag{6.12}$$

and

$$Q^-(r\underline{\xi},\underline{\omega}) = \frac{1}{\sigma_{m-1}}\left[\sum_{k=1}^{\infty}\frac{1}{r^{m+k-2}}P^{(k)}(\underline{\omega}^{-1}\underline{\xi}) - \sum_{k=-\infty}^{-1}\frac{m+|k|-2}{|k|}\frac{1}{r^{m+|k|-2}}P^{(k)}(\underline{\omega}^{-1}\underline{\xi})\right] - \tilde{N}(r\underline{\xi},\underline{\omega}), \tag{6.13}$$

where $\tilde{N}$ is the canonical harmonic conjugate outside the unit ball of the double layer potential N, where

$$N(r\underline{\xi}) = \frac{1}{\sigma_{m-1}}\frac{1}{r^{m-2}}$$

and

$$\tilde{N}(r\underline{\xi},\underline{\omega}) = \frac{1}{\sigma_{m-1}}\frac{m-2}{r^{m-2}}\int_0^{\infty}\frac{\rho^{m-2}}{|\rho\underline{\xi}-\underline{\omega}|^m}d\rho\ \underline{\xi}\wedge\underline{\omega}, \qquad \text{a.e. } r > 1. \tag{6.14}$$

Acknowledgement

The author wishes to specially thank A. Axelsson. In a meeting in Finland Axelsson first explained to the author the definition of Hilbert transform on curves and surfaces that is closely related to the author's Ph.D. thesis work. The idea of the present work was strongly influenced by Axelsson. Some ingredients, including computation of the scalar part of the singular Cauchy integral on the unit sphere and the operator equations, were directly taken from him. He also generously shares with the author the forthcoming related work [2]. The author also sincerely thanks A. McIntosh, K.I. Kou, Y. Yang and Y.C. Lin for helpful discussions. X.M. Li read a draft version of the article and gave valuable comments and a list of suggested corrections. The author wishes to thank him as well.

References

[1] A. Axelsson, *Transmission problems and boundary operators,* Integral Equation and Operator Theory, **50** (2004), 147–164.

[2] A. Axelsson, K.I. Kou and T. Qian, *Hilbert transforms and the Cauchy integral in Euclidean space*, preprint.

[3] S. Bell, *The Cauchy transform, potential theory and conformal mappings,* 1992.

[4] F. Brackx, B. De Knock, H. De Schepper and D. Eelbode, *On the interplay between the Hilbert transform and conjugate harmonic functions,* Mathematical Methods in the Applied Sciences, **29** (2006), 1435–1450.

[5] F. Brackx and H. De Schepper, *Conjugate harmonic functions in Euclidean space: a spherical approach,* Computational Methods and Function Theory, to appear in CMFT 2006.

[6] F. Brackx, H. De Schepper and D. Eelbode, *A new Hilbert transform on the unit sphere in* $\mathbf{R}^m$, Complex Variables and Elliptic Equations, **51** (2006), 453–462.

[7] F. Brackx and N. Van Acker, *A conjugate Poisson kernel in Euclidean space,* Simon Stevin, **67** (1993), 3–14.

[8] D. Constales, *A conjugate harmonic to the Poisson kernel in the unit ball of* $\mathbf{R}^n$, Simon Stevin, **62** (1988), 289–291.

[9] R. Coifman, A. McIntosh and Y. Meyer, *L'intégrale de Cauchy définit un opérateur borné sur* L^2 *pour les courbes lipschitziennes,* Ann. Math. **116** (1982), 361–387.

[10] R. Delanghe, F. Sommen and V. Soucek, *Clifford Algebra and Spinor-Valued Functions,* **53**, Kluwer Academic Publishers, Dordrecht, Boston, London, 1992.

[11] E. Fabes, M. Jodeit and N. Riviére, *Potential techniques for boundary value problems on* C^1 *domains,* Acta Math., **141** (1978), 165–186.

[12] J.B. Garnett, *Bounded Analytic Functions*, Academic Press, 1987.

[13] J. Gilbert and M. Murray, *Clifford Algebra and Dirac Operator in Harmonic Analysis,* Cambridge University Press, Cambridge, MA 1991.

[14] C.E. Kenig, *Weighted Hardy spaces on Lipschitz domains,* Amer. J. Math., **102** (1980), 129–163.

[15] C.E. Kenig, *Harmonic analysis techniques for second order elliptic boundary value problems,* CBMS, Regional Conference Series in Mathematics, **83**, 1991.

[16] C. Li, A. McIntosh and S. Semmes, *Convolution singular integrals on Lipschitz surfaces,* J. Amer. Math. Soc., **5** (1992), 455–481.

[17] C. Li, A. McIntosh and T. Qian, *Clifford algebras, Fourier transforms, and singular convolution operators on Lipschitz surfaces,* Revista Mathemática Iberoamericana, **10** (1994), 665–721.

[18] T. Qian, *Singular integrals with holomorphic kernels and H^∞ Fourier multipliers on star-shaped Lipschitz curves,* Studia Mathematica, **123** (1997), 195–216.

[19] T. Qian, *Fourier Analysis on starlike Lipschitz surfaces,* Journal of Functional Analysis, **183** (2001), 370–412.

[20] T. Qian and Y. Yang, *Cauchy Singular Integral and Hilbert Transforms on the Unit Sphere,* preprint.

[21] G. Verchota, *Layer potentials and regularity for the Dirichlet problem for Laplace's equation,* J. Funct. Anal., **59** (1984), 572–611.

Tao Qian
Department of Mathematics
University of Macau
Macao, China SAR
e-mail: fsttq@umac.mo

Quaternionic and Clifford Analysis
Trends in Mathematics, 277–289

n-Dimensional Bloch Classes

L.F. Reséndis O. and L.M. Tovar S.

Abstract. In this paper we give a new generalization for the unit ball in $\mathbb{R}^n$, of Bloch space. We justify our definition by showing the connection between our proposal with the analytic, quaternionic and monogenic cases.

Mathematics Subject Classification (2000). Primary 30G35; Secondary 30C45.

Keywords. Subharmonic function, Bloch and $\mathcal{Q}_p$ classes.

1. Introduction

Let $\mathbb{D}$ be the open unit disk in the complex plane $\mathbb{C}$ and T its boundary. Let $\mathcal{A}$ be the space of analytic functions $f : \mathbb{D} \to \mathbb{C}$. The well-known Bloch space is defined as follows

$$\mathcal{B} = \{ \ f \in \mathcal{A} \ : \sup_{z\in\mathbb{D}}(1 - |z|^2)|f'(z)| < \infty \ \} \ .$$

According to R. Remmert (see [22], p. 229), the auxiliary function $|f'(z)|(1-|z|^2)$ was first introduced in 1929 by Landau (see [19], p. 83). Since then, the Bloch space has been intensively studied (see, e.g., [3],[12], [16], [23]). In this paper we give a generalization of the Bloch space for real-valued subharmonic functions defined in the open unit ball of $\mathbb{R}^n$. Let $\phi_a : \mathbb{C} \to \mathbb{C}$ be the Möbius transformation,

$$\phi_a(z) = \frac{a-z}{1-\overline{a}z}, \qquad |a| < 1, \tag{1.1}$$

with pole at $z = 1/\overline{a}$. Observe that $\phi_a^{-1} = \phi_a$ and

$$1 - |\phi_a(z)|^2 = \frac{(1-|a|^2)(1-|z|^2)}{|1-\overline{a}z|^2} = (1-|z|^2)|\phi_a'(z)|$$

For $z, \ a \in \mathbb{D}$, consider a Green's function of $\mathbb{D}$, with logarithmic singularity at a, defined by

$$g(z,a) = \ln\frac{|1-\overline{a}z|}{|z-a|} = \ln\frac{1}{|\phi_a(z)|} \ . \tag{1.2}$$

This work was completed with a partial support from CONACYT, COFAA-IPN and UAM-2230302.

So, $g(z,a)$ is given by the composition of the Möbius transformation ϕ_a and the fundamental solution of the two-dimensional real Laplacian. There are several spaces related with the Bloch space, for instance the Dirichlet space (see, e.g., [1], [26])

$$D = \left\{ f \in \mathcal{A} : \iint_{\mathbb{D}} |f'(z)|^2 \, dx \, dy < \infty \right\} .$$

We say that $f \in \mathcal{A}$ belongs to the space $BMOA(T)$ (Bounded Mean Oscillation Analytic functions) if

$$\sup_{|I|} \frac{1}{|I|} \int_I |f(e^{i\theta}) - f_I| \, d\theta < \infty$$

where

$$f_I = \frac{1}{|I|} \int_I f(e^{i\theta}) \, d\theta$$

and I is a subarc of T. After F. John and L. Nirenberg [20], A. Baernstein [8] gave the following characterization.

Theorem 1.1. *Let $f \in \mathcal{A}$. Then $f \in BMOA$ if and only if*

$$\sup_{a \in \mathbb{D}} \iint_{\mathbb{D}} |f'(z)|^2 g(z,a) \, dx \, dy .$$

R. Aulaskari and P. Lapan introduced in their seminal paper [2], the $\mathcal{Q}_p$ spaces for $1 \leq p < \infty$, as the set of functions $f \in \mathcal{A}$ such that

$$\sup_{a \in \mathbb{D}} \iint_{\mathbb{D}} |f'(z)|^2 g^p(z,a) \, dx \, dy < \infty .$$

In this paper the Bloch space appears in a very natural way. More precisely

Theorem 1.2 (Proposition 1, [2]). *If $f \in \mathcal{A}$ and $0 < p < \infty$, then*

$$(1 - |a|^2)^2 |f'(a)|^2 \leq \frac{1}{\pi} \left(\frac{2e}{p} \right)^p \iint_{\mathbb{D}} |f'(z)|^2 g^p(z,a) \, dx \, dy \tag{1.3}$$

for all $a \in D$.

They estimated the integral in (1.3) by the left hand and obtained the expression defining the Bloch space. Thus $\mathcal{Q}_p \subset B$. Moreover, they proved $B = \mathcal{Q}_p$ for $1 < p < \infty$. By Baernstein [8], $\mathcal{Q}_1 = BMOA(T) \cap H^1$, where H^1 is the Hardy space on $\mathbb{D}$. Then the $\mathcal{Q}_p$ spaces for $0 \leq p < 1$ were introduced by R. Aulaskari, J. Xiao and R. Zhao in [6], where the main fact of this work was that these spaces fill the gap between the Dirichet space and the Bloch space. Further generalizations were given also by R. Zhao in [27], where he defines for $0 < p < \infty$, $-2 < q < \infty$, $0 \leq s < \infty$, the $F(p,q,s)$ weighted spaces as the set of analytic functions $f \in \mathcal{A}$ such that

$$\sup_{a \in \mathbb{D}} \iint_{\mathbb{D}} |f'(z)|^p (1 - |z|^2)^q g^s(z,a) \, dx \, dy < \infty .$$

2. The Bloch space in the quaternionic case

In 1999, K. Gürlebeck et al. [17] generalized the $\mathcal{Q}_p$ spaces for hyperholomorphic functions defined in the unit ball of $\mathbb{R}^3$. More precisely, consider the set of real quaternions $\mathbb{H}$, this means elements of the form

$$a = \sum_{k=0}^{3} a_k e_k, \quad \text{where} \quad a_k \in \mathbb{R},\ k = 0, 1, 2, 3,$$

e_0 is the unit and e_1, e_2, e_3 are called imaginary units satisfying, $e_k^2 = -e_0$, $e_1e_2 = -e_2e_1 = e_3$, $e_2e_3 = -e_3e_2 = e_1$ and $e_3e_1 = -e_1e_3 = e_2$.
The natural operations for addition and multiplication make $\mathbb{H}$ a skew-field.The quaternion conjugation in $\mathbb{H}$, is defined for the basic elements as

$$\overline{e_0} := e_0, \quad \bar{e}_k := -e_k, \quad k = 1,\ 2,\ 3$$

and it extends onto $\mathbb{H}$ by linearity. Note that we have the property

$$a\bar{a} = \bar{a}a = |a|_{\mathbb{H}}^2 = a_0^2 + a_1^2 + a_2^2 + a_3^2 = |a|_{\mathbb{R}^4}\ .$$

Therefore, for $a \in \mathbb{H} - \{0\}$ the quaternion

$$a^{-1} := \frac{1}{|a|^2}\bar{a}$$

is an inverse to a. Observe that $\overline{ab} = \bar{b}\bar{a}$ for $a,\ b \in \mathbb{H}$. Let $\Omega \subset \mathbb{R}^3$ be a domain. We shall consider $\mathbb{H}$-valued functions defined in Ω, depending on $x = (x_0, x_1, x_2)$:

$$f : \Omega \mapsto \mathbb{H}\ .$$

On $C^1(\Omega, \mathbb{H})$ it is considered a generalized Cauchy-Riemann operator

$$D(f) = e_0\frac{\partial f}{\partial x_0} + e_1\frac{\partial f}{\partial x_1} + e_2\frac{\partial f}{\partial x_2}\ .$$

Here, D is a right-linear operator with respect the scalars in $\mathbb{H}$. The operator $\overline{D}$

$$\overline{D}(f) = e_0\frac{\partial f}{\partial x_0} - e_1\frac{\partial f}{\partial x_1} - e_2\frac{\partial f}{\partial x_2}$$

is the adjoint Cauchy-Riemann operator. The solutions of $D(f)(x) = 0,\ x \in \Omega$ are called (left) hyperholomorphic (or monogenic) functions and generalize the class of analytic functions in the one-dimensional complex function theory. Let Δ the three-dimensional Laplace operator $\Delta = \sum_{k=0}^{2} \frac{\partial^2}{\partial x_k^2}$. Then on $C^2(\Omega, \mathbb{H})$ – in analogy to the complex case – we have the factorization $\Delta = D\overline{D} = \overline{D}D$.
Using the adjoint generalized Cauchy-Riemann operator $\overline{D}$ instead of the derivative $f'(z)$, the quaternionic Möbius transformation $\phi_a = (a-x)(1-\bar{a}x)^{-1}$, and the modified fundamental solution $g(x) = \frac{1}{4\pi}(\frac{1}{|x|} - 1)$ of the real Laplacian, quaternionic $\mathcal{Q}_p$-spaces are defined by

$$\mathcal{Q}_p(\mathcal{H}) = \left\{ f \in \ker D\ :\ \mathcal{Q}_p(f) := \sup_{a\in B} \int_B |\overline{D}f(x)|^2 g^p(\phi_a(x))\, dB_x < \infty \right\}. \quad (2.1)$$

The generalizations of the Green function and the higher-dimensional Möbius transformation seem to be natural, where $-\frac{1}{2}\overline{D}$ plays the role of a derivative, as it is shown in [21] [25] for dimension four, and for arbitrary dimension in [18]. The introduction of the Bloch space in the quaternionic case is motivated by the analogous of Proposition 1 of Aulaskari and Lappan [2].

To be more precise, Gürlebeck et al. proved:

Proposition 2.1 (Proposition 4.1, [17]). *Let f be hyperholomorphic and $0 < p < 3$, then we have*

$$(1-|a|^2)^3|\overline{D}f(a)|^2 \le C_1 \int_B |\overline{D}f(x)|^2 \left(\frac{1}{|\phi_a(x)|}-1\right)^p dB_x \,,$$

where the constant C_1 does not depend on a and f.

From the previous proposition is natural to define the quaternionic Bloch space as the set of monogenic functions $f : B \to \mathbb{H}$ such that

$$\sup_{a\in B}(1-|a|^2)^{\frac{3}{2}}|\overline{D}f(a)| < \infty \,.$$

In the same paper they also proved:

Proposition 2.2 (Theorem 4.1, Theorem 5.1 [17]). *Let f be hyperholomorphic in the unit ball. Then the following conditions are equivalent:*

1. $f \in B$.
2. $\mathcal{Q}_p(f) < \infty$ *for all* $2 < p < 3$.
3. $\mathcal{Q}_p(f) < \infty$ *for some* $2 < p < 3$.
4. *For all* $2 < p < 3$.
$$\sup_{a\in B}\int_B |\overline{D}f(x)|^2(1-|\phi(x)|^2)^p\, dB_x < \infty \,.$$
5. *For some* $2 < p < 3$.
$$\sup_{a\in B}\int_B |\overline{D}f(x)|^2(1-|\phi(x)|^2)^p\, dB_x < \infty \,.$$

3. Bloch spaces in $\mathbb{R}^n$

Several generalizations of Bloch spaces have been done for higher dimensions and for different classes of functions. For the unit ball in $\mathbb{R}^n$, J. Cnops et al. in [14], generalize $\mathcal{Q}_p$ and Bloch spaces by using Clifford Analysis and recently S. Bernstein in [11] with harmonic functions. In this section we present a generalization of Bloch and $\mathcal{Q}_p$ spaces for subharmonic functions in the unit ball of $\mathbb{R}^n$, where we follow the approach presented by A.F. Beardon in [9], for Möbius transformations in $\mathbb{R}^n$. In a forthcoming paper we will continue the development of this approach, by considering the general case $n \ge 4$ for $(1-|\phi_a(x)|^2)^p$ and for $n \ge 3$ with the Green function obtained as the composition of ϕ_a with the fundamental solution of the n-dimensional real Laplacian.

The motivation to do it in that way is, that maintaining the usual notation of $\mathbb{R}^n$, the proof of several properties result simple and it is possible to obtain several important results.

The ball $B(a,r)$ in $\mathbb{R}^n$ is defined by

$$B(a,r) = \{\ x \in \mathbb{R}^n\ : |x-a| < r\}$$

and its boundary, the sphere $S(a,r)$ in $\mathbb{R}^n$ is given by

$$S(a,r) = \{\ x \in \mathbb{R}^n\ : |x-a| = r\}$$

where $a \in \mathbb{R}^n$ and $r > 0$. We will denote by $B = B(0,1)$ and $S^{n-1} = S(0,1)$. Denote by $\hat{\mathbb{R}}^n := \mathbb{R}^n \cup \{\infty\}$. The reflection (or inversion) in $S(a,r)$ is given by the function $\phi_{a,r} : \hat{\mathbb{R}}^n \to \hat{\mathbb{R}}^n$ defined for $x \neq a$ as

$$\phi_{a,r}(x) = a + \frac{r^2}{|x-a|^2}(x-a) \tag{3.1}$$

and $\phi_{a,r}(a) = \infty$. It is clear that $\phi_{a,r}^{-1} = \phi_{a,r}$. Let $P(a,t)$ be the plane in $\mathbb{R}^n$ given by

$$P(a,t) = \{\ \ x \in \mathbb{R}^n\ :\ (x,a) = t\ \}$$

where $a \in \mathbb{R}^n$, $a \neq 0$ and (x,a) is the usual scalar product and $t \in \mathbb{R}$.
A Möbius transformation in $\hat{\mathbb{R}}^n$ is defined as a finite composition of reflections (in spheres or planes).

Clearly, each Möbius transformation is a homeomorphism of $\mathbb{R}^n \cup \{\infty\}$ onto itself. Composition and inverse of Möbius transformations are again Möbius transformations and the identity map in $\hat{\mathbb{R}}^n$ is a Möbius transformation too. Therefore the set of Möbius transformations form a group, which is called the General Möbius group $GM(\hat{\mathbb{R}}^n)$.

Define for $0 \neq a \in \mathbb{R}^n$ the Kelvin inverse (see [7], p. 59)

$$a^* = \frac{1}{|a|^2} a\ . \tag{3.2}$$

We have the following characterization of Möbius transformations.

Theorem 3.1 (Theorem 3.5.1 [9]**).** *Let ϕ be a Möbius transformation. If $\phi(B) = B$ then*

$$\phi(x) = (\sigma x)A$$

where σ is a reflection in some sphere orthogonal to S^{n-1} and A is an orthogonal matrix.

Theorem 3.2 (Theorem 3.5.1 [9]**).** *Let $\phi_{a,r}$ be a reflection in $S(a,r)$. Then the following conditions are equivalent:*

1. *$S(a,r)$ and S^{n-1} are orthogonal;*
2. *$1 + r^2 = |a|^2$ (see (3.2.2), [9]);*
3. *$\phi_{a,r}(a^*) = 0$ (equivalently, $\phi_{a,r}(0) = a^*$);*
4. *$\phi_{a,r}(B) = B$.*

If the conditions of the previous theorem are satisfied, for instance, $1+r^2 = |a|^2$, we write for the reflection $\phi_{a,r} = \phi_a$. Of course, by (3.2) we have

$$\phi_a(x) = \phi_{a^*}(x) = \frac{1}{|a^*|^2}a^* + \frac{1-|a^*|^2}{||a^*|^2x - a^*|^2}(|a^*|^2x - a^*)$$

and $a^* \in B$. By (3.2) we present our results using a and a^*. Also, for $|a| > 1$, it is well known the next formula ([9] 3.4.2)

$$\frac{1-|\phi_a(x)|^2}{|a|^2-1} = \frac{1-|x|^2}{|x-a|^2} . \tag{3.3}$$

Theorem 3.3. *Let* $0 < R < 1$ *and* $a \in \mathbb{R}^n$ *be with* $|a| > 1$. *Let* $\phi_a : \hat{\mathbb{R}}^n \to \hat{\mathbb{R}}^n$ *be the reflection in* $S(a, \sqrt{|a|^2-1})$. *Then the pseudo-hyperbolic ball* $U(a,R) = \phi(B(0,R))$ *is a Euclidean ball with center and radius*

$$c = \frac{1-R^2}{|a|^2-R^2}a \,, \qquad \tilde{R} = \frac{(|a|^2-1)R}{|a|^2-R^2}$$

respectively.

Proof. Observe first that,

$$\phi_a(\frac{R}{|a|}a) = a + \frac{|a|^2-1}{(R-|a|)|a|}a .$$

Since the reflection acts radially, the diameter of the pseudo-hyperbolic ball is

$$\begin{aligned} \left|\phi_a(\frac{R}{|a|}a) - \phi_a(-\frac{R}{|a|}a)\right| &= \left|\frac{|a|^2-1}{(R-|a|)|a|}a - \frac{|a|^2-1}{(-R-|a|)|a|}a\right| \\ &= \frac{2R(|a|^2-1)}{R^2-|a|^2} . \end{aligned}$$

The center is given by

$$\begin{aligned} \frac{1}{2}\left(\phi_a(\frac{R}{|a|}a) + \phi_a(-\frac{R}{|a|}a)\right) &= \frac{1}{2}\left(a + \frac{|a|^2-1}{(R-|a|)|a|}a + a + \frac{|a|^2-1}{(-R-|a|)|a|}a\right) \\ &= \frac{1}{2}\left(2a + \frac{|a|^2-1}{(R-|a|)|a|}a - \frac{|a|^2-1}{(R+|a|)|a|}a\right) \\ &= a + \frac{|a|^2-1}{R^2-|a|^2}a = \frac{1-R^2}{|a|^2-R^2}a . \end{aligned}$$

□

Remark 3.4. If $a \in \mathbb{R}^n$ verifies $|a| > 1$, then $a^* = \dfrac{a}{|a|^2} \in B$ and the pseudo-hyperbolic ball $U(a,R)$ has as center

$$c = \frac{1-R^2}{|a|^2-R^2}a = \frac{1-R^2}{\left(1-\frac{R^2}{|a|^2}\right)}\frac{a}{|a|^2} = \frac{1-R^2}{1-|a^*|^2R^2}a^*$$

and radius

$$r = \frac{(|a|^2 - 1)R}{|a|^2 - R^2} = \frac{\left(1 - \frac{1}{|a|^2}\right)}{\left(1 - \frac{R^2}{|a|^2}\right)} R = \frac{(1 - |a^*|^2)R}{1 - |a^*|^2 R^2} .$$

Observe that the form of these formulae is similar to the correspondent ones in the complex case (see [24]).

As a result of a straightforward calculation we obtain:

Proposition 3.5. *Let* $0 < R < 1$ *and* $a \in \mathbb{R}^n$ *be with* $|a| > 1$. *Let* $\phi_a : \hat{\mathbb{R}}^n \to \hat{\mathbb{R}}^n$ *be the reflection in* $S(a, \sqrt{|a|^2 - 1})$. *Then*

$$a^* = \frac{a}{|a|^2} \in \left[\frac{1 - R^2}{|a|^2 - R^2} a, \phi_a(-\frac{R}{|a|} a)\right] = \left[\frac{1 - R^2}{|a|^2 - R^2} a, \frac{1 + R|a|}{|a|^2 + R|a|} a\right]$$

and

$$\left|\phi_a(-\frac{R}{|a|} a) - a^*\right| = \frac{(1 - R)(|a|^2 - 1)}{|a|(|a| + R)} = R^* .$$

As a consequence of the previous result we get

Corollary 3.6. *For* $0 < R < 1$ *and* $a \in \mathbb{R}^n$ *with* $|a| > 1$, *let* $\phi_a : \hat{\mathbb{R}}^n \to \hat{\mathbb{R}}^n$ *be the reflection in* $S(a, \sqrt{|a|^2 - 1})$. *Then*

$$B(a^*, R^*) \subset B(c, r) .$$

We say that $u : B \to \mathbb{R}$ belongs to the class $\mathcal{SH}(B)$ if u is a subharmonic function, that is, for each $b \in B$ and $0 < r < 1 - |b|$

$$u(b) \leq \int_S u(b + r\zeta)\, d\sigma(\zeta)$$

where $d\sigma(\zeta)$ is the normalized surface area measure on S (see [7], p. 224, [15], p. 264).

Proposition 3.7. *Let* $u \in \mathcal{SH}(B)$ *and* $1 \leq p < \infty$. *Then* $|u|$, u^p *and* $|u|^p \in \mathcal{SH}(B)$.

Proof. It follows from an easy estimation that $|u| \in \mathcal{SH}(B)$. Now, as the function $x \mapsto x^p$ is a convex function, by Jensen's inequality we get the result. □

Corollary 3.8. *Let* $u_i : B \to \mathbb{R}$ *be harmonic functions. Then* $\sum_{i=1}^n c_i u_i^2(x)$, *with* $0 \leq c_i < \infty$ *is a subharmonic function.*

Recall that for a Borel measurable integrable function f on $\mathbb{R}^n$

$$\frac{1}{nV(B)} \int_{\mathbb{R}^n} f\, dV = \int_0^\infty r^{n-1} \int_S f(r\zeta)\, d\sigma(\zeta)\, dr .$$

We say that the function u belongs to the Dirichlet subharmonic class $D^{sh}_{p,\phi}$, if $u \in \mathcal{SH}(B)$ and

$$\int_B u^2(x) \left(1 - |x|^2\right)^p dB(x) < \infty .$$

We say that the function u belongs to the subharmonic class $\mathcal{Q}_p^{sh}$, if $u \in \mathcal{SH}(B)$ and

$$\sup_{\phi} \int_B u^2(x) \left(1 - |\phi(x)|^2\right)^p dB(x) < \infty ,$$

where ϕ is a Möbius transformation of B. By Theorem 3.1 and the definition of $\mathcal{Q}_p^{sh}$, we can omit the orthogonal transformation and consider only reflections on $S(a, \sqrt{|a|^2 - 1})$ or the identity, that is, a subharmonic function u belongs to the subharmonic class $\mathcal{Q}_{p,\phi}^{sh}$ if and only if

$$\sup_{|a|>1} \int_B u^2(x) \left(1 - |\phi_a(x)|^2\right)^p dB(x) < \infty .$$

If we consider the Möbius transformation $\phi(x) = x$, we obtain immediately $\mathcal{Q}_{p,\phi}^{sh} \subset D_p^{sh}$.

Theorem 3.9. *Let $0 < R < 1$ and $a \in \mathbb{R}^n$, with $|a| > 1$ Let $0 < p < \infty$ and $u \in \mathcal{Q}_p^{sh}$. Then*

$$u^2(a^*) \frac{(1-R)^n (|a|^2 - 1)^n}{|a|^n (|a| + R)^n} \leq \frac{1}{V(B)\,(1 - R^2)^p} \int_B u^2(x) \left(1 - |\phi_a(x)|^2\right)^p dB_x .$$

Proof. Let $0 < R < 1$ and $U(a, R)$ be the pseudohyperbolic ball with radius R. Then by Corollary 3.6

$$\begin{aligned}
&\int_B u^2(x) \left(1 - |\phi_a(x)|^2\right)^p dB_x \\
&\quad \geq \int_{U(a,R)} u^2(x) \left(1 - |\phi_a(x)|^2\right)^p dB_x \\
&\quad \geq \left(1 - R^2\right)^p \int_{B(a*,R^*)} u^2(x)\, dB_x \\
&\quad = \left(1 - R^2\right)^p nV(B) \int_0^{R^*} r^{n-1} \int_S u^2(a^* + r\zeta)\, d\sigma(\zeta)\, dr \\
&\quad = \left(1 - R^2\right)^p nV(B) \left[\frac{r^n}{n}\right]_0^{R*} u^2(a^*) \\
&\quad = \left(1 - R^2\right)^p V(B) \left(\frac{(1-R)(|a|^2 - 1)}{|a|(|a| + R)}\right)^n u^2(a^*) .
\end{aligned}$$

□

Observe how simple is the previous proof (see Proposition 4.1 in [17] and Lemma 3.2 in [13]).

Theorem 3.10. *Let $u : B^n \to \mathbb{R}$ be a subharmonic function and $0 < R < 1$. Then the following conditions are equivalent:*

a) $$\sup_{|a|>1} \frac{(1-R)^{\frac{n}{2}} (|a|^2 - 1)^{\frac{n}{2}}}{|a|^{\frac{n}{2}} (|a| + R)^{\frac{n}{2}}} |u(a^*)| < \infty.$$

b)
$$\sup_{|a|>1} \frac{(|a|^2-1)^{\frac{n}{2}}}{|a|^n} |u(a^*)| < \infty.$$

c)
$$\sup_{|b|<1} (1-|b|^2)^{\frac{n}{2}} |u(b)| < \infty.$$

Proof. If $|a| > 1$ and $0 < R < 1$, then that a) holds if and only if b) does, it follows from $|a| < |a| + R < 2|a|$. Taking $a^* = \frac{a}{|a|^2}$, we get b) if and only if c) holds. □

From the previous theorem is natural to define the respective Bloch class as the set of functions $u \in \mathcal{SH}(B)$ such that

$$B(u) = \sup_{|b|<1} (1-|b|^2)^{\frac{n}{2}} |u(b)| < \infty .$$

We denote this class as B^{sh}.

Remark 3.11. If $u \in \mathcal{SH}(B)$, then $-u$ is superharmonic, so we do not have a space. If we restrict to the family of harmonic functions, then we get a space.

As a consequence of the previous definition and Theorems 3.9 and 3.10 we have the next inclusion

Corollary 3.12. *For all* $0 \le p < \infty$, $Q^{sh}_{p,\phi} \subset B^{sh}$.

4. Applications

As an immediate application, we consider the case $n = 3$. Let $a \in \mathbb{R}^3$ with $|a| > 1$ and $a^* \in B \subset \mathbb{R}^3$. It is a straightforward calculation to prove the following equality:

$$\frac{|a^*|^2(1-|a^*|^2)}{||a^*|^2 x - a^*|^2} = \frac{|a|^2-1}{|x-a|^2} . \tag{4.1}$$

Proposition 4.1. *Let* $B \subset \mathbb{R}^3$. *If* $u : B \to \mathbb{R}$ *is a subharmonic function in* B^{sh} *and* $2 < p < \infty$, *then for all* $|a| > 1$

$$\int_B u^2(x)(1-|\phi_a(x)|^2)^p \, dB_x \le 2\pi B^2(f)\beta\left(p-2, \frac{1}{2}\right) ,$$

where β *denotes the beta function.*

Proof. Since $u \in B^{sh}$, $u(x) \le \frac{B(u)}{(1-|x|^2)^{3/2}}$, the following estimation follows from the change of variable formula

$$\begin{aligned}
\int_B u^2(x)(1-|\phi_a(x)|^2)^p dB_x &\le B^2(u) \int_B \frac{1}{(1-|x|^2)^3} (1-|\phi_a(x)|^2)^p dB_x \\
&= B^2(u) \int_B \frac{1}{(1-|x|^2)^3} (1-|\phi_a(x)|^2)^p dB_x \\
&= B^2(u) \int_B \frac{1}{(1-|\phi_a(x)|^2)^3} (1-|x|^2)^p \frac{(|a|^2-1)^3}{|x-a|^6} dB_x .
\end{aligned}$$

Here we used the fact that the Jacobian determinant of ϕ_a is given by

$$\frac{(1-|\phi_a(x)|^2)^3}{(1-|x|^2)^3}=\frac{(|a|^2-1)^3}{|x-a|^6}\,.$$

(See [17] and (4.1).) Now after (3.3) we get

$$\begin{aligned}\int_B u^2(x)(1-|\phi_a(x)|^2)^p\,dB_x &\le B^2(u)\int_B(1-|x|^2)^{p-3}\,dB_x\\ &= B^2(u)3V(B)\int_0^1(1-r^2)^{p-3}r^2\int_S d\sigma\,dr\\ &= 2\pi B^2(u)\beta(p-2,\frac{1}{2})\,.\end{aligned}$$

□

Theorem 4.2. *Let $B\subset\mathbb{R}^3$. If $u:B\to\mathbb{R}$ is a subharmonic function then the following conditions are equivalent.*

1. $u\in B^{sh}$.
2. $u\in \mathcal{Q}^{sh}_{p,\phi}$, *for all* $2<p<\infty$.
3. $u\in \mathcal{Q}^{sh}_{p,\phi}$, *for some* $2<p<\infty$.

Proof. The implication $(1\Rightarrow 2)$ follows from Proposition 4.1. It is obvious that $(2\Rightarrow 3)$. From Corollary 3.12 we have that $(3\Rightarrow 1)$. □

The previous theorem means that for the dimension 3, all the $\mathcal{Q}^{sh}_{p,\phi}$ classes for $2<p<\infty$ coincide and are identical to the Bloch class. Compare the previous result with Proposition 2.2. So we are obtaining some kind of generalization.

5. Examples

Now we present some particular examples where we apply Corollary 3.8.

Example. Let $\mathcal{H}(B,\mathbb{R})$ the space of real valued n-dimensional harmonic functions. Then we can define the space $\mathcal{Q}_p(\mathcal{H}(\mathbb{R}))$ as

$$\sup_{|a|>1}\int_B|\nabla f(x)|^2\left(1-|\phi_a(x)|^2\right)^p\,dB_x<\infty\,,$$

with its corresponding Bloch space

$$\sup_{x\in B}(1-|x|^2)^{\frac{n}{2}}|\nabla f(x)|<\infty\,.$$

Remark 5.1. Note that in [7] (see p. 43, Ex. 10, 11) the Bloch space is defined as

$$\sup_{x\in B}(1-|x|^2)|\nabla f(x)|<\infty\,,$$

that is, a simple generalization of the complex case (n=2). In [10] the corresponding Dirichlet D_p, $-1 < p < \infty$ spaces were studied and defined by

$$\int_B |\nabla f(x)|^2 (1 - |x|^2)^p \, dB_x \, .$$

Therefore $\mathcal{Q}_p(\mathbb{H}(\mathbb{R})) \subset D_p$.

Example. Let $\mathcal{H}(B, \mathbb{C})$ be the space of complex-valued n-dimensional harmonic functions. Then $f(x) = \operatorname{Re} f(x) + i \operatorname{Im} f(x)$ and each component is a harmonic function. Then we can define the space $\mathcal{Q}_p(\mathcal{H}(\mathbb{C}))$ as

$$\sup_{|a|>1} \int_B |f(x)|^2 \left(1 - |\phi_a(x)|^2\right)^p \, dB_x < \infty \, ,$$

with its corresponding Bloch space

$$\sup_{x \in B} (1 - |x|^2)^{\frac{n}{2}} |f(x)| < \infty \, .$$

Note that this a Bergman-type space.

Example. For notation of this example, the main references are [10], [13] [14]. For $-1 < p$, consider the fractional *Dirichlet space* $\mathbf{D}_p(\mathcal{M}, \mathbf{Cl}_{0,n}, B)$ (see [10]) of left monogenic functions, defined by

$$\mathbf{D}_p(\mathcal{M}, \mathbf{Cl}_{0,n}, B) = \left\{ f \in \mathcal{M}(\mathbf{Cl}_{0,n}) : \int_B |\overline{D} f(x)|_0^2 \left(1 - |x|^2\right)^p dB_x < \infty \right\},$$

where $B \subset \mathbb{R}^m$, $m \leq n$, is the unit ball and $\mathbf{Cl}_{0,n}$ the 2^n-dimensional universal Clifford algebra over $\mathbb{R}$. The corresponding $\mathcal{Q}_p$ spaces (see [14]) are

$$\mathcal{Q}_p(\mathcal{M}, \mathbf{Cl}_{0,n}, B) = \left\{ f \in \mathcal{M}(\mathbf{Cl}_{0,n}) : \int_B |\overline{D} f(x)|_0^2 \left(1 - |\phi_a(x)|^2\right)^p dB_x < \infty \right\}.$$

Of course for $0 < p < \infty$, $\mathcal{Q}_p(\mathcal{M}, \mathbf{Cl}_{0,n}, B) \subset \mathbf{D}_p(\mathcal{M}, \mathbf{Cl}_{0,n}, B)$. It is well known that $D\overline{D} f = \Delta f = 0$, then each $f_A : B \to \mathbb{R}$ is a harmonic function.
The associated Bloch space is

$$\sup_{x \in B} (1 - |x|^2)^{\frac{m}{2}} |\overline{D} f(x)|_0 < \infty \, .$$

Acknowledgment

We are grateful to the referee for all his remarks and suggestions.

References

[1] A. Aleman, *Hilbert spaces of analytic functions between the Hardy and Dirichlet spaces*, Proc. Amer. Math. Soc. **115** (1992), 97–104.

[2] R. Aulaskari and P. Lappan, *Criteria for an analytic function to be Bloch and a harmonic or meromorphic funtion to be normal*, Complex Analysis and its Applications, Pitman Research Notes in Math. **305**, Longman Scientific and Technical, Harlow (1994), 136–146.

[3] J. Anderson, J. Clunie and Ch. Pomerenke, *On Bloch Functions and Normal Functions*, J. Reine Angew. **270** (1974), 12–37.

[4] R. Aulaskari, D. Stegenga and J. Xiao, *Some subclasses of BMOA and their characterization in terms of Carleson measures*, Rocky Mountain J. Math. **26** (1996), 485–506.

[5] R. Aulaskari, L.F. Reséndis O. and L.M. Tovar S., *$\mathcal{Q}_p$ spaces and Harmonic Majorants*, Complex Variables **49** (2004), 241–256.

[6] R. Aulaskari, J. Xiao and R. Zhao *On Subspaces and Subsets of BMOA and UBC*, Analysis, **15**, (1995), 101–121.

[7] S. Axler, P. Bourdon and W. Ramey *Harmonic Function Theory*, Graduate Texts in Mathematics **137**, (Second Edition), 2001.

[8] A. Baernstein *Analytic functions of Bounded Mean Oscillation*, Aspects of Contemporary Complex Analysis, Academic Press, 1980, 2–26.

[9] A.F. Beardon *The Geometry of Discrete Groups*, Graduate Texts in Mathematics **91**, 1983.

[10] S. Bernstein, K. Gürlebeck, L.F. Reséndis O. and L.M. Tovar S. *Dirichlet and Hardy Spaces of Harmonic and Monogenic Functions*, Journal for Analysis and its Applications, Volume **24**, (2005), No. 4, 763–789.

[11] S. Bernstein, *Harmonic $\mathcal{Q}_p$ spaces and functions of Bounded Mean Oscillation*, preprint.

[12] J.C. Bishop, *Bounded functions in the little Bloch space*, Pacific J. Math. **142** (1990), 209–225.

[13] J. Cnops, R. Delanghe. *Möbius Invariant Spaces in the Unit Ball*, Applicable Analysis. Vol. **73**, (1-2), 45–64.

[14] J. Cnops, R. Delanghe, K. Gürlebeck and M. Shapiro, *$\mathcal{Q}_p$ spaces in Clifford Analysis*, Advances in Applied Clifford Algebras **11** (S), Ed. Universidad Nacional Autónoma de México, 2001, 201–218.

[15] J.B. Conway *Functions of One Complex Variable*, Graduate Texts in Mathematics, 1973, (1st ed.).

[16] N. Danikas *Some Banach spaces of Analytic Functions*, Function spaces and Complex Analysis, Summer school Ilomantsi, Finland, August 25–29, Joensuu 1997, Re. Ser. 2 (1999), 9–35.

[17] K. Gürlebeck, U. Kähler, M. Shapiro and L.M. Tovar*On $\mathcal{Q}_p$ spaces of quaternion-valued functions*, Complex Variables, **64**, (2001), 33–50.

[18] K. Gürlebeck and H.R. Malonek, *On stricted inclusions of weighted Dirichlet spaces of monogenic functions*, Bulletin of the Australian Mathematical Society, **39** (1999), 115–135.

[19] E. Landau, *Über die Blochsche Konstante und zwei verwandte Weltkonstanten*, Math. Zeitschr. **30**, 608-634 (1929), Coll. Works 9, 75–101.

[20] F. John and L. Nirenberg *On functions of bounded mean oscillation*, Comm. Pure Appl. Math. **14** (1961), 415–426.

[21] I.M. Mitelman and M.V. Shapiro, *Differentiation of the Martinelli-Bochner integrals and the notion of hyperderivability*, Mathematische Nachrichten, **172**, 1995, 211–238.

[22] R. Remmert. *Classical Topics in Complex Function Theory*, Springer Verlag, Graduate Texts in Mathematics, **172**.

[23] K. Stephenson, *Construction of an inner function in the little Bloch Space*, Trans. Amer. Math. Soc. **1988**, 713–720.

[24] K. Stroethoff, *Besov-Type Characterization for the Bloch Space*, Bull. Austral. Math. Soc. Vol. **39**, (1989), 405–420.

[25] A. Sudbery *Quaternionic Analysis*, Mathematical Proceedings of the Cambridge Philosophical Society, **85**, 199–225.

[26] S. Yamashita, *Hyperbolic Hardy Classes and Hyperbolically Dirichlet-finite Functions*, Hokkaido Math. J. **10**, (1981), 700–722.

[27] R. Zhao, *On α-Bloch functions and $VMOA$*, Acta Math.Sci. **16**, (3) (1996), 349-3-60.

L.F. Reséndis O.
Universidad Autónoma Metropolitana-Azcapotzalco
Av. San Pablo # 180, D.F., C.P. 02200
México
e-mail: `lfro@correo.azc.uam.mx`

L.M. Tovar S.
Escuela Superior de Física y Matemáticas del IPN
Edif. 9, Unidad ALM, Zacatenco del IPN., C.P. 07300, D.F.
México
e-mail: `tovar@esfm.ipn.mx`